Electrical Principles for Higher TEC

D.W. Tyler

*Senior Lecturer
Reading College of Technology*

BSP PROFESSIONAL BOOKS
OXFORD LONDON EDINBURGH
BOSTON PALO ALTO MELBOURNE

Copyright © D. W. Tyler 1982

All rights reserved. No part of this
publication may be reproduced, stored
in a retrieval system, or transmitted,
in any form or by any means, electronic,
mechanical, photocopying, recording
or otherwise without the prior
permission of the copyright owner.

First published in Great Britain by
Granada Publishing 1982
Reprinted 1984
Reprinted 1985 by
Collins Professional and Technical Books
Reprinted 1988 by
BSP Professional Books

British Library
Cataloguing in Publication Data
Tyler, D. W. (David W)
   Electrical principles for higher TEC
   1. Electrical engineering—Questions &
   Answers—For technicians
   I. Title
   621.3'076

ISBN 0-632-02362-7

BSP Professional Books
A division of Blackwell Scientific
   Publications Ltd
Editorial Offices:
Osney Mead, Oxford OX2 0EL
   (Orders: Tel. 0865 240201)
8 John Street, London WC1N 2ES
23 Ainslie Place, Edinburgh EH3 6AJ
3 Cambridge Center, Suite 208, Cambridge
   MA 02142, USA
667 Lytton Avenue, Palo Alto, California
   94301, USA
107 Barry Street, Carlton, Victoria 3053,
   Australia

Printed and bound in Great Britain by
Mackays of Chatham PLC, Kent

Contents

Preface
1 Symbolic notation ... 1
2 Network theorems ... 49
3 Complex waveforms ... 88
4 A.c. bridges and potentiometers ... 130
5 Electrostatics ... 161
6 Electromagnetism ... 204
7 Permanent magnets ... 245
8 Two-port networks ... 259
9 Resonance, Q-factor and coupled a.c. circuits ... 298
10 Laplace transforms and the solution of circuit transients ... 346
11 Transmission lines ... 391
Appendix 1 Star/delta transformation ... 420
Appendix 2 Delta/star transformation ... 423
Appendix 3 Mutual inductance ... 426
Index ... 431

Preface

This book provides good coverage of the electrical principles required by undergraduates and Higher TEC students at levels IV and V.

A sound grasp of the principles involved in electrical and electronic engineering is of considerable value. To appreciate this one has only to look back and view the extremely rapid development of equipment which has taken place. We have moved from thermionic valves to p-n-p transistors, on to n-p-n transistors, integrated circuits, NMOS, PMOS, CMOS, etc. The speed at which changes take place seems ever accelerating. In power engineering, control, instrumentation and supervision methods have been revolutionised, many of the older electro-mechanical devices having been replaced by integrated electronics and the omnipresent computer.

All such equipment conforms to common principles and a good knowledge of these enables the new technology, as it progresses, to be mastered. For example, the steady-state and transient response of circuits, mutual inductance as a means of obtaining improved bandwidth for power matching and transfer; electro-magnetism and electro-statics, transmission systems and the effects of harmonics.

In many courses the principles (and often the mathematics) seem to be given second rate attention, being sacrificed to a specialist study of particular devices. When these become obsolete, re-training is that much more difficult without a thorough grasp of fundamentals.

I hope that this book will go at least some way towards a greater appreciation of 'Principles'.

                                                                         D.W.T.

1 Symbolic Notation

1.1 PHASOR MANIPULATION

This section is a brief revisionary introduction to the circuit theory which should have been covered fully in previous work.

Consider the two voltages shown phasorially in Fig. 1.1.

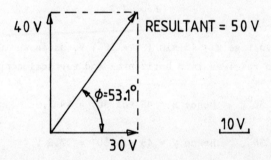

Fig. 1.1

$$v_1 = 30 \sin \omega t \text{ V} \qquad v_2 = 40 \sin (\omega t + 90°) \text{ V}.$$

The peak values of the two voltages are 30 V and 40 V respectively. v_2 leads v_1 by $90°$.

To add these two voltages together the parallelogram is constructed using the peak values of voltage. By scale drawing or by the use of Pythagoras' theorem the resultant V_R = 50 V. The angle between V_1 and V_R = $53.1°$.

$$V_R = \sqrt{(30^2 + 40^2)} = 50 \text{ V}; \qquad \tan \phi = \frac{40}{30} = 1.33, \text{ hence } \phi = 53.1°.$$

The resultant voltage is therefore given by the equation

$$v_R = 50 \sin(\omega t + 53.1°) \text{ V}.$$

In this case the two voltages have been added together to find a resultant. It is possible to resolve a single voltage into two component voltages which are at right angles.

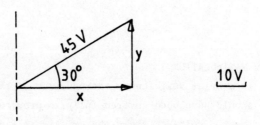

Fig. 1.2

Given a voltage $v = 45 \sin(\omega t + 30°)$ V, as shown in Fig. 1.2, this is to be resolved into horizontal and vertical components.

$$\frac{x}{45} = \cos 30°, \quad \text{hence } x = 45 \cos 30° = 39 \text{ V}.$$

$$\frac{y}{45} = \sin 30°, \quad \text{hence } y = 45 \sin 30° = 22.5 \text{ V}.$$

Therefore $v_x = 39 \sin \omega t$ V, and $v_y = 22.5 \sin(\omega t + 90°)$ V.

This resolution of the original voltage into quadrature voltages facilitates the addition and subtraction of phasors as demonstrated in worked example 1.1.

Worked example 1.1   Add together the following three voltages:

$$v_1 = 60 \sin(\omega t + 45°) \text{ V}$$
$$v_2 = 75 \sin(\omega t + 80°) \text{ V}$$
$$v_3 = 10 \sin \omega t \text{ V}.$$

Express the answer in the same form.

SYMBOLIC NOTATION

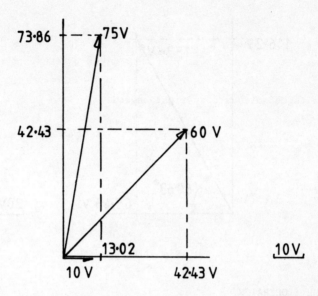

Fig. 1.3

Resolve each of the voltages into horizontal (x) and vertical (y) components.

$V_{x1} = 60 \cos 45° = 42.43$ V $V_{y1} = 60 \sin 45° = 42.43$ V

$V_{x2} = 75 \cos 80° = 13.02$ V $V_{y2} = 75 \sin 80° = 73.86$ V

$V_{x3} = 10$ V.   There is no component of V_3 in the y direction.

Adding horizontal components: $V_{x(total)} = 42.43 + 13.02 + 10 = 65.45$ V.

Adding vertical components: $V_{y(total)} = 42.43 + 73.86 = 116.29$ V.

The resultant voltage has a modulus (size) of $\sqrt{(65.45^2 + 116.29^2)} = 133.4$ V.

$$\tan \phi = \frac{116.29}{65.45} = 1.77, \quad \text{hence } \phi = 60.63°.$$

Hence $v_R = 133.4 \sin(\omega t + 60.63°)$ V (see Fig. 1.4).

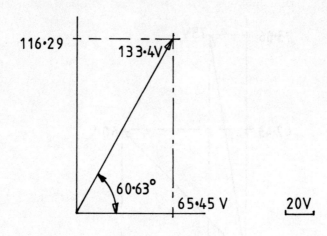

Fig. 1.4

1.2 THE j OPERATOR

The use of the j operator can reduce considerably the amount of work involved in solving electrical circuits containing inductance and capacitance, especially when these are in parallel or series parallel combination.

By definition, multiplying a phasor by j moves its position 90° anti-clockwise without changing its magnitude.

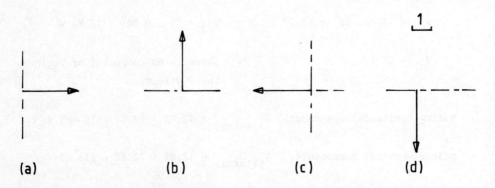

Fig. 1.5

In Fig. 1.5(a) a phasor 3 units long is drawn horizontally to the

SYMBOLIC NOTATION

right. It represents an electrical quantity to scale. On a graph this is the usual +x direction. If we write (j3), by definition this is the same phasor turned through 90° anti-clockwise. This is shown in Fig. 1.5(b). A further 90° shift is shown in Fig. 1.5(c) and this is described as (j × j3) or ($j^2$3). This is the original phasor reversed, i.e. (-3). This means that

$$j^2 = -1, \quad \text{hence } j = \sqrt{-1}.$$

Since there is no numerical solution for this quantity (-1 × -1 = 1 and +1 × +1 = 1), the numbers involving j are often called imaginary numbers. This can be very misleading since they are in no way imaginary as will be seen as the chapter develops.

In Fig. 1.5(d) a further j operation moves the phasor to the vertically downward position where it becomes ($j^3$3). Now

$$j^3 = j \times j^2 = j \times -1 = -j.$$

This axis is the (-j) axis.

Multiplying by j for the last time brings the phasor back to the original position so that ($j^4$3) must be equal to the original value (3).

$$j^4 = j^2 \times j^2 = -1 \times -1 = 1.$$

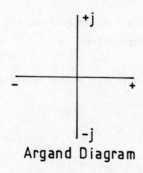

Argand Diagram

Fig. 1.6

Figure 1.6 combines these axes in one diagram. This is known as the Argand diagram.

Two quantities, OP = (4 + j5) and OQ = (-6 + j3), are shown plotted on an Argand diagram in Fig. 1.7.

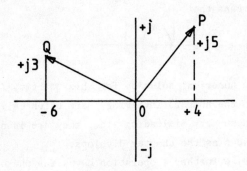

Fig. 1.7

1.3 RECTANGULAR CO-ORDINATES

The position of a phasor representing an electrical quantity can be described by quoting its horizontal and vertical components. By returning to Fig. 1.2, in which $v = 45 \sin(\omega t + 30°)$ V, the horizontal component $x = 39$ V, while the vertical component $y = 22.5$ V. This voltage may be expressed as

(39 + j22.5) V.

Such a quantity is called a complex number, the values 39 and j22.5 being its rectangular co-ordinates.

<u>Worked example 1.2</u>   Resolve the phasor shown in Fig. 1.8 into its rectangular co-ordinates.

$\frac{x}{5} = \cos 60°$ $\qquad\qquad$ $x = 5 \cos 60° = 2.5$

$\frac{y}{5} = \sin 60°$ $\qquad\qquad$ $y = 5 \sin 60° = 4.33$

The horizontal component is from O, the origin, extending to the right

SYMBOLIC NOTATION

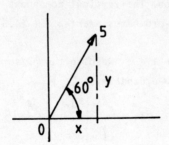

Fig. 1.8

and is in the +x direction. The vertical component is from the origin upwards in the +j direction.

The rectangular co-ordinates are (2.5 + j4.33).

Worked example 1.3   Resolve the phasor shown in Fig. 1.9 into its rectangular co-ordinates.

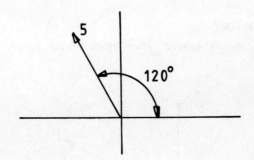

Fig. 1.9

Horizontal component = 5 cos 120°   (cos 120° = -cos 60°)
                          = 5 × (-0.5)
                          = -2.5

Vertical component   = 5 sin 120°   (sin 120° = sin 60°)
                          = 4.33

The horizontal component is 2.5 measured to the left of the origin, i.e. in the -x direction. The vertical component is again upwards.

The rectangular co-ordinates are (-2.5 + j4.33).

Self-assessment example 1.4   Resolve the phasor shown in Fig. 1.10 into its rectangular co-ordinates.

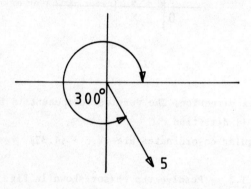

Fig. 1.10

1.4   THE POLAR FORM

Let z denote a complex number represented by the phasor OP in Fig. 1.11.

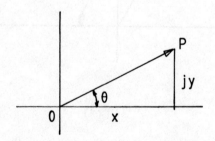

Fig. 1.11

In rectangular form it has been shown that z may be expressed as (x + jy).

SYMBOLIC NOTATION

By Pythagoras' theorem, OP = $\sqrt{(x^2 + y^2)}$, and $\tan \theta = \frac{y}{x}$, hence $\theta = \arctan \frac{y}{x}$.

If the length of OP = r, the phasor may be described as $r\underline{/\theta}$ (read as 'r angle theta'). r is called the modulus of the complex number. θ is called the argument of the complex number.

Mod z or $|z|$ can be written as an abbreviation for the modulus, i.e. r together with arg z for the angle the phasor makes with the +x direction, where:

$z = (x + jy) = r\underline{/\theta}$, mod z or $|z| = \sqrt{(x^2 + y^2)}$, and arg z = $= \arctan \frac{y}{x}$.

$r\underline{/\theta}$ is known as the polar form of the complex number.

The phasor in Fig. 1.8 could be described as $5\underline{/60^o}$. The size or modulus of the phasor is 5 and the angle it makes with the +x direction or argument is 60°. The phasor in Fig. 1.9 would be described as $5\underline{/120^o}$, and that in Fig. 1.10 as $5\underline{/300^o}$. The angles may be measured in a clockwise direction when they become negative. The phasor in Fig. 1.10 could equally well be described as $5\underline{/-60^o}$.

When using symbolic phasor notation it is sound practice always to draw the triangle concerned roughly to scale so that it is possible to see the angle involved and in which quadrant of the Argand diagram one is working.

1.5 THE USE OF PEAK AND r.m.s. VALUES

When expressing a voltage as $v = V_m \sin(\omega t + \phi)$ volts,

    v = instantaneous value at time t seconds,
    V_m = maximum value of the wave,
    $\omega (= 2\pi f)$ is the angular velocity in radians per second.

A phasor drawn to scale may be used to represent this voltage and this in turn may be expressed in either rectangular or polar forms.
For example: $v = 100 \sin(\omega t + 30^o)$.
In polar form this is $100\underline{/30^o}$ V.
In rectangular form, $100(\cos 30^o + j \sin 30^o) = (86.6 + j50)$ V.

Mod V_m or $|V_m|$ = 100 V.   The argument = $+30°$.

One can also work in r.m.s. values. The r.m.s. value of a sinusoid with maximum value 100 V = 100 × 0.707 = 70.7 V.
In polar form this is 70.7$\underline{/30°}$ V.
In rectangular form, 70.7(cos 30° + j sin 30°) = (61.2 + j35.35) V, these voltages being r.m.s. values.

Mod V or $|V|$ = 70.7 V.   The argument = $+30°$.

Ammeters and voltmeters which are r.m.s. indicating are usually employed and Fig. 1.12 shows such meters connected to measure current and voltage in a single-phase circuit. These indicate r.m.s. moduli and have no means of resolving angle.

$$\frac{|V|}{|I|} = |Z|.$$

Nothing may be deduced about the phase angle of the load unless either a wattmeter is connected in the circuit to measure power, or other forms of ammeter and voltmeter are used which are capable of indicating phase.

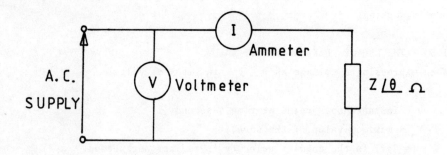

Fig. 1.12

<u>Worked example 1.5</u>   Express (1 + j7) in the polar form.
Drawing the quantities approximately to scale (Fig. 1.13) shows that the resultant lies in the first quadrant of the Argand diagram. The angle θ lies between 0° and 90°. By Pythagoras' theorem,

SYMBOLIC NOTATION

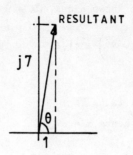

Fig. 1.13

the resultant $= \sqrt{(1^2 + 7^2)} = 7.07$

$\tan \theta = \frac{7}{1}$,   hence $\theta = 81.9°$.

The polar form is $7.07\underline{/81.9°}$.

<u>Worked example 1.6</u>   Express $(-7 - j9)$ in the polar form.

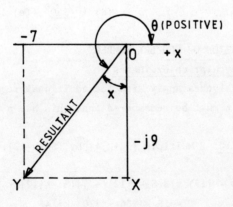

Fig. 1.14

Drawing the rectangular co-ordinates in their correct positions as shown in Fig. 1.14 shows that the resultant lies in the third quadrant and that θ lies between $180°$ and $270°$. In the figure, work in triangle OXY and for the moment ignore the (j) and negative signs. OXY is a triangle with two known sides, OX = 9 and XY = 7.

The resultant = $\sqrt{(9^2 + 7^2)}$ = 11.4

To find the angle θ,

$\tan x = \frac{7}{9}$ = 0.777,   hence x = 37.9°.

Therefore the resultant phasor lies at 90° + 37.9° = 127.9° from the +x direction (clockwise). Hence θ = 360° - 127.9° = 232.1° (anti-clockwise).

The resultant may be described as (i) 11.4/-127.9°
                                    or (ii) 11.4/232.1°.

Self-assessment examples
1.7 Express (3 - j4) in the polar form.
1.8 Convert to the polar form (a) (10 + j2), (b) (-4 - j6),
                              (c) (-9 + j4), (d) (5 - j12).
1.9 Convert to the rectangular form (a) 28/85°, (b) 4/220°, (c) 6/165°,
                                    (d) 10/350°, (e) 14/-75°.

1.6   MULTIPLICATION OF COMPLEX NUMBERS
1.6.1 Using rectangular co-ordinates
All the rules of algebra apply with the additional requirement that where j^2 occurs it must be remembered that this has a value of -1.

Worked example 1.10   Multiply (3 + j4) by (5 - j12).

(3 + j4) × (5 - j12) = 3(5 - j12) + j4(5 - j12)
                    = 15 - j36 + j20 - $j^2$48

Now j^2 = -1, so that -$j^2$48 = -(-1) × 48 = +48.

The solution is 15 + 48 - j36 + j20
              = (63 - j16).

SYMBOLIC NOTATION

1.6.2 Using the polar form

Consider multiplying two polar quantities, $A\underline{/\theta}$ and $B\underline{/\phi}$, together. Firstly consider these two quantities in their rectangular form.

$$A\underline{/\theta} = A(\cos\theta + j\sin\theta) \qquad B\underline{/\phi} = B(\cos\phi + j\sin\phi).$$

Therefore $A\underline{/\theta} \times B\underline{/\phi} = A(\cos\theta + j\sin\theta) \times B(\cos\phi + j\sin\phi)$
$$= AB(\cos\theta\cos\phi + j\cos\theta\sin\phi + j\sin\theta\cos\phi - \sin\theta\sin\phi).$$

Using the double-angle formulae: $\cos(\theta + \phi) = \cos\theta\cos\phi - \sin\theta\sin\phi$
$\sin(\theta + \phi) = \cos\theta\sin\phi + \sin\theta\cos\phi$

$$A\underline{/\theta} \times B\underline{/\phi} = AB(\cos(\theta + \phi) + j\sin(\theta + \phi)).$$

Converted to the polar form this is $AB\underline{/\theta + \phi}$.

Hence to multiply two quantities in the polar form, multiply the two moduli together and add their arguments.

To demonstrate the method the figures from worked example 1.10 will be used.

(3 + j4) in the polar form = $5\underline{/53.13°}$.
(5 - j12) in the polar form = $13\underline{/-67.38°}$.

The solution for this product in rectangular form has been shown to be (63 - j16). In polar form this is $65\underline{/-14.25°}$.

Working in polar form: $5\underline{/53.13°} \times 13\underline{/-67.38°}$
$= 5 \times 13\underline{/53.13° + (-67.38°)}$
$= 65\underline{/-14.25°}$

which is the required result.

Worked example 1.11 Multiply (3 + j2) by (-2 - j4) using both rectangular and polar forms.

(a) Rectangular

$$(3 + j2) \times (-2 - j4) = -6 - j12 - j4 - j^2 8$$
$$= -6 + 8 - j16$$
$$= (2 - j16) \quad \text{(In polar form } 16.1\underline{/-82.9^o}\text{)}$$

(b) Polar

$$(3 + j2) = 3.6\underline{/33.7^o} \qquad (-2 - j4) = 4.47\underline{/-116.6^o}.$$
$$\text{Product} = 3.6 \times 4.47\underline{/33.7^o + (-116.6^o)} = 16.1\underline{/-82.9^o}$$

which agrees with the result obtained using rectangular co-ordinates.

<u>Self-assessment example 1.12</u>   Express the answers to the following products in both rectangular and polar forms.
(a) $(10 - j15) \times 6\underline{/225^o}$, (b) $(-5 + j6) \times (3 - j10)$.

1.7   DIVISION OF COMPLEX NUMBERS
1.7.1 Using rectangular co-ordinates
Consider performing the calculation $\frac{(x + jy)}{(a - jb)}$.
The division cannot be carried out while the expression is in this form. It is possible to divide by real numbers alone or by j terms alone but not simultaneously by a combination of both.

To turn the expression into a suitable form it is necessary to multiply both the numerator and denominator by a quantity which will turn the denominator into a pure number. This process is called rationalisation. The quantity involved is called the complex conjugate of the denominator.

The complex conjugate of the denominator is the denominator itself with the sign of the j term changed.

The original quotient becomes $\frac{(x + jy)}{(a - jb)} \times \frac{(a + jb)}{(a + jb)}$.

Note that the operation has not changed the value of the expression since the two (a + jb) terms could be cancelled.

Multiply out the numerator: $ax + jbx + jay - by$
$$= (ax - by) + j(bx + ay).$$

SYMBOLIC NOTATION

Multiply out the denominator: $a^2 + jab - jab + b^2$
$$= a^2 + b^2.$$

Thus multiplying the denominator by the complex conjugate has eliminated the j term. Therefore

$$\frac{(x + jy)}{(a - jb)} = \frac{(ax - by) + j(bx + ay)}{a^2 + b^2}$$

$$= \frac{(ax - by)}{a^2 + b^2} + \frac{j(bx + ay)}{a^2 + b^2}.$$

<u>Worked example 1.13</u>   Divide $(3 + j2)$ by $(2 + j4)$.

$\frac{(3 + j2)}{(2 + j4)}$ becomes $\frac{(3 + j2)(2 - j4)}{(2 + j4)(2 - j4)}$ after rationalising.

Multiplying out both numerator and denominator

$$\frac{6 - j12 + j4 + 8}{2^2 + 4^2} = \frac{(14 - j8)}{20} = \frac{14}{20} - j\frac{8}{20}.$$

Thus $\frac{(3 + j2)}{(2 + j4)} = \frac{14}{20} - j\frac{8}{20} = 0.7 - j0.4$

<u>Worked example 1.14</u>   Evaluate $\frac{(-7 - j9)}{(5 - j12)}$.

Rationalise the expression by multiplying numerator and denominator by $(5 + j12)$. This gives

$$\frac{(-7 - j9)(5 + j12)}{5^2 + 12^2} = \frac{-35 - j84 - j45 + 108}{169}$$

$$= \frac{(73 - j129)}{169}$$

$$= (0.43 - j0.76).$$

1.7.2 Using the polar form
Consider dividing $A\underline{/\theta}$ by $B\underline{/\phi}$.
Converting to rectangular form gives

$$\frac{A\underline{/\theta}}{B\underline{/\phi}} = \frac{A(\cos \theta + j \sin \theta)}{B(\cos \phi + j \sin \phi)}.$$

Rationalise the expression by multiplying numerator and denominator by $(\cos \phi - j \sin \phi)$. This gives

$$\frac{A(\cos \theta + j \sin \theta)(\cos \phi - j \sin \phi)}{B(\cos^2 \phi + \sin^2 \phi)}$$

But $\cos^2 \phi + \sin^2 \phi = 1$. Hence the expression becomes

$$\frac{A}{B}(\cos \theta \cos \phi - j \cos \theta \cos \phi + j \sin \theta \cos \phi + \sin \theta \sin \phi)$$
$$= \frac{A}{B}((\cos \theta \cos \phi + \sin \theta \sin \phi) + j(\sin \theta \cos \phi - \cos \theta \sin \phi)).$$

Using the double-angle formulae: $\cos(\theta - \phi) = \cos \theta \cos \phi + \sin \theta \sin \phi$
$\sin(\theta - \phi) = \sin \theta \cos \phi - \cos \theta \sin \phi$,

$$\frac{A\underline{/\theta}}{B\underline{/\phi}} = \frac{A}{B}(\cos(\theta - \phi) + j \sin(\theta - \phi))$$

which in the polar form is $\frac{A}{B}\underline{/\theta - \phi}$.

Division in the polar form is accomplished by dividing the moduli and subtracting the argument of the denominator from that of the numerator, having due regard for the signs.

Using the quantities from worked example 1.14:

$(-7 - j9) = 11.4\underline{/-127.87^\circ}$ $\quad (5 - j12) = 13\underline{/-67.38^\circ}.$

Therefore $\dfrac{(-7 - j9)}{(5 - j12)} = \dfrac{11.4\underline{/-127.87^\circ}}{13\underline{/-67.38^\circ}} = \dfrac{11.4}{13}\underline{/-127.87^\circ - (-67.38^\circ)}$

$$= 0.877\underline{/-60.49^\circ}.$$

Converting this to rectangular form yields $(0.43 - j0.76)$, which is the result previously obtained.

<u>Worked example 1.15</u> Divide $50\underline{/-126^\circ}$ by $7.5\underline{/36^\circ}$.

$$\frac{50\underline{/-126^\circ}}{7.5\underline{/36^\circ}} = \frac{50}{7.5}\underline{/-126^\circ - 36^\circ} = 6.67\underline{/-162^\circ}.$$

SYMBOLIC NOTATION

<u>Self-assessment example 1.16</u>   Divide $(125 + j70)$ by $14\underline{/65^\circ}$. Express your answer in both rectangular and polar forms.

From the analyses performed so far, it should be observed that:

(a) to ADD or SUBTRACT complex numbers they have to be in the rectangular $(a + jb)$, form
(b) when MULTIPLYING or DIVIDING complex numbers it is far simpler to work in the polar form. Although $\frac{(a + jb)(x + jy)}{(p + jq)}$ can be evaluated, it is better expressed as $\frac{c\underline{/\theta} \times d\underline{/\phi}}{e\underline{/\alpha}}$ when the solution is $\frac{c \times d}{e}\underline{/\theta + \phi - \alpha}$.

1.8   THE SIGNIFICANCE OF j IN a.c. CIRCUITS
1.8.1 Resistance

When an alternating voltage is applied to a resistance, the current flowing is in phase with the voltage (see Fig. 1.12 in which, for this case, the load is considered to be purely resistive).

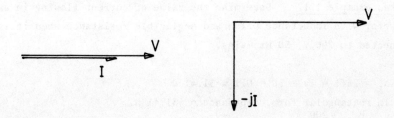

Fig. 1.15 *Fig. 1.16*

As shown in Fig. 1.15, both voltage and current phasors are in the +x direction. The voltage can be expressed as $(V + j0)$ and the current as $(I + j0)$

$$R = \frac{(V + j0)}{(I + j0)} \quad \text{or simply} \quad \left|\frac{V}{I}\right| = R \; \Omega.$$

There are no j terms involved.

1.8.2 Inductance

When an alternating voltage is applied to an inductance with negligible resistance the current flowing lags the voltage by $90°$ (take Fig. 1.12 to have such an inductance as its load). The phasors are shown in Fig. 1.16.

$$X_L = 2\pi f L = \frac{|V|}{|I|}.$$

In rectangular form, impedance $= \dfrac{(V + j0)}{(0 - jI)} = \dfrac{|V|}{-j|I|} = \dfrac{X_L}{-j}.$

Multiplying the numerator and the denominator by j, the impedance becomes

$$\frac{jX_L}{-j^2} = jX_L \ \Omega.$$

In rectangular form the impedance of an inductive reactance with negligible resistance may be represented as jX_L Ω. In polar form this is $X_L \underline{/90°}$ Ω.

<u>Worked example 1.17</u>   Determine the value of current flowing in an inductor with inductance 0.1 H and negligible resistance when it is connected to 200 V, 50 Hz mains.

$X_L = 2\pi f L = 2\pi \times 50 \times 0.1 = 31.41 \ \Omega.$
In rectangular form, impedance $= j31.41 \ \Omega.$
$I = \dfrac{V}{Z} = \dfrac{200}{j31.41}$.

Multiplying numerator and denominator by -j gives

$$I = \frac{-j200}{-j^2 31.41} = \frac{-j200}{31.41} = -j6.37 \ A.$$

OR working in polar form

$$I = \frac{200\underline{/0°}}{31.41\underline{/90°}} = \frac{200}{31.41}\underline{/0° - 90°}$$

$$= 6.37\underline{/-90°} \ A.$$

SYMBOLIC NOTATION

Both methods place the current in the correct phasor position.

Further to section 1.5, although peak or r.m.s. values for voltage and current can equally well be used it is more normal to use r.m.s. values, which is the case in this example.

1.8.3 Capacitance

In a capacitor the current leads the applied voltage by $90°$ (in Fig. 1.12 the load is now considered to be a capacitor). As shown in Fig. 1.17, the voltage is again $(V + j0)$. The current phasor is vertically upwards and is therefore jI amperes.

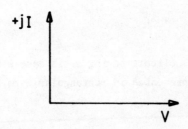

Fig. 1.17

$$X_C = \frac{1}{2\pi f C} = \left|\frac{V}{I}\right| \; \Omega.$$

In rectangular form, impedance $= \dfrac{(V + j0)}{(0 + jI)} = \dfrac{|V|}{+j|I|} = \dfrac{X_C}{j}.$

Multiplying numerator and denominator by $-j$ gives

$$\text{impedance} = \frac{-jX_C}{-j^2} = -jX_C \; \Omega.$$

The impedance of a capacitive reactance may be represented as $-jX_C$ in rectangular form. In polar form this is $X_C\underline{/-90°}$.

<u>Worked example 1.18</u> Determine the value of current flowing in a 40 μF capacitor connected to 200 V, 50 Hz mains.

$$X_C = \frac{1}{2\pi \times 50 \times 40 \times 10^{-6}} = 79.58 \; \Omega.$$

In rectangular form, impedance = $-j79.58$ Ω.

$$I = \frac{V}{Z} = \frac{200}{-j79.58}.$$

Multiplying numerator and denominator by j gives

$$I = \frac{j200}{-j^2 79.58} = j2.07 \text{ A}.$$

OR in polar form

$$I = \frac{200\underline{/0^o}}{79.58\underline{/-90^o}} = \frac{200}{79.58}\underline{/0^o - (-90^o)} = 2.51\underline{/90^o} \text{ A}.$$

1.9 SERIES CIRCUITS

The impedance of a series circuit having resistance R Ω and inductive reactance X_L Ω may be represented in rectangular form as

$$Z = (R + jX_L) \text{ Ω}.$$

Similarly for a series circuit having resistance R Ω and capacitive reactance X_C Ω, the impedance may be written as

$$Z = (R - jX_C) \text{ Ω}.$$

For a circuit with resistance, inductive reactance and capacitive reactance in series, the impedance in rectangular form will be

$$Z = (R + jX_L - jX_C) = (R + j(X_L - X_C)) \text{ Ω}.$$

Worked example 1.19

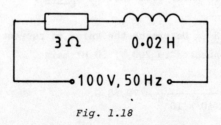

Fig. 1.18

SYMBOLIC NOTATION

A circuit comprises a 3 Ω resistor in series with an inductor having 0.02 H inductance and negligible resistance. It is connected to a 100 V, 50 Hz supply. Calculate (a) the value of the circuit current, (b) the phase angle between current and voltage.

Inductive reactance = $2\pi \times 50 \times 0.02 = 6.28$ Ω.

Using the correct j notation this is written as j6.28 Ω.

Total impedance of the circuit = $(3 + j6.28)$ Ω.

$$I = \frac{V}{Z} = \frac{100}{(3 + j6.28)} = \frac{100(3 - j6.28)}{(3^2 + 6.28^2)} = \frac{300 - j628}{48.44}$$

$$= (6.19 - j12.96) \text{ A}.$$

Converting to polar form

$$I = 14.37 \underline{/-64.46^\circ} \text{ A}$$

OR working directly in polar form

$$I = \frac{100}{(3 + j6.28)} = \frac{100}{6.96 \underline{/64.46^\circ}} = 14.37 \underline{/-64.46^\circ} \text{ A}.$$

Modulus of the current = 14.37 A (answer (a)) (this is the value of current indicated on an a.c. ammeter included in the circuit).

The current lags the voltage by 64.46° (answer (b)).

<u>Worked example 1.20</u> The current in a circuit is $(4.5 + j12)$ A, when the applied voltage is $(100 + j150)$ V. Calculate the value of the circuit impedance.

$$I = \frac{V}{Z}, \quad \text{re-arranging gives } Z = \frac{V}{I}.$$

$$Z = \frac{(100 + j150)}{(4.5 + j12)} = \frac{180.27 \underline{/56.3^\circ}}{12.82 \underline{/69.4^\circ}}$$

$$= 14.06 \underline{/-13.1^\circ} \text{ Ω}$$

$$= (13.7 - j3.2) \text{ Ω}.$$

The circuit comprises a resistance of value 13.7 Ω in series with a capacitor with capacitive reactance 3.2 Ω at the particular frequency being used.

Worked example 1.21   A circuit comprises a resistance of value 120 Ω in series with a capacitor with capacitive reactance 250 Ω at the supply frequency. The current is $0.9\underline{/0°}$ A. Calculate (a) the circuit voltage, (b) the voltage across the resistor, (c) the voltage across the capacitor, (d) the phase angle between the circuit current and the supply voltage.

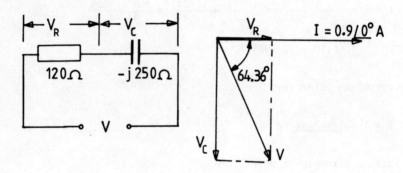

Fig. 1.19

(a) $V = IZ = 0.9(120 - j250) = (108 - j225)$ V $= 249.6\underline{/-64.36°}$ V.
(b) V_R is in phase with the circuit current and is equal to 108 V.
(c) V_C lags the circuit current and is equal to 225 V.
(d) The phase angle between the circuit current and the voltage is 64.36°.

Self-assessment example 1.22   A current of 2.5 A flows in a circuit comprising a resistance of value 50 Ω in series with an inductor with inductance 0.2 H when connected to 50 Hz mains. Determine (a) the modulus of the supply voltage, (b) the phase angle between the circuit current and the supply voltage, (c) the value of the circuit current and its phase angle with respect to the supply voltage if the inductor is replaced by a capacitor with capacitance 80 μF, the supply voltage meanwhile remaining unchanged.

SYMBOLIC NOTATION

Worked example 1.23

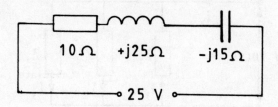

Fig. 1.20

Calculate the value of the current flowing in the circuit shown in Fig. 1.20.

$$Z = 10 + j25 - j15$$
$$= 10 + j10$$
$$= 14.14\underline{/45^\circ}.$$

$$I = \frac{V}{Z} = \frac{25}{14.14\underline{/45^\circ}} = 1.77\underline{/-45^\circ} \text{ A}.$$

Self-assessment example 1.24

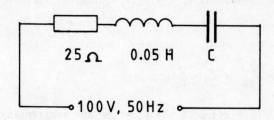

Fig. 1.21

Calculate the value of capacitance necessary to cause 1 A to flow in the circuit shown in Fig. 1.21. (Hint: What must be the value of the total circuit impedance Z?)

1.10 PARALLEL CIRCUITS

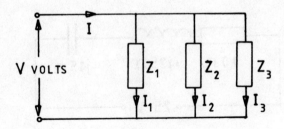

Fig. 1.22

Figure 1.22 shows three impedances Z_1, Z_2 and Z_3 connected in parallel across a supply of V volts. The potential difference across each impedance is the same, namely the supply voltage V.

$$I_1 = \frac{V}{Z_1} \quad I_2 = \frac{V}{Z_2} \quad I_3 = \frac{V}{Z_3}$$

The total supply current, $I = I_1 + I_2 + I_3$

$$I = \frac{V}{Z_1} + \frac{V}{Z_2} + \frac{V}{Z_3}$$

$$= V\left(\frac{1}{Z_1} + \frac{1}{Z_2} + \frac{1}{Z_3}\right).$$

Replacing these impedances by a single equivalent impedance Z_{eq} gives

$$I = \frac{V}{Z_{eq}}.$$

Hence $\dfrac{V}{Z_{eq}} = V\left(\dfrac{1}{Z_1} + \dfrac{1}{Z_2} + \dfrac{1}{Z_3}\right).$

Cancelling V on both sides of the equation and rearranging gives

$$Z_{eq} = \frac{1}{\frac{1}{Z_1} + \frac{1}{Z_2} + \frac{1}{Z_3}}.$$

If there are only two parallel impedances

SYMBOLIC NOTATION

$$Z_{eq} = \frac{1}{\frac{1}{Z_1} + \frac{1}{Z_2}} = \frac{1}{\frac{Z_2 + Z_1}{Z_1 Z_2}} = \frac{Z_1 Z_2}{Z_1 + Z_2}.$$

These formulae for equivalent impedance will be recognised from previous work involving resistors. However when complex impedances are involved, the rectangular or polar form must always be used when substituting in the formulae to take account of the angles involved. Worked examples 1.25 and 1.27 will make this clear.

Worked example 1.25

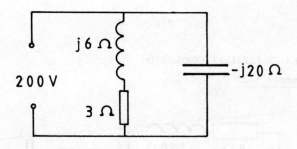

Fig. 1.23

Calculate the value of an impedance equivalent to that of the parallel combination shown in Fig. 1.23. Hence determine the value of the circuit current.

$$Z_{eq} = \frac{Z_1 Z_2}{Z_1 + Z_2} = \frac{(3 + j6)(-j20)}{3 + j6 - j20} = \frac{6.71 \underline{/63.4^\circ} \times 20 \underline{/-90^\circ}}{(3 - j14)}$$

$$Z_{eq} = \frac{134.2 \underline{/-26.6^\circ}}{14.32 \underline{/-77.9^\circ}}$$

$$= 9.37 \underline{/51.3^\circ} \; \Omega.$$

$$I = \frac{V}{Z} = \frac{200 \underline{/0^\circ}}{9.37 \underline{/51.3^\circ}}$$

$$= 21.34 \underline{/-51.3^\circ} \; A.$$

ELECTRICAL PRINCIPLES FOR HIGHER TEC

Self-assessment example 1.26   Determine the value of a single impedance equivalent to the parallel combination shown in Fig. 1.24.

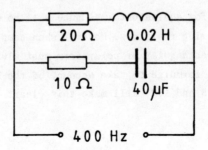

Fig. 1.24

1.11   SERIES PARALLEL CONNECTION
Worked example 1.27

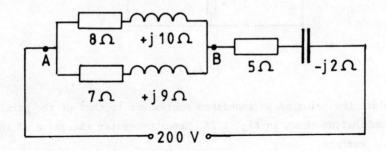

Fig. 1.25

Determine the value of the circuit current for the series parallel combination of impedances shown in Fig. 1.25.

It is firstly necessary to reduce the parallel pair of impedances to a single equivalent impedance. This may be added to (5 - j2) using the rectangular form.

$$Z_{eq} = \frac{(8 + j10)(7 + j9)}{(8 + j10 + 7 + j9)} = \frac{12.81 \underline{/51.3^\circ} \times 11.4 \underline{/52.12^\circ}}{(15 + j19)}$$

$$= \frac{164.2 \underline{/103.42^\circ}}{24.2 \underline{/51.7^\circ}}$$

26

SYMBOLIC NOTATION

$$= 6.03\underline{/51.7^o}\ \Omega = (3.72 + j4.75)\ \Omega.$$

This is added to $(5 - j2)\ \Omega$

$$Z_{total} = (3.72 + j4.75) + (5 - j2) = (8.72 + j2.75)\ \Omega$$
$$= 9.15\underline{/17.4^o}\ \Omega$$

$$I = \frac{V}{Z} = \frac{200}{9.15\underline{/17.4^o}} = 21.85\underline{/-17.4^o}\ A.$$

<u>Worked example 1.28</u>   For the circuit shown in Fig. 1.25, determine the value of the current in each of the two parallel impedances.

From worked example 1.27, the total circuit current = $21.85\underline{/-17.4^o}$ A. The equivalent impedance Z_{eq} of the parallel pair of impedances = $6.03\underline{/51.76^o}\ \Omega$.

The potential difference developed across $Z_{eq} = IZ_{eq}$
(V_{AB} in Fig. 1.25)
$$V_{AB} = 21.85\underline{/-17.4^o} \times 6.03\underline{/51.76^o}$$
$$= 131.8\underline{/34.36^o}\ V$$

Current in $(8 + j10)\ \Omega = \dfrac{131.8\underline{/34.36^o}}{(8 + j10)} = \dfrac{131.8\underline{/36.36^o}}{12.81\underline{/51.3^o}}$

$$= 10.28\underline{/-17.04^o}\ A.$$

Current in $(7 + j9)\ \Omega = \dfrac{131.8\underline{/34.36^o}}{(7 + j9)} = \dfrac{131.8\underline{/34.36^o}}{11.4\underline{/52.12^o}}$

$$= 11.56\underline{/-17.76^o}\ A.$$

<u>Self-assessment example 1.29</u>   Determine the value of the current in the $(10 + j30)\ \Omega$ impedance in Fig. 1.26.

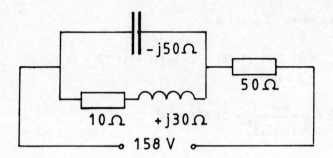

Fig. 1.26

1.12 ADMITTANCE, CONDUCTANCE AND SUSCEPTANCE

For the circuit shown in Fig. 1.27, the impedance $Z = (R + jX_L)$ Ω. The admittance Y is defined as $1/Z$. The unit of Y is the siemen (S).

$$Y = \frac{1}{(R + jX_L)} \text{ S}$$

Rationalising

$$Y = \frac{(R - jX_L)}{R^2 + X_L^2} = \frac{(R - jX_L)}{Z^2}$$

$$= \frac{R}{Z^2} - \frac{jX_L}{Z^2}$$

$$= G - jB$$

where $G = \dfrac{R}{Z^2}$, the conductance of the circuit

$B = \dfrac{X_L}{Z^2}$, the inductive susceptance of the circuit.

SYMBOLIC NOTATION

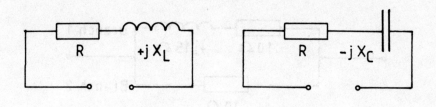

Fig. 1.27   Fig. 1.28

Similarly for the resistive capacitive circuit in Fig. 1.28;

$$Y = \frac{1}{(R - jX_C)} \text{ S}$$

$$= \frac{(R + jX_C)}{R^2 + X_C^2} = \frac{(R + jX_C)}{Z^2}$$

$$= \frac{R}{Z^2} + \frac{jX_C}{Z^2}$$

$$= G + jB$$

where $G = \frac{R}{Z^2}$, the circuit conductance, as in the inductive case

$B = \frac{X_C}{Z^2}$, the capacitive susceptance of the circuit.

Notice that the sign of the j term is reversed when considering admittance as opposed to impedance.

Admittance is particularly useful where circuits have several parallel arms as the next two worked examples will show.

<u>Worked example 1.30</u>   For the circuit shown in Fig. 1.29, determine (a) the total admittance, (b) the total circuit current when the supply voltage is 25 V.

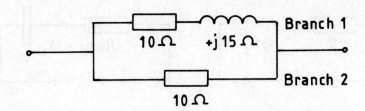

Fig. 1.29

Branch 1: $Y_1 = \dfrac{1}{(10 + j15)} = \dfrac{(10 - j15)}{10^2 + 15^2} = \dfrac{10}{325} - j\dfrac{15}{325}$

$= (0.031 - j0.046)$ S.

Branch 2: $Y_2 = \dfrac{1}{10} = 0.1$ S.

Now since $\dfrac{1}{Z_{eq}} = \dfrac{1}{Z_1} + \dfrac{1}{Z_2}$ and $Y = \dfrac{1}{Z}$

$Y_{eq} = Y_1 + Y_2 = (0.031 - j0.046) + 0.1$
$= (0.131 - j0.046)$ S.

Also $I = \dfrac{V}{Z_{eq}} = V\left(\dfrac{1}{Z_{eq}}\right) = VY_{eq}$, hence

$I = 25(0.131 - j0.046) = (3.27 - j1.154)$ A
$= 3.47\underline{/-19.4^o}$ A.

<u>Worked example 1.31</u> For the circuit shown in Fig. 1.30, determine (a) the total conductance, (b) the total susceptance, (c) the total admittance in polar form, (d) the current drawn from the supply, and (e) the circuit impedance.

SYMBOLIC NOTATION

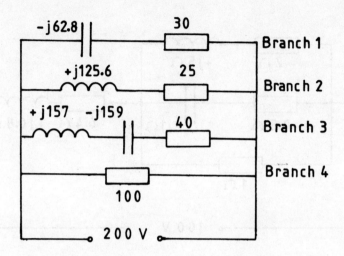

Fig. 1.30

Branch 1: $Y_1 = \dfrac{1}{(30 - j62.8)} = \dfrac{(30 + j62.8)}{30^2 + 62.8^2} = 0.0062 + j0.0129$

Branch 2: $Y_2 = \dfrac{1}{(25 + j125.6)} = \dfrac{(25 - j125.6)}{25^2 + 125.6^2} = 0.0015 - j0.0076$

Branch 3: $Y_3 = \dfrac{1}{40 + j157 - j159} = \dfrac{(40 + j2)}{40^2 + 2^2} = 0.0249 + j0.0012$

Branch 4: $Y_4 = \dfrac{1}{100} = 0.01$

 Adding: $\quad Y_{total} = 0.0426 + j0.0065$

Answers

(a) $G = 0.0426$ S.
(b) $B = 0.0065$ S (+j, so that the circuit overall is capacitive).
(c) $Y = (0.0426 + j0.0065) = 0.043\underline{/8.4^\circ}$ S in the polar form.
(d) $I = VY = 200 \times 0.043\underline{/8.4^\circ} = 8.6\underline{/8.4^\circ}$ A.
(e) $Z = \dfrac{1}{Y}$
$ = \dfrac{1}{0.043\underline{/8.4^\circ}} = 23.26\underline{/-8.4^\circ}\ \Omega$
$\phantom{(e) Z = \dfrac{1}{0.043}} = (23 - j3.4)\ \Omega.$

ELECTRICAL PRINCIPLES FOR HIGHER TEC

Self-assessment example 1.32

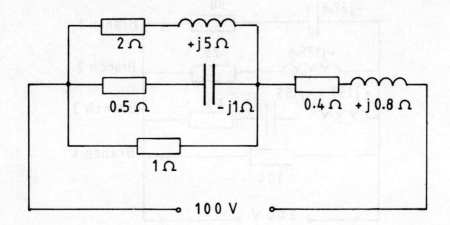

Fig. 1.31

For the circuit shown in Fig. 1.31, calculate (a) the admittance of the three parallel arms, (b) the equivalent impedance of the three parallel arms, (c) the total impedance of the circuit, (d) the circuit current.

1.13   POWER IN a.c. CIRCUITS
1.13.1 Resistive inductive circuit

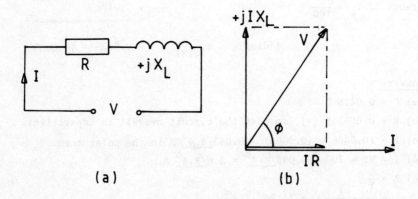

Fig. 1.32

SYMBOLIC NOTATION

In the RL circuit shown in Fig. 1.32(a), the current lags the voltage by an angle ϕ as shown in the phasor diagram Fig. 1.32(b).

$$v = V_m \sin(\omega t + \phi) \text{ volts}$$
$$i = I_m \sin \omega t \text{ amperes.}$$

The instantaneous power, p, in a circuit is the product of instantaneous voltage and current.

$$p = vi$$
$$= V_m \sin(\omega t + \phi) \times I_m \sin \omega t \text{ W.} \quad (1.1)$$

Using the double-angle formulae:

$$\cos(A - B) = \cos A \cos B + \sin A \sin B \quad (1.2)$$
$$\cos(A + B) = \cos A \cos B - \sin A \sin B \quad (1.3)$$

subtract equation 1.3 from equation 1.2

$$\cos(A - B) - \cos(A + B) = 2\sin A \sin B$$
$$\text{Hence, } \sin A \sin B = \tfrac{1}{2}(\cos(A - B) - \cos(A + B)) \quad (1.4)$$

Let $A = (\omega t + \phi)$ and $B = \omega t$ in equation 1.1, which gives

$$\text{instantaneous power} = V_m \sin A \times I_m \sin B$$
$$= \tfrac{1}{2} V_m I_m (\cos(A - B) - \cos(A + B))$$
$$(\text{From equation (1.4)})$$

or in terms of $(\omega t + \phi)$ and ωt

$$\text{instantaneous power} = \tfrac{1}{2} V_m I_m (\cos(\omega t + \phi - \omega t) - \cos(\omega t + \phi + \omega t))$$
$$= \tfrac{1}{2} V_m I_m (\cos \phi - \cos(2\omega t + \phi)) \text{ W.}$$

Now ϕ is constant for a particular circuit so that $\tfrac{1}{2} V_m I_m \cos \phi$ is constant throughout the cycle. The mean value of $\tfrac{1}{2} V_m I_m \cos(2\omega t + \phi)$ is zero over any number of complete cycles since this produces equal areas above and below the time axis. Both these quantities are shown in Fig. 1.33(a).

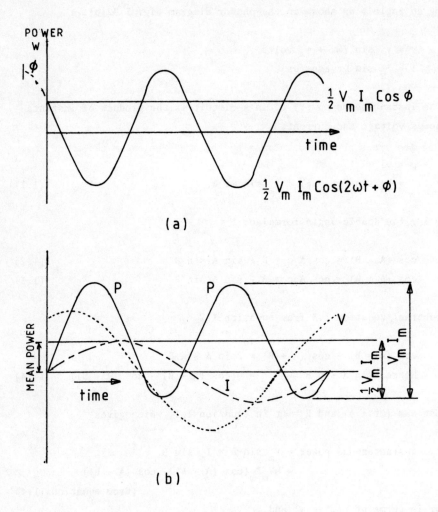

Fig. 1.33

The power curve P in Fig. 1.33(b) is derived by adding $-\frac{1}{2}V_m I_m \cos(2\omega t + \phi)$ to the constant term $\frac{1}{2}V_m I_m \cos \phi$. The voltage and current waves have also been added.

The power to the circuit can be seen to be a pulsating quantity. It pulsates at twice the supply frequency. The areas of the power curve above the time axis represent power supplied to the circuit, while the small lobes beneath the axis represent power being returned

SYMBOLIC NOTATION

to the supply from the circuit inductance as the magnetic field collapses. The difference between the areas above and below the time axis represents the heat loss due to the circuit resistance.

The mean or average value of the difference between the power supplied and the power returned is $\frac{1}{2}V_m I_m \cos\phi$ watts.

Figure 1.33 has been drawn for $\phi = 60°$. The shapes are typical however for all lagging phase angles, the power curve remaining a sinusoid but moving downward with respect to the time axis so decreasing the areas above and increasing the areas below as the phase angle increases. At $\phi = 90°$, the areas above and below are equal, so that no power is dissipated in the circuit. Now since

$$\frac{V_m}{\sqrt{2}} = V \text{ (r.m.s.)} \quad \text{and} \quad \frac{I_m}{\sqrt{2}} = I \text{ (r.m.s.)}$$

$$VI = \frac{V_m I_m}{\sqrt{2}\sqrt{2}} = \frac{1}{2}V_m I_m.$$

The mean power is therefore given by

$$P = VI \cos\phi \text{ watts} \qquad 1.5)$$

(mean power = modulus of V × modulus of I × cosine of the angle between them, V and I being r.m.s. values).

1.13.2 The resistive capacitive circuit

In an RC circuit the voltage lags the current by an angle ϕ as shown in Fig. 1.34. The mean power may be deduced in the same manner as for the RL circuit but using the angle $-\phi$ throughout. This results in the equation

$$\text{mean power} = VI \cos(-\phi) \text{ watts}$$

but since $\cos(-\phi) = \cos\phi$, the result is as before.

In Fig. 1.33, interchanging the voltage and current curves produces an identical power curve with areas above and below the time axis. The areas below the axis represent power being returned to the supply from the charged capacitor.

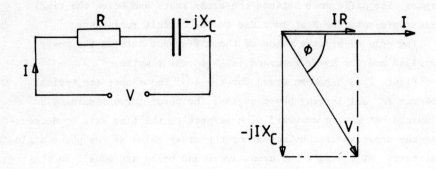

Fig. 1.34

1.13.3 Resistance only

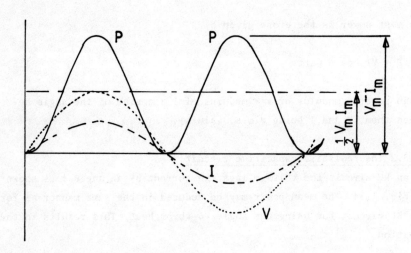

Fig. 1.35

Using the formula

$$P = VI \cos \phi \; W$$

For resistance only, $\phi = 0$, hence

SYMBOLIC NOTATION

P = VI W.

The curves of voltage, current and power are shown in Fig. 1.35.

1.14 WATTS, VOLT-AMPERES AND VOLT-AMPERES REACTIVE (W, VA and VA_R)

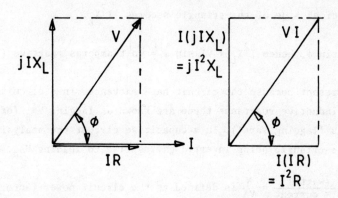

Fig. 1.36

Consider the RL circuit and its phasor diagram in Fig. 1.32. The phasor diagram is reproduced in Fig. 1.36(a). It has been shown that

$$\text{power} = VI \cos \phi \text{ W.} \qquad \text{(equation 1.5)}$$

In Fig. 1.36(a),

$$\frac{IR}{V} = \cos \phi.$$

Transposing gives

$$IR = V \cos \phi \qquad (1.6)$$

Substituting 1.6 into 1.5 gives

$$\text{power} = I \times IR = I^2 R \text{ W.}$$

Since IR and I are in phase, the circuit power may be determined knowing only the modulus of the current and the circuit resistance. Multiplying each of the quantities in Fig. 1.36(a) by I results in the diagram shown in Fig. 1.36(b).

The base = I^2R = circuit power (W).
The hypotenuse = VI = volt-amperes (VA).
The vertical side of the triangle becomes jI^2X_L.

$\dfrac{I^2X_L}{VI} = \sin\phi$, hence $I^2X_L = VI \sin\phi$ = volt-amperes reactive (VA_R).

These are present because the circuit has reactance. In a circuit possessing inductive reactance these are known as lagging VA_R (often spoken of as 'lagging vars'). In a capacitive circuit the analysis is similar, the triangle being inverted giving rise to leading VA_R.

$\dfrac{\text{power (watts)}}{\text{voltage} \times \text{current}} = \dfrac{W}{VA}$ is defined as the circuit power factor.
(1.7)

Phasors are used to represent sinusoidal quantities so that in Fig. 1.36

$\dfrac{W}{VA} = \dfrac{I^2R}{VI} = \cos\phi$.

For other waveforms the definition of power factor is still applicable but since non-sinusoidal waveforms comprise a series of different sine waves, no single angle is involved and $\cos\phi$ cannot be used. (See further work on complex waveforms, chapter 3.)

Now the two methods of power determination developed so far will be examined using the circuit shown in Fig. 1.37.

Worked example 1.33

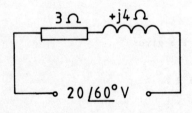

Fig. 1.37

SYMBOLIC NOTATION

Determine the power in the circuit shown in Fig. 1.37 using
(a) $P = VI \cos \phi$ W, (b) $P = I^2 R$ W.

The impedance, $Z = (3 + j4) = 5\underline{/53.13^\circ}$ Ω.

$$I = \frac{V}{Z} = \frac{20\underline{/60^\circ}}{5\underline{/53.13^\circ}} = 4\underline{/6.87^\circ} \text{ A.}$$

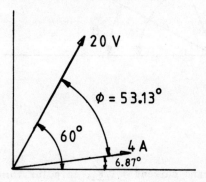

Fig. 1.38

The modulus of the current is 4 A. The modulus of the voltage is 20 V. The angle between them is 53.13°, which is the impedance angle. (Note that in Fig. 1.38 the impedance angle is the angle between the voltage and current phasors.)

(a) Power = $VI \cos \phi$ = 20 × 4 × cos 53.13° = 48 W.
(b) Power = $I^2 R$ = 4^2 × 3 = 48 W ((modulus of current)^2 × resistance).

1.15 DETERMINATION OF POWER USING COMPLEX NUMBERS

If an attempt is made to calculate the power in an a.c. circuit as a VI product, one of the limitations of using complex numbers will be discovered. Almost invariably it is best to determine the power developed in an alternating current circuit by using the formula: $P = I^2 R$ watts.

This section is introduced to make the reader aware of the pitfalls of indiscriminately multiplying voltage by current believing that the outcome will be the value of the circuit power in every case.

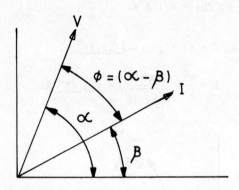

Fig. 1.39

In Fig. 1.39,

voltage = $V\underline{/\alpha^\circ}$, and current = $I\underline{/\beta^\circ}$, in polar form.

The product, voltage × current = $V\underline{/\alpha}\ I\underline{/\beta} = VI\underline{/(\alpha + \beta)}$
$= VI(\cos(\alpha + \beta) + j\sin(\alpha + \beta))$.

However, power = $VI\cos(\alpha - \beta)$, and reactive volt-amperes = $VI\sin(\alpha - \beta)$,

so that the result obtained by multiplying the two complex quantities is invalid. By reversing the sign of one of the angles in the polar forms above, the power and reactive volt-amperes may be determined. Thus by changing $\underline{/\beta}$ to $\underline{/-\beta}$

$V\underline{/\alpha}\ I\underline{/-\beta} = VI\underline{/(\alpha - \beta)} = VI\underline{/\phi}$
$= VI\cos\phi + jVI\sin\phi$.
   (power)   (VA_R)

The same effect may be achieved by reversing one of the signs of the j terms when multiplying in the rectangular form.

Let $V = (a + jb)$ and $I = (x + jy)$.

SYMBOLIC NOTATION

Reversing the sign of the y term the product becomes

$$(a + jb)(x - jy) = ax - jay + jbx + by$$
$$= (ax + by) + j(bx - ay).$$
$$\text{(power)} \qquad (VA_R)$$

<u>Worked example 1.34</u>   Using the circuit in Fig. 1.37, determine the circuit power using both rectangular and polar forms of voltage and current.

$$V = 20\underline{/60^\circ} = (10 + j17.32) \text{ V}$$
$$I = 4\underline{/6.87^\circ} = (3.97 + j0.478) \text{ A (polar forms from worked example 1.33)}.$$

In polar form: reverse the sign of the current angle.

$$20\underline{/60^\circ} \times 4\underline{/-6.87^\circ} = 80\underline{/53.13^\circ} = 48 + j64$$

The power is 48 watts. (64 VA_R are also present but this result was not called for.)

In rectangular form: reverse the sign of the j term in the current expression.

$$(10 + j17.32)(3.97 - j0.478) = 39.72 - j4.78 + j68.78 - j^2 8.28$$
$$= 48 + j64$$

Again the power is 48 W.

Generally it is to be recommended that power be calculated as $I^2 R$ or $VI \cos \phi$. Worked example 1.34 is carried out so that the reader will not, in ignorance of the shortcomings of the method, simply multiply voltage and current in the complex form with the expectancy that the result will be power.

<u>Worked example 1.35</u>   Calculate the value of power developed in each arm of the circuit shown in Fig. 1.40.

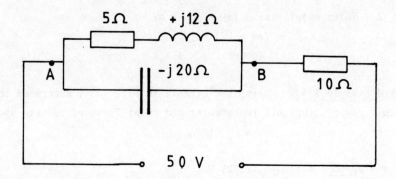

Fig. 1.40

Equivalent impedance of the two parallel arms,

$$Z_{eq} = \frac{(5 + j12)(-j20)}{5 + j12 - j20}$$

$$= \frac{13\underline{/67.4^o} \times 20\underline{/-90^o}}{9.43\underline{/-58^o}}$$

$$= 27.57\underline{/35.4^o}$$

$$= (22.47 + j15.97) \ \Omega.$$

Total impedance of the circuit = 22.47 + j15.97 + 10
$$= (32.47 + j15.97) = 36.2\underline{/26.2^o} \ \Omega.$$

Circuit current = $\frac{V}{Z} = \frac{50}{36.2\underline{/26.2^o}} = 1.38\underline{/-26.2^o}$ A.

Power developed in the 10 Ω resistor = I^2R = $1.38^2 \times 10$ = 19.04 W.

Potential difference developed across the 10 Ω resistor

$$= IR$$
$$= 1.38\underline{/-26.2^o} \times 10$$
$$= 13.8\underline{/-26.2^o}$$
$$= (12.38 - j6.1) \ V.$$

SYMBOLIC NOTATION

Potential difference across the two parallel arms (V_{AB} in Fig. 1.40) = supply voltage - voltage drop across the 10 Ω resistor

$= 50 - (12.38 - j6.1)$

$= (37.62 + j6.1) = 38.1\underline{/9.21^{\circ}}$ V.

Current in $(5 + j12)$ Ω impedance $= \dfrac{V_{AB}}{(5 + j12)} = \dfrac{38.1\underline{/9.21^{\circ}}}{13\underline{/67.4^{\circ}}}$

$= 2.93\underline{/-58.19^{\circ}}$ A.

Power developed in $(5 + j12)$ Ω impedance $= I^2R = 2.93^2 \times 5$

$= 42.9$ W.

No power is developed in the capacitive branch. Power is only developed in resistance.

<u>Self-assessment example 1.36</u>   Determine the power developed in each arm of the circuit shown in Fig. 1.41.

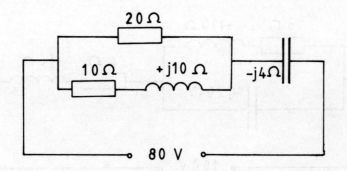

Fig. 1.41

<u>Further problems</u>

1.37 Perform the following calculations involving complex numbers expressing your answers in both polar and rectangular form.

(a) $\dfrac{(4 + j6)}{(10 - j8)}$,   (b) $\dfrac{25\underline{/60^{\circ}}}{10\underline{/-45^{\circ}}}$,   (c) $\dfrac{(100 + j60)}{50\underline{/45^{\circ}}}$,

(d) $65\underline{/85^{\circ}} \times 40\underline{/-10^{\circ}}$,   (e) $(5 + j20)(10 - j6)$,

(f) $(12 - j14) \times 50\underline{/25^\circ}$.

1.38. Determine the value of the circuit current in each of the following cases:

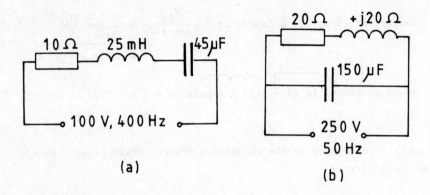

Fig. 1.42   Fig. 1.43

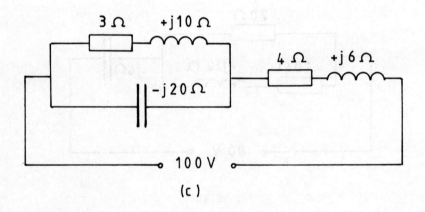

Fig. 1.44

SYMBOLIC NOTATION

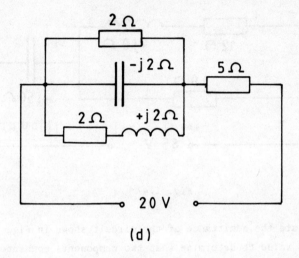

Fig. 1.45

1.39. Determine the power in each branch of the circuits in Figs. 1.46 and 1.47.

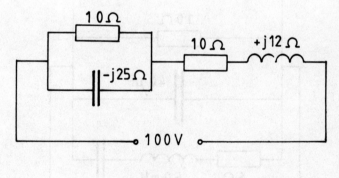

Fig. 1.46

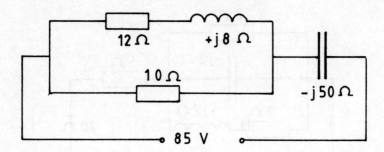

Fig. 1.47

1.40. Calculate the admittance of the circuit shown in Fig. 1.48.
Use this value to determine what two components connected in series could be used to replace the given combination of impedances. (Answer in ohms and farads or ohms and henrys.)

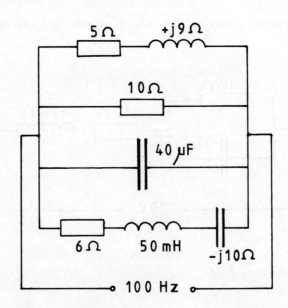

Fig. 1.48

1.41. For the circuit shown in Fig. 1.49, calculate (a) the circuit

SYMBOLIC NOTATION

current, (b) the total circuit power, (c) the power developed in the 20 Ω resistor, (d) the reactive volt-amperes developed in the (4 + j6) Ω impedance, and (e) the overall circuit power factor.

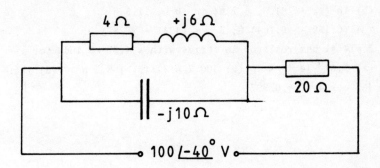

Fig. 1.49

CHAPTER 1   ANSWERS

1.4. (2.5 - j4.33)
1.7. 5/-53.1°
1.8. (a) 10.2/11.3°   (b) 7.2/-123.7°   (c) 9.85/156°
      (d) 13/-67.38°
1.9. (a) (2.18 + j24.9)   (b) -3.06 - j2.57)   (c) (-5.8 + j1.55)
      (d) (9.84 - j1.74)   (e) (3.62 - j13.5)
1.12. (a) 108.2/168.7°; (-106.1 + j21.2)   (b) 81.54/56.5°; (45 + j68)
1.16. 10.23/-35.75°; (8.3 - j5.98)
1.22. (a) 200.7 V   (b) 51.5°   (c) I = 3.14 A, leading V by 38.5°
1.24. Z = 100 Ω; |25 + j15.71 - jX$_C$| = 100; X$_C$ = 112.53 Ω;
      C = 28.29 μF
1.26. 15.17/-29.86° Ω; (13.16 - j7.55) Ω
1.29. I$_{total}$ = 1/-18.4° A; current in (10 + j30) Ω = 2.24/-44.97° A
1.32. (a) (1.47 + j0.63) S   (b) Z$_{eq}$ = (0.57 - j0.25) Ω
      (c) (0.97 + j0.55) Ω   (d) 89.7/-29.8° A; (77.8 - j44.6) A
1.36. 400 W; 400 W; zero
1.37. (a) 0.56/94.95°; (-0.048 + j0.558)   (b) 2.5/105°;
      (-0.65 + j2.4)   (c) 2.33/-14°; (2.26 - j0.56)
      (d) 2600/75°; (673 + j2511)   (e) 240.3/45°; (170 + j170)
      (f) 922/-24.4°; (839.6 - j380.8)

ELECTRICAL PRINCIPLES FOR HIGHER TEC

1.38. (a) $1.82\underline{/-79.5^{\circ}}$ A; $(0.33 - j1.79)$ A (b) $9.2\underline{/48.54^{\circ}}$ A; $(6.09 + j6.89)$ A (c) $3.68\underline{/-56.54^{\circ}}$ A; $(2.02 - j3.07)$ A (d) $3.22\underline{/3.69^{\circ}}$ A; $(3.2 + j0.207)$ A

1.39. (a) In $(10 + j12)$ Ω, 238.4 W; in 10 Ω, 205.7 W
 (b) In $(12 + j8)$ Ω, 6.7 W; in 10 Ω, 11.5 W

1.40. $Y = (0.159 - j0.104)$ S; $Z_{eq} = (4.41 + j2.87)$ Ω; $Z = 4.41$ ohm resistor in series with a 4.57 mH inductor

1.41. (a) $3.07\underline{/-44.4^{\circ}}$ A (b) 306.2 W (c) 188.5 W (d) 176.4 VA$_R$
 (e) $\cos 4.4^{\circ} = 0.997$

2 Network Theorems

The solution of problems involving several closed loops is possible using Kirchhoff's laws. However as the number of closed loops increases, the number of unknown currents increases and the solutions of the equations become successively more difficult. To assist in such solutions a number of circuit theorems and transforms have been evolved. These are applicable to both a.c. and d.c. circuits with linear response provided that in a.c. circuits due regard is taken of the phase of the a.c. supplies and the impedances are expressed in complex form. Linear response means that the circuit components have the same value of impedance irrespective of the magnitude and direction of the current in them. A pure resistor is an example of a linear device. An iron-cored coil is a non-linear device in that at a certain value of current it becomes magnetically saturated when its impedance falls.

In the succeeding sections the following will be considered:

(a) the superposition theorem
(b) Thevenin's theorem
(c) Norton's theorem
(d) The star/delta and delta/star transformations.

One further theorem, (e), is the maximum power transfer theorem. This concerns the conditions to be satisfied in order that the maximum amount of power is transferred from a source of e.m.f., possibly through a transfer network, into a load.

2.1 THE SUPERPOSITION THEOREM

In a network supplied by more than one source of e.m.f., each e.m.f. produces the same effect as if it were acting alone, all other e.m.f.s meanwhile being replaced by their respective internal impedances.

This means that in a network supplied by several sources of e.m.f. the current can be calculated in each branch of the network due to one of the e.m.f.s with all the other e.m.f.s removed, leaving only their internal impedances. This will be repeated for each of the e.m.f.s in turn. Finally the individual currents in each branch are added algebraically to give the actual branch current. The word 'superposition' comes from superimpose which is adding one current to another in the same conductor. Worked example 2.1 should make the method clear.

Worked example 2.1 In Fig. 2.1 two sources of e.m.f. E_1 and E_2 are connected in parallel to supply current to a 3 Ω load resistor. Source E_1 has an e.m.f. of 18 V and an internal resistance of 1 Ω. Source E_2 has an e.m.f. of 27 V and an internal resistance of 2 Ω and is connected in series with a 4 Ω resistor. The two e.m.f.s are in phase. Determine the value of current in each branch of the circuit and the potential difference developed across the 3 Ω load resistor.

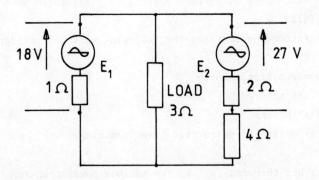

Fig. 2.1

Since each source of e.m.f. acts as though it were alone, the circuit can be considered firstly as in Fig. 2.2(a) and then as in Fig. 2.2(b).

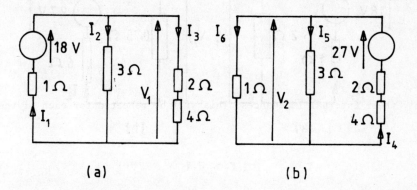

Fig. 2.2

Using Kirchhoff's first law:

In Fig. 2.2(a) $I_1 = I_2 + I_3$ and in Fig. 2.2(b) $I_4 = I_5 + I_6$.

The circuit in Fig. 2.2(a) may be further simplified by determining the equivalent resistance of 3 Ω and (4 + 2) Ω in parallel.

$$R_{eq} = \frac{3 \times 6}{3 + 6} = 2 \ \Omega.$$

Likewise for the circuit in Fig. 2.2(b), 3 Ω in parallel with 1 Ω

$$R_{eq} = \frac{3 \times 1}{3 + 1} = 0.75 \ \Omega.$$

The circuits then appear as in Fig. 2.3(a) and (b).

Considering the source E_1 alone, as in Fig. 2.3(a):

$$I_1 = \frac{18}{(1 + 2)} = 6 \ A.$$

$$V_1 = I_1 \times 2 = 6 \times 2 = 12 \ V.$$

In Fig. 2.2(a):

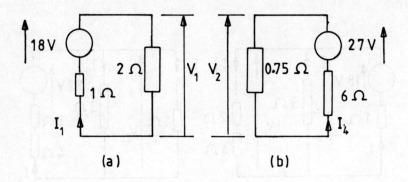

Fig. 2.3

$$I_2 = \frac{V_1}{3} = \frac{12}{3} = 4 \text{ A.}$$

$$I_3 = \frac{V_1}{6} = \frac{12}{6} = 2 \text{ A.}$$

Considering the source E_2 alone, as in Fig. 2.3(b):

$$I_4 = \frac{27}{(6 + 0.75)} = 4 \text{ A.}$$

$$V_2 = I_4 \times 0.75 = 4 \times 0.75 = 3 \text{ V.}$$

In Fig. 2.2(b):

$$I_5 = \frac{V_2}{3} = \frac{3}{3} = 1 \text{ A.}$$

$$I_6 = \frac{V_2}{1} = \frac{3}{1} = 3 \text{ A.}$$

Recombining the two circuits to form the original once more, there are six currents to consider (Fig. 2.4).

Adding the currents in each limb with due regard to direction gives:

Current in source E_1 = 6 A (up) + 3 A (down) = <u>3 A (up)</u>.
Current in the load resistor = 4 A (down) + 1 A (down)
= <u>5 A (down)</u>.

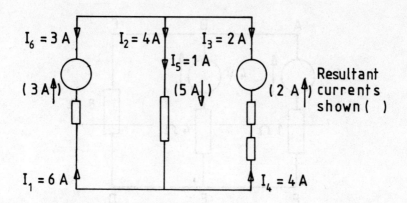

Fig. 2.4

Current in source E_2 = 2 A (down) + 4 A (up) = <u>2 A (up)</u>.

These results are shown in brackets in Fig. 2.4. Checking, it can be seen that the two source currents sum to give the current in the load resistor.

Finally, the potential difference developed across the 3 Ω load resistor is

IR = 5 × 3 = 15 V.

<u>Self-assessment example 2.2</u> For the circuit shown in Fig. 2.5, determine the current in each branch and the power developed in the 8 Ω resistor.

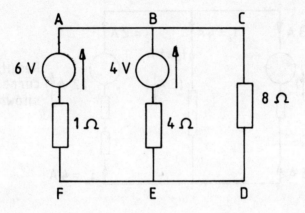

Fig. 2.5

2.2 THEVENIN'S THEOREM

Consider the circuit shown in Fig. 2.6. Two sources of e.m.f. feed into a network which has provision for feeding a load at terminals A and B. Any number of sources of e.m.f. may be involved and the network comprises any arrangement of linear components.

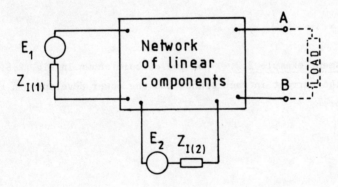

Fig. 2.6

Thevenin's theorem states that the whole circuit up to load terminals AB may be replaced by a single source of e.m.f. in series with a single impedance, as shown in Fig. 2.7.

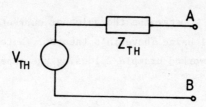

Fig. 2.7

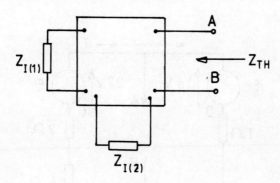

Fig. 2.8

The value of the single e.m.f., the Thevenin generator voltage V_{TH}, is the voltage which appears at the load terminals AB in the original network with any load removed. The value of the single impedance, Z_{TH}, is that of the internal impedance of the complete network and generators viewed from the terminals AB, the sources of e.m.f. being suppressed, leaving only their internal impedances. This is shown in Fig. 2.8.

The load impedance may be connected to the terminals A and B of the Thevenin equivalent circuit (Fig. 2.7) when the current will be given by:

$$I_{load} = \frac{V_{TH}}{Z_{TH} + Z_{load}} \text{ A.}$$

It is therefore only necessary to calculate or measure V_{AB} and the internal impedance in order to be able to determine the current which would flow in any load connected to AB.

Worked example 2.3   Determine the value of current in the 3 Ω resistor in Fig. 2.1 using Thevenin's theorem. (Note: the answer is already known from worked example 2.1 using the superposition theorem.)

The first step is to remove the load resistor (Fig. 2.9). There are 18 V attempting to drive current in a clockwise direction round the circuit and 27 V acting in the opposite direction.

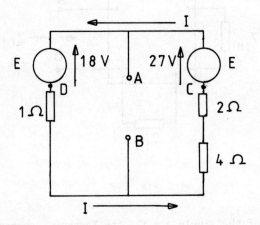

Fig. 2.9

The nett e.m.f. = 27 - 18 = 9 V (anti-clockwise).
The total resistance of the circuit = 1 + 2 + 4 = 7 Ω.
Hence I = $\frac{9}{7}$ = 1.286 A (anti-clockwise: see Fig. 2.9).

Consider point B and the bottom line in the diagram to be at reference potential. From B to C there will be a voltage drop of 1.286 × 6 = 7.714 V. Point C is therefore at a negative potential of 7.714 V with respect to B. From C to A the potential is raised by 27 V by source E_2. Hence the potential of A is -7.714 + 27 = 19.286 V above that of B. This is then the value of the Thevenin voltage.

As a check, continuing round the circuit, from A to D there is an opposing e.m.f. of 18 V so that point D is at a potential of 19.286 - 18 = 1.286 V above that of B. By virtue of the current flowing from point D back to the bottom line there is a voltage drop of 1.286 × 1 = 1.286 V, which brings us back to datum level. Now the internal impedance will be considered.

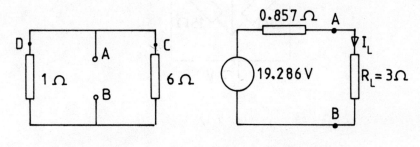

Fig. 2.10   Fig. 2.11

In Fig. 2.10 the e.m.f.s have been suppressed, leaving only their internal impedances. From point A to point B there are two parallel routes, one through D and 1 Ω and the other through C and 6 Ω.

The effective resistance of 6 Ω in parallel with 1 Ω = $\frac{6 \times 1}{6 + 1}$ = 0.875 Ω.

The Thevenin circuit may now be drawn as in Fig. 2.11 up to points A and B. The 3 Ω load resistor is added and the load current becomes:

$$I_L = \frac{19.286}{0.875 + 3} = 5 \text{ A}.$$

This agrees with the result obtained using the superposition theorem in worked example 2.1.

Worked example 2.4   In the bridge circuit shown in Fig. 2.12, determine the value of the current in the 40 Ω resistor using Thevenin's theorem. (The voltage source has negligible internal resistance.)

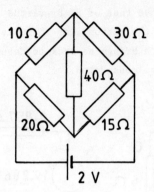

Fig. 2.12

The first step in the solution is to remove the resistor in which the current is to be found. The circuit is then as shown in Fig. 2.13.

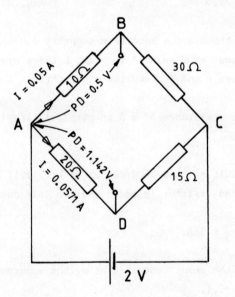

Fig. 2.13

To find the Thevenin voltage:

NETWORK THEOREMS

The current from the battery divides into two parts at point A.

The current in the top arms = $\frac{2}{10 + 30}$ = 0.05 A.

Hence the potential difference between A and B = IR = 0.05 × 10 = 0.5 V.

The current in the bottom arms = $\frac{2}{20 + 15}$ = 0.0571 A.

Hence the potential difference between A and D = IR = 0.0571 × 20 = 1.142 V.

The potential difference between B and D = $V_{AD} - V_{AB}$ = 1.142 - 0.5 = 0.642 V.

This is the Thevenin voltage. Since the voltage drop from A to B is less than that from A to D, point B is at a higher potential than point D and current will flow from B to D in the load resistor when it is re-connected.

Next the internal resistance of the network must be found. The voltage of the battery is suppressed and since it has negligible internal resistance, points A and C are effectively connected together. The circuit then becomes as shown in Fig. 2.14.

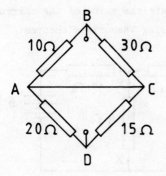 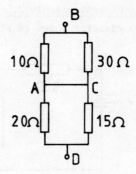

Fig. 2.14

The effective resistance of 30 Ω in parallel with 10 Ω =

$$\frac{30 \times 10}{30 + 10} = 7.5 \; \Omega.$$

The effective resistance of 20 Ω in parallel with 15 Ω =
$$\frac{20 \times 15}{20 + 15} = 8.57 \; \Omega.$$

The resistance from B to D = 7.5 + 8.57 = 16.07 Ω.

This is the Thevenin internal resistance.

The Thevenin circuit becomes as shown in Fig. 2.15 and reconnecting the load resistor results in a load current given by

$$I_L = \frac{0.642}{16.07 + 40} = 0.0115 \; A.$$

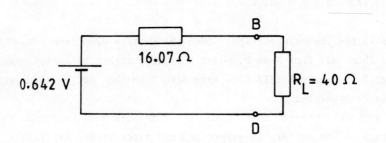

Fig. 2.15

Self-assessment example 2.5   Calculate the value of the current in the 8 Ω resistor shown in Fig. 2.16 using Thevenin's theorem.

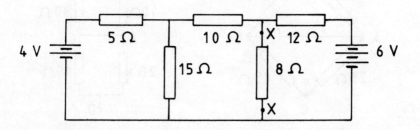

Fig. 2.16

(One method is to reduce the circuit to the left of XX to a Thevenin

NETWORK THEOREMS

equivalent when, adding it to that to the right of XX, the circuit becomes as in worked example 2.1.)

In the preceding examples only resistive circuits have been considered. Thevenin's theorem is now applied to a circuit containing reactance.

Worked example 2.6   Obtain the Thevenin equivalent circuit for the network shown in Fig. 2.17 and hence determine the value of current which would flow in an impedance of (60 + j300) Ω connected across AB.
What power will be dissipated in this impedance?

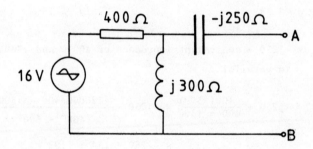

Fig. 2.17

Thevenin voltage:
Since there is no current in the capacitor with AB open-circuited, the potential difference across terminals AB is the same as that across the pure inductor, j300 Ω.

$$\text{Potential difference across the inductor} = \frac{16}{400 + j300} \times j300 \text{ volts}$$

$$= \frac{j4800(400 - j300)}{400^2 + 300^2}$$

$$= \frac{1.44 \times 10^6 + j1.92 \times 10^6}{250 \times 10^3}$$

Thevenin voltage = 5.76 + j7.68 V

Internal (Thevenin) impedance:
Suppressing the source e.m.f. gives the circuit shown in Fig. 2.18.

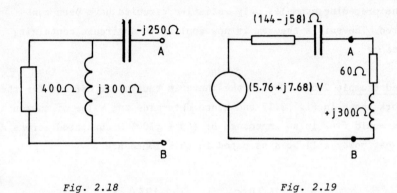

Fig. 2.18 Fig. 2.19

$Z_{internal}$ = -j250 + equivalent impedance of 400 Ω and j300 Ω in parallel

$$= -j250 + \frac{400(j300)}{400 + j300} = -j250 + \frac{j120000(400 - j300)}{400^2 + 300^2}$$

$$= -j250 + 144 + j192$$

Thevenin impedance = (144 - j58) Ω.

The Thevenin circuit together with the load impedance of (60 + j300) Ω is shown in Fig. 2.19.

$$\text{Current in the load impedance} = \frac{5.76 + j7.68}{(144 - j58) + (60 + j300)}$$

$$= \frac{5.76 + j7.68}{(204 + j242)}$$

$$= \frac{9.6 \underline{/53.13°}}{316.5 \underline{/49.87°}}$$

$$= 0.03 \underline{/3.26°} \text{ A}.$$

The current in the load has a modulus of 0.03 A and it leads the supply voltage by 3.26°.

$$\text{Power in load impedance} = I^2 R = (0.03)^2 \times 60 = 0.054 \text{ W}.$$

NETWORK THEOREMS

<u>Self-assessment example 2.7</u> By the use of Thevenin's theorem determine the value of current in a 46.74 Ω resistor connected to terminals AB in Fig. 2.20. The two sources of e.m.f. have outputs which are in phase.

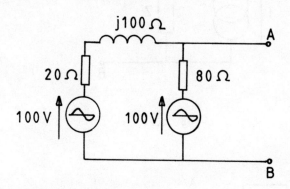

Fig. 2.20

2.3   NORTON'S THEOREM

Referring back to the network and sources of e.m.f. shown in Fig. 2.6, Norton's theorem states that the whole circuit up to the load terminals AB may be replaced by a single constant current source shunted by a single impedance as shown in Fig. 2.21. The value of current delivered by the current generator is that which would be delivered into a short-circuit across the terminals AB in the original network. The value of the single impedance is that of the internal impedance of the network, the sources of e.m.f. being suppressed leaving only their internal impedances, which is precisely the same as for Thevenin's theorem.

Thevenin circuits may be simply transformed into Norton circuits in the following manner. Starting with a Thevenin circuit as shown in Fig. 2.22, short-circuiting terminals AB results in a current given by:

$$I = \frac{V_{TH}}{Z_I}.$$

This is the Norton current. Since Z_I is the same for both theorems,

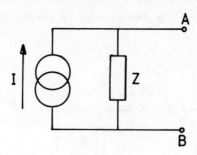

Fig. 2.21

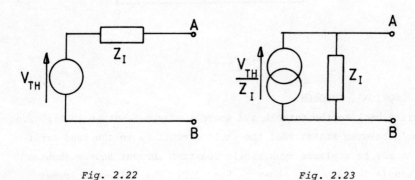

Fig. 2.22   *Fig. 2.23*

the Norton circuit may be drawn as in Fig. 2.23.

It may now be demonstrated that the two circuits are equivalent. Connecting a load impedance Z_L to the Thevenin circuit in Fig. 2.22 will result in a current in the load impedance of:

$$I = \frac{V_{TH}}{Z_I + Z_L} \text{ A}.$$

Now considering the Norton circuit in Fig. 2.23, connecting the same impedance across the terminals AB puts Z_L in parallel with Z_I.

The equivalent impedance $Z_{eq} = \dfrac{Z_I Z_L}{Z_I + Z_L}$.

NETWORK THEOREMS

Potential difference $V_{AB} = IZ_{eq} = \dfrac{V_{TH}}{Z_I}\left[\dfrac{Z_I Z_L}{Z_I + Z_L}\right]$

Current in $Z_L = \dfrac{V_{AB}}{Z_L} = \dfrac{V_{TH}}{Z_I}\left[\dfrac{Z_I Z_L}{Z_I + Z_L}\right]\dfrac{1}{Z_L}$

$= \dfrac{V_{TH}}{Z_I + Z_L}$ A as before.

<u>Worked example 2.8</u>   Determine, by the use of Norton's theorem, the value of current in the 10 Ω resistor connected to terminals AB of the network shown in Fig. 2.24. (The source has negligible internal resistance.)

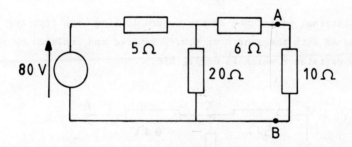

Fig. 2.24

Shorting terminals A and B effectively connects the 6 Ω resistor in parallel with the 20 Ω resistor and eliminates the 10 Ω resistor from the calculations.

The equivalent resistance of 20 Ω and 6 Ω in parallel
$= \dfrac{20 \times 6}{20 + 6} = 4.6$ Ω.

Total resistance of the circuit = 4.6 + 5 = 9.6 Ω.

Current delivered by the source = $\dfrac{V}{R_{total}} = \dfrac{80}{9.6} = 8.33$ A.

At junction X in Fig. 2.25, the current will divide between the 20 Ω and 6 Ω resistors.

$$I_{S.C.} = \frac{V_{XY}}{6} = \frac{8.33 \times 4.6}{6} = 6.39 \text{ A.}$$

This is the Norton current.

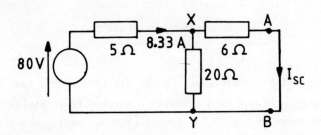

Fig. 2.25

The internal resistance is now needed looking back from the terminals AB with the source of e.m.f. removed and replaced by its internal resistance which is negligible.

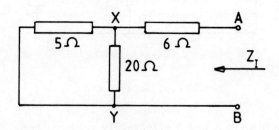

Fig. 2.26

In Fig. 2.26, looking left from terminals XY, 5 Ω is in parallel with 20 Ω.

$$Z_{eq} = \frac{20 \times 5}{20 + 5} = 4 \text{ Ω} \qquad Z_I = 6 + 4 = 10 \text{ Ω.}$$

The Norton circuit is as shown in Fig. 2.27

NETWORK THEOREMS

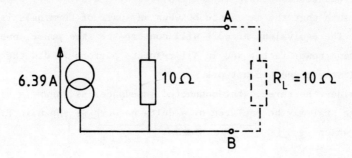

Fig. 2.27

With the load resistor of 10 Ω connected to AB, one half of the supply current flows in Z_I and the other half in Z_L.

Current in Z_L = 0.5 × 6.39 = 3.195 A.

Self-assessment example 2.9   Deduce the Norton equivalent to the circuit shown in Fig. 2.28.

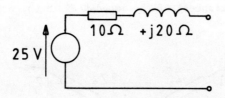

Fig. 2.28

2.4 THE STAR/DELTA AND DELTA/STAR TRANSFORMATIONS

Some networks and bridges are best solved using these transforms possibly followed by the application of Thevenin's theorem. An application of the star/delta transformation is in the solution of three-phase, star-connected load problems where only a three-wire supply is available. In this case, if the load is not balanced across the phases the star point voltage is difficult to calculate. With the delta-equivalent network this is not required.

ELECTRICAL PRINCIPLES FOR HIGHER TEC

2.4.1 The star/delta transformation

Any star-connected network may be replaced with a delta (π)-connected network such that the impedance between any pair of terminals is unchanged. The equivalent network will consume the same power, operate at the same power factor and in all respects perform as did the original star-connected network.

Consider the three, star-connected impedances as shown in Fig. 2.29. The star may be replaced by a delta network as shown in Fig. 2.30 in which:

$$Z_A = Z_1 Z_2 \left(\frac{1}{Z_1} + \frac{1}{Z_2} + \frac{1}{Z_3}\right), \quad Z_B = Z_2 Z_3 \left(\frac{1}{Z_1} + \frac{1}{Z_2} + \frac{1}{Z_3}\right), \quad \text{and}$$

$$Z_C = Z_1 Z_3 \left(\frac{1}{Z_1} + \frac{1}{Z_2} + \frac{1}{Z_3}\right).$$

To calculate the value of Z_A, first multiply together Z_1 and Z_2. These are the two impedances in the original star connected to the same terminals as Z_A, i.e. terminals 1 and 2. For Z_B, the terminals are 2 and 3 so that the product $Z_2 Z_3$ is required. Finally for Z_C, this is connected to terminals 1 and 3 so that the product $Z_1 Z_3$ is used.

In each case the product is multiplied by the sum of the reciprocals (sum of the admittances) of the impedances in the original star.

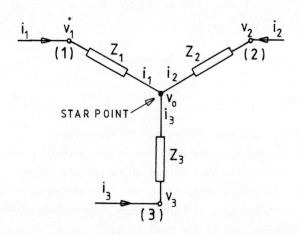

Fig. 2.29

NETWORK THEOREMS

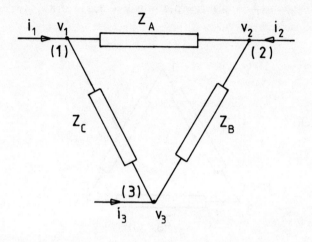

Fig. 2.30

This transformation may be extended for use with any number of impedances connected to a common point.

For a full proof of the transformation see appendix 1.

<u>Worked example 2.10</u> Deduce the equivalent delta to the star given in Fig. 2.31.

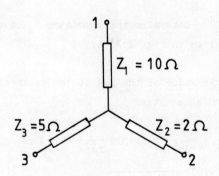

Fig. 2.31

Firstly evaluate

$$\frac{1}{Z_1} + \frac{1}{Z_2} + \frac{1}{Z_3} = \frac{1}{10} + \frac{1}{5} + \frac{1}{2} = 0.1 + 0.5 + 0.2 = 0.8$$

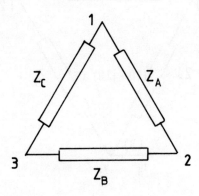

Fig. 2.32

Z_A = 0.8 × (product of the two resistances connected to terminals 1 and 2 in the original star)

Z_A = 0.8 × 10 × 2 = 16 Ω.

Similarly Z_B = 0.8 × 5 × 2 = 8 Ω
and Z_C = 0.8 × 5 × 10 = 40 Ω.

<u>Worked example 2.11</u> Determine the impedance Z_A in the equivalent delta to the star given in Fig. 2.33(a).

Firstly determine the value of the sum of the reciprocals of the impedances in the original star.

$$\frac{1}{20} + \frac{1}{(10 + j10)} + \frac{1}{-j25} \left(\text{multiply } \frac{1}{-j25} \text{ by } \frac{j}{j}\right)$$

$$= 0.05 + \frac{(10 - j10)}{100 + 100} + j0.04$$

$$= 0.05 + 0.05 - j0.05 + j0.04$$

$$= 0.1 - j0.01$$

NETWORK THEOREMS

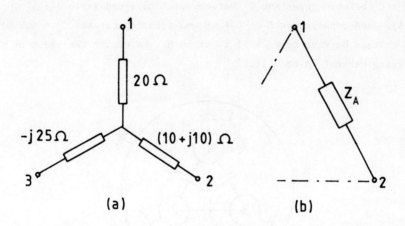

Fig. 2.33

$Z_A = 20 \times (10 + j10) \times (0.1 - j0.01)$
$= 20 \times 14.14\underline{/45°} \times 0.1005\underline{/-5.71°}$
$= 28.42\underline{/39.29°}$
$= (22 + j18)\ \Omega.$

<u>Self-assessment example 2.12</u> Convert the star given in Fig. 2.34 into an exact equivalent delta.

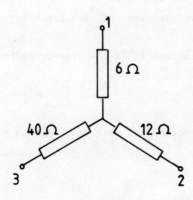

Fig. 2.34

71

Worked example 2.13   A three-phase, three-core cable has capacitance C_c between cores and C_s between each core and earth (Fig. 2.35). Each capacitance C_c = 0.1 µF and each capacitance C_s = 0.2 µF. The voltage between cores = 11 kV at 50 Hz. Determine the value of charging current in each line.

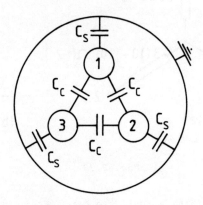

Fig. 2.35

The capacitances C_s are effectively connected in star since they are formed between each core and a common point, namely earth. One method of solving the problem is to convert the capacitive reactances due to capacitances C_s into an equivalent delta when they can be added to those due to capacitances C_c.

Let the capacitive reactance of each capacitor C_s be X_{C_s} Ω.

The sum of the reciprocals of the star-connected reactances

$$= 3\left[\frac{1}{X_{C_s}}\right].$$

The capacitive reactance of each delta branch $= X_{C_s} \times X_{C_s} \times 3\left[\frac{1}{X_{C_s}}\right]$

$$= 3X_{C_s}.$$

For a capacitor to have three times the reactance of the original capacitance, C_s, it must have a capacitance equal to one third of C_s. Hence each branch of the delta equivalent has the value of $\frac{1}{3}C_s$. In

NETWORK THEOREMS

Fig. 2.36 this is shown together with C_c to make up the total capacitance between one pair of cores.

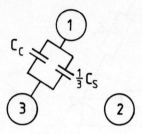

Fig. 2.36

The total capacitance between cores = $C_c + \frac{1}{3}C_s = \left(0.1 + \frac{0.2}{3}\right)$ μF

= 0.1667 μF.

Capacitive reactance, $X_c = \frac{1}{\omega C} = \frac{1}{2\pi \times 50 \times 0.1667 \times 10^{-6}} = 19099$ Ω.

Since the cable is symmetrical, each pair of cores will yield a similar result.

The line current in conductor 1 will be the sum of the currents due to the capacitances between conductor 1 and conductor 2 and between conductor 1 and conductor 3. From previous studies on balanced, three-phase, delta-connected systems it should be remembered that the line current $I_L = \sqrt{3} \times$ (phase current). The currents I_{12} and I_{13} are phase currents in such a system so that the current in each line is $\sqrt{3}\,I_{12}$ amperes.

$$I_{12} = \frac{V_{line}}{X_C} = \frac{11000}{19099} = 0.576 \text{ A}.$$

Line charging current = $\sqrt{3} \times 0.576 = 0.997$ A.

2.4.2 The delta/star transformation

Any delta-connected network may be replaced by a star-connected network which is exactly electrically equivalent.

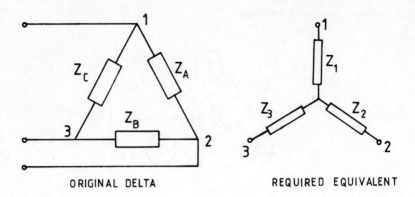

Fig. 2.37

In Fig. 2.37:

$$Z_1 = \frac{Z_A Z_C}{Z_A + Z_B + Z_C} \quad Z_2 = \frac{Z_A Z_B}{Z_A + Z_B + Z_C} \quad \text{and} \quad Z_3 = \frac{Z_B Z_C}{Z_A + Z_B + Z_C}.$$

The branches of the equivalent star are found by multiplying together the two impedances in the original delta connected to the relevant terminal and then dividing by the sum of the delta-connected impedances. (Compare this with the result of the star/delta transform and take particular note that $\frac{1}{Z_A + Z_B + Z_C}$ is NOT the same as the sum of the individual reciprocals.)

For a proof of the transform see appendix 2.

Worked example 2.14   Determine the value of current in the 40 Ω resistor in the Wheatstone bridge arrangement shown in Fig. 2.38 using the delta/star transformation. (Note: this current has already been deduced using Thevenin's theorem in worked example 2.4 so it will be interesting to compare the complexity of the solutions and indeed to see whether the same result is achieved.)

First the values of the impedances must be calculated which form an equivalent star to the delta ABD in Fig. 2.38. Let Z_1 be connected to A, Z_2 to B, and Z_3 to D.

The sum of the three delta impedances = 10 + 20 + 40 = 70 Ω

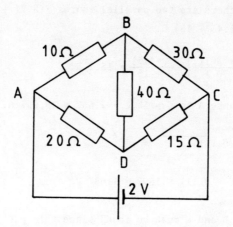

Fig. 2.38

Therefore: $Z_1 = \dfrac{10 \times 20}{70} = 2.857\ \Omega$

$Z_2 = \dfrac{10 \times 40}{70} = 5.71\ \Omega$

$Z_3 = \dfrac{20 \times 40}{70} = 11.43\ \Omega.$

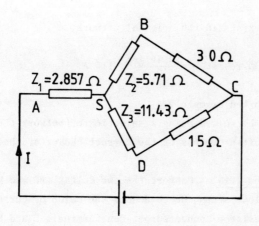

S = STAR POINT

Fig. 2.39

The circuit is redrawn as in Fig. 2.39.

From S to C there are two parallel routes: $(5.71 + 30)$ Ω in parallel with $(11.43 + 15)$ Ω.

$$Z_{eq} = \frac{(5.71 + 30)(11.43 + 15)}{5.71 + 30 + 11.43 + 15} = 15.2 \text{ Ω}.$$

Total resistance A to C = $15.2 + 2.857 = 18.057$ Ω.

$$I = \frac{2}{18.057} = 0.111 \text{ A}.$$

Hence p.d. SC = $0.111 \times 15.2 = 1.687$ V.

The p.d. between B and D must be found since this p.d. exists across the same two points and the 40 Ω resistor in the original network.

Current in top branch SBC = $\frac{1.687}{35.71} = 0.0472$ A.
P.d. SB = $0.0472 \times 5.71 = 0.27$ V.

Current in bottom branch SDC = $\frac{1.687}{26.43} = 0.064$ A.
P.d. SD = $0.064 \times 11.43 = 0.73$ V.

Therefore p.d. BD = $0.73 - 0.27 = 0.46$ V.

Since this p.d. also exists in the original network,

$$\text{current in the 40 Ω resistor} = \frac{0.46}{40} = 0.0115 \text{ A}.$$

This is as found in worked example 2.4.

Most problems can be solved using a selection of network theorems. The ease of solution often depends on the correct choice of theorem.

<u>Self-assessment problem 2.15</u>   Convert the two deltas ABC and DEF in Fig. 2.40 into equivalent stars. Hence deduce the value of current flowing in the 10 Ω resistor connected between terminals C and E.

NETWORK THEOREMS

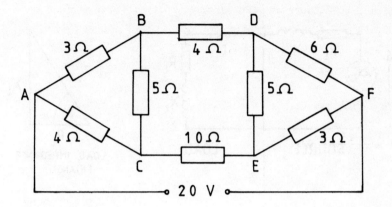

Fig. 2.40

2.5 THE MAXIMUM POWER TRANSFER THEOREM

When considering equipment supplied with power from the mains it is never necessary to consider the conditions for maximum power transfer. The power supply is treated as being infinitely great compared with the demand being made, and an impedance to suit the user rather than to suit the supply network is connected.

However when a small source of power is involved, such as the output from an audio-amplifier or a telephone system, it is often important that an impedance of the correct magnitude, and where possible with the correct phase angle, be used in order that the maximum use of the available power is made by the load.

The conditions for maximum power transfer into an impedance with fixed phase angle will now be considered, followed by a discussion as to the effects of variable phase angle.

Consider the source shown in Fig. 2.41 with internal impedance $(R_I + jX_I)$ Ω supplying a load with impedance $(R_L + jX_L)$ Ω, which may be varied in magnitude while maintaining the phase angle constant. This would be the case if the load comprised a tapped coil, a variable wire-wound rheostat or a fixed impedance being fed through a variable-ratio transformer.

From the impedance triangle for the load:

77

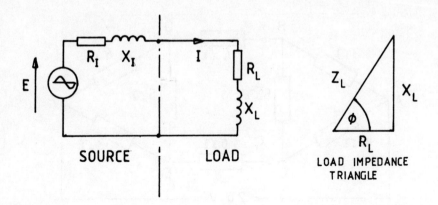

Fig. 2.41

$|Z_L| = \sqrt{(R_L^2 + X_L^2)}, \quad R_L = Z_L \cos \phi, \quad X_L = Z_L \sin \phi.$

The total circuit resistance $= R_I + R_L = R_I + Z_L \cos \phi.$

The total circuit reactance $= X_I + X_L = X_I + Z_L \sin \phi.$

Hence, total circuit impedance $= \sqrt{[(R_I + Z_L \cos \phi)^2 + (X_I + Z_L \sin \phi)^2]}.$
$$(2.1)$$

Load power $= I^2 R$ and $I = \dfrac{E}{Z}$

Therefore,

$$\text{load power} = \left\{ \dfrac{E}{\sqrt{[(R_I + Z_L \cos \phi)^2 + (X_I + Z_L \sin \phi)^2]}} \right\}^2 \times Z_L \cos \phi.$$

Squaring the denominator eliminates the root sign and multiplying out the brackets yields:

$$\text{power} = \dfrac{E^2 Z_L \cos \phi}{R_I^2 + 2R_I Z_L \cos \phi + Z_L^2 \cos^2 \phi + X_I^2 + 2X_I Z_L \sin \phi + Z_L^2 \sin^2 \phi}$$

Divide numerator and denominator by Z_L and collect $\cos^2 \phi$ and $\sin^2 \phi$.

NETWORK THEOREMS

$$\text{Power} = \frac{E^2 \cos \phi}{\frac{R_I^2}{Z_L} + 2R_I \cos \phi + \frac{X_I^2}{Z_L} + 2X_I \sin \phi + Z_L(\cos^2 \phi + \sin^2 \phi)}.$$

But $\cos^2 \phi + \sin^2 \phi = 1$, hence

$$\text{power} = \frac{E^2 \cos \phi}{\frac{R_I^2}{Z_L} + 2R_I \cos \phi + \frac{X_I^2}{Z_L} + 2X_I \sin \phi + Z_L} \text{ watts}$$

For the power to be a maximum, the denominator (= y, say) must be made into a minimum. $|Z_L|$ is the only variable, E, cos ϕ, R_I and X_I all being constant. Differentiating the denominator with respect to Z_L gives

$$\frac{dy}{dZ_L} = -\frac{R_I^2}{Z_L^2} - \frac{X_I^2}{Z_L^2} + 1$$

$\frac{d^2 y}{dZ_L^2}$ is positive which indicates a minimum value.

Equating $\frac{dy}{dZ_L}$ to zero gives

$$-\frac{R_I^2}{Z_L^2} - \frac{X_I^2}{Z_L^2} + 1 = 0$$

Transposing gives

$$Z_L^2 = R_I^2 + X_I^2$$

$$|Z_L| = \sqrt{(R_I^2 + X_I^2)}$$

Therefore, maximum power transfer is achieved when the modulus of the load impedance is equal to the modulus of the source impedance.

<u>Worked example 2.16</u>   For the circuit shown in Fig. 2.42 calculate:

(a) the value of load impedance to give maximum power transfer,
(b) the value of load resistance, (c) the power developed in the load, (d) the power loss in the source.

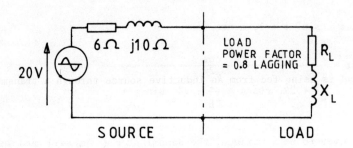

Fig. 2.42

(a) For maximum power transfer $|Z_L| = \sqrt{(R_I^2 + X_I^2)} = \sqrt{(6^2 + 10^2)}$

$$= 11.66 \, \Omega.$$

For power factor 0.8, $\cos \phi = 0.8$, hence $\phi = 36.87°$. (Fig. 2.41.)

(b) $R_L = Z_L \cos \phi = 9.33 \, \Omega.$
$X_L = Z_L \sin \phi = 7 \, \Omega.$

The total impedance of the circuit $= \sqrt{[(R_I + R_L)^2 + (X_I + X_L)^2]}$
(equation 2.1)
$= \sqrt{[(6 + 9.33)^2 + (10 + 7)^2]}$
$= 22.9 \, \Omega.$

Therefore the circuit current $I = \dfrac{V}{Z} = \dfrac{20}{22.9} = 0.873$ A.

(c) Power in the load $= I^2 R_L = 0.873^2 \times 9.33 = 7.12$ W.
(d) Power loss in the source $= I^2 R_I = 0.873^2 \times 6 = 4.57$ W.

Three other possibilities will now be considered:
(a) Where the load is purely resistive.
Maximum power transfer will occur when $R_L = |Z_I|$.

NETWORK THEOREMS

(b) Where the load comprises a variable resistance in series with a fixed inductive reactance X_F Ω, maximum power transfer will occur when

$$R_L = \sqrt{[(R_I)^2 + (X_I + X_F)^2]}.$$

The load resistance has to be made equal to the modulus of the impedance of the rest of the circuit. It could be envisaged that a resistive load is being fed from an inductive source through a transmission line or network which is also inductive. The load resistance would then be made equal to the modulus of the impedance of the rest of the system complete.

(c) Where a completely free choice may be made, the maximum possible power transferred into the load is obtained when the load impedance is the complex conjugate of the impedance of the source. For example, if the source has impedance $(R_I + jX_I)$ Ω, then maximum power transfer occurs when the load impedance is $(R_I - jX_I)$ Ω.

Total circuit resistance = $2R_I$ Ω.
With an applied e.m.f. of E volts, $I = \dfrac{E}{2R_I}$ A.

$$\text{Load power} = I^2 R_I = \left(\frac{E}{2R_I}\right)^2 R_I$$

$$= \frac{E^2}{4R_I} \text{ watts.} \tag{2.2}$$

With all the other conditions envisaged, the power transfer into the load is less than this but is the best that can be achieved under the constraints detailed. However there is a price to pay since such a circuit will be selective, i.e. it exhibits resonance which implies that this maximum power transfer occurs at one particular frequency only. Where a reasonable bandwidth is required, such as in audio work, this method cannot be used. It could be employed in circuits required to perform a single function, for example, closing a relay.

Worked example 2.17

(a) A 5 V, 50 Hz source has an internal impedance which is equivalent to a 600 Ω resistance in series with a 3.183 µF capacitor. Determine the load components required for maximum power transfer when the load is (i) variable in reactance and resistance, (ii) a variable resistance only.

(b) Determine the load power developed in (a) (i) and (ii) above.

(a) 3.183 μF at 50 Hz gives capacitive reactance,

$$X_C = \frac{1}{2\pi \times 50 \times 3.183 \times 10^{-6}}$$

$$= 1000 \ \Omega \ (-j1000 \ \Omega).$$

(i) Variable reactance allows tuning of the circuit.
A coil is required with inductive reactance 1000 Ω (+j1000 Ω)

$$L = \frac{1000}{2\pi \times 50} = 3.183 \ H.$$

Maximum power transfer occurs when the load comprises a 600 Ω resistor in series with a coil with inductance 3.183 H.

(ii) For resistance only,

$$R_L = Z_I$$

$$= \sqrt{(600^2 + 1000^2)}$$

$$= 1166 \ \Omega.$$

(b) (i) Power $= \dfrac{E^2}{4R_I}$ (equation 2.2)

$$= \frac{5^2}{4 \times 600} = 0.0104 \ W.$$

(ii) Circuit impedance $= \sqrt{[(1166 + 600)^2 + 1000^2]} = 2029 \ \Omega.$

$$I = \frac{5}{2029} \ A.$$

Load power $= I^2 R_L = \left(\dfrac{5}{2029}\right)^2 1166 = 0.00708 \ W.$

NETWORK THEOREMS

<u>Self-assessment example 2.18</u>   A source with e.m.f. 10 V and internal impedance $(3 + j4)$ Ω is connected to a load through a section of line with inductive reactance $+j5$ Ω and negligible resistance.
(a) Calculate the value of load impedance to give maximum power transfer (i) if the load is purely resistive, (ii) if the load has a power factor of 0.9 lagging. (b) for (a) (i) and (ii), calculate the value of power developed in the load.

Further problems
2.19. With respect to the terminals AB, determine the equivalent (a) Thevenin voltage, (b) Norton current. With what internal resistance is each of these generators associated. (Fig. 2.43.)

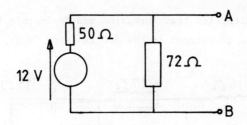

Fig. 2.43

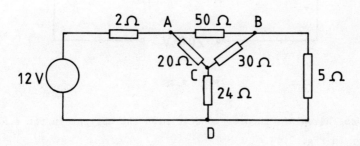

Fig. 2.44

2.20. Replace the delta ABC in Fig. 2.44 with an equivalent star. Hence determine the power dissipated in the 5 Ω resistor.
2.21. Using the delta/star transformation in delta XYZ, followed by

an application of Thevenin's theorem, determine the value of a component to be connected to terminals AB which will cause maximum power to be transferred into it. (Fig. 2.45.)

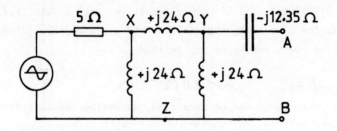

Fig. 2.45

2.22. Using the delta/star transformation and Thevenin's theorem determine the value of current in the 5 Ω resistor in Fig. 2.46.

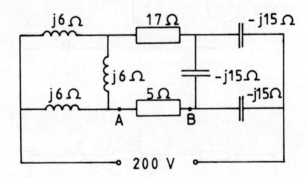

Fig. 2.46

2.23. Use Thevenin's theorem to determine the current in the 4.58 Ω resistor in Fig. 2.47.

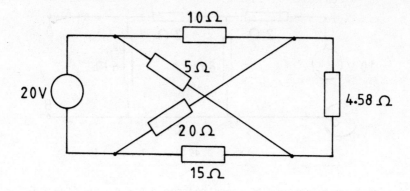

Fig. 2.47

2.24. Determine the value of current in the 3 Ω resistor when terminals X and Y in Fig. 2.48 are connected directly together.

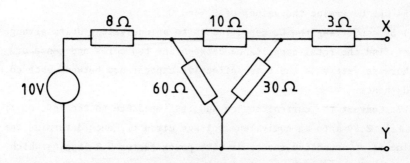

Fig. 2.48

2.25. (a) A load of $(R + j20)$ Ω is fed from a generator with an e.m.f. of 100 V and internal impedance $(25 + j50)$ Ω. (i) What value of R in the load would cause maximum power transfer to be accomplished? (ii) What is the value of the power in (i)?
(b) Determine (i) the equivalent voltage generator (Thevenin), (ii) the equivalent current generator (Norton), which may be used to represent the network given in Fig. 2.49. With what impedance are these two generators associated?

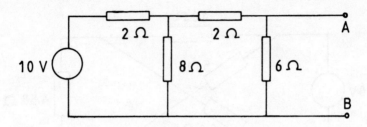

Fig. 2.49

2.26. (a) A three-core, metal-sheathed cable has a capacitance C_1 between any two conductors and C_2 between each conductor and sheath. When the three conductors are shorted together, the total capacitance between conductors and sheath is found to be 0.48 μF. When two conductors are connected together and the third is connected to the sheath, the capacitance between the pair and sheath is found to be 0.54 μF. Determine the values of C_1 and C_2.

(b) By converting the C_2 capacitances to an equivalent delta arrangement find the total capacitance between any two cores and hence or otherwise determine the total effective capacitance between each core and sheath.

2.27. Convert the current source and its impedance to the left of XY in Fig. 2.50 into an equivalent voltage circuit. Hence determine the value of an inductive reactance with power factor 0.8 lagging which should be connected to terminals AB to get maximum power transfer into that impedance. What is the value of this power?

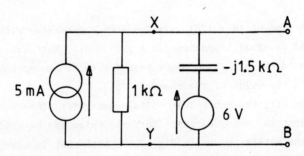

Fig. 2.50

NETWORK THEOREMS

CHAPTER 2   ANSWERS

2.2. I_{AF} = 0.908 A; I_{BE} = 0.272 A; I_{CD} = 0.636 A; Power = 3.24 W

2.5. V_{TH} = 4.6 V; Z_I = 6.407 Ω; I = 0.319 A

2.7. V_{TH} = 100 V; Z_I = (48 + j32) Ω; I = $1 \underline{/-18.66^o}$ A

2.9. I = $1.118 \underline{/-63.43^o}$ A; Z_I = (10 + j20) Ω

2.12. $\Sigma \frac{1}{Z}$ = 0.275; Z_{12} = 19.8 Ω; Z_{23} = 132 Ω; Z_{31} = 66 Ω

2.15. I_{source} = 2.87 A; I in 10 Ω = 1.054 A

2.18. (a)(i) 9.486   (ii) R_L = 8.537 Ω; X_L = 4.135 Ω   (b)(i) 4 W
(ii) 2.79 W

2.19. (a) 11.8 V   (b) 0.4 A; Z_I = 29.5 Ω

2.20. 0.45 W

2.21. 1.14 Ω pure resistor

2.22. V_{TH} (V_{AB}) = $191.5 \underline{/9.44^o}$ V; Z_I = (0.47 - j5.834) Ω;
I = $23.94 \underline{/56.3^o}$ A

2.23. 0.111 A

2.24. 0.405 A

2.25. (a)(i) 74.3 Ω   (ii) 50.4 W   (b) V_{TH} = 5 V; I_{Norton} = 2.22 A;
Z_I = 2.25 Ω

2.26. (a) Joining three together results in $3C_2$. A common pair to sheath = $2C_2 + 2C_1$. C_2 = 0.16 μF; C_1 = 0.11 μF
(b) Between cores: 0.1633 μF. Core/sheath: 0.49 μF

2.27. R_L = 665.6 Ω. X_L = j499.2 Ω.   Power = 0.01025 W.

3 Complex Waveforms

So far it has been assumed that alternating voltages and currents have been sinusoidal in form and that each could therefore be represented by a single phasor and the required calculations carried out using complex numbers.

The waveforms met in practice however often differ from the sinusoidal form, and even when this difference is relatively small, ignoring it could result in quite large errors when calculating values of circuit current and power for example.

All such non-sinusoidal waveforms met in electrical engineering under steady-state conditions repeat regularly, in other words each positive half cycle is identical and each negative half cycle is identical. Positive and negative half cycles may however be different in shape. Such waves are called 'complex' and cannot be represented by a single phasor.

Two typical complex waves are shown in Fig. 3.1.

3.1 SYNTHESIS OF COMPLEX WAVES

All complex waves are made up of a number of pure sine waves of different frequencies and amplitudes. The first sine wave has a frequency equal to that of the complex wave and is known as the fundamental. The other sine waves have frequencies which are integral multiples of the fundamental, e.g. twice, four times, six times, etc., which are known as EVEN harmonics or three times, five times, seven times, etc., which are known as ODD harmonics.

A complex voltage waveform may be described by the equation:

COMPLEX WAVEFORMS

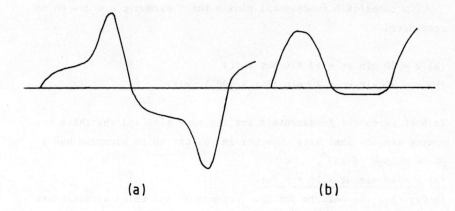

Fig. 3.1

$$v = V_{m(1)} \sin(\omega t \pm \phi_1) + V_{m(2)} \sin(2\omega t \pm \phi_2) +$$
$$V_{m(3)} \sin(3\omega t \pm \phi_3) + \ldots + V_{m(n)} \sin(n\omega t \pm \phi_n)$$

where $V_{m(1)}$, $V_{m(2)}$, ..., $V_{m(n)}$ are the maximum values of the various sine waves and ϕ_1, ϕ_2, ..., ϕ_n are their phase angles. The fundamental has angular velocity ω rad/s, the second harmonic, 2ω rad/s, and the nth harmonic, $n\omega$ rad/s.

Generally as the frequency of the harmonic increases, its amplitude decreases. Not all harmonics are present in a single complex wave, most being made up of a fundamental plus odd harmonics only or a fundamental plus even harmonics only. The shape of the complex waveform is affected by the amplitude of the harmonics present and their respective phase angles.

It is possible to resolve any complex wave into its fundamental and harmonics by the use of the Fourier analysis. This analysis yields both the magnitude and the phase of all components of the wave. However the number of different wave shapes met in electrical engineering is limited and here synthesis of waves (making complex waves by adding individual sine waves) will be considered rather than analysis. Practically, waveform analysis is often carried out using a waveform analyser which accepts the complex wave and gives a direct readout of all the waves present.

Two cases of a fundamental plus a third harmonic are now to be considered.

(a) $v = 20 \sin \omega t + 10 \sin 3\omega t$ volts
(b) $v = 20 \sin \omega t + 10 \sin (3\omega t + 180°)$ volts.

In both cases the fundamentals are the same size and the third harmonics are the same size. However in (b) the third harmonic has a phase change of $180°$.

(a) $v = 20 \sin \omega t + 10 \sin 3\omega t$

In Fig. 3.2 the phasors for the fundamental and third harmonic are shown for $t = 0$.

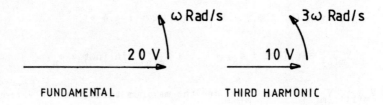

Fig. 3.2

At time $t = 0$, both waves start at zero and are travelling in a positive sense. For each complete cycle of the fundamental the third harmonic will complete three cycles. The waves are plotted to scale in Fig. 3.3. The fundamental and third harmonic are added together at regular intervals. For example, when the third harmonic reaches its first maximum of 10 V, the fundamental has reached 10 V also. Adding the two instantaneous values gives 20 V. Repeating the process at each zero and each maximum of the third harmonic, the complex wave can be plotted as shown in Fig. 3.3.

(b) $v = 20 \sin \omega t + 10 \sin (3\omega t + 180°)$

In this case the third harmonic voltage has already completed one half of a cycle (+ $180°$) and is about to go negative as the fundamental is about to rise from zero going positive. The phasors are shown in Fig. 3.4 at $t = 0$.

COMPLEX WAVEFORMS

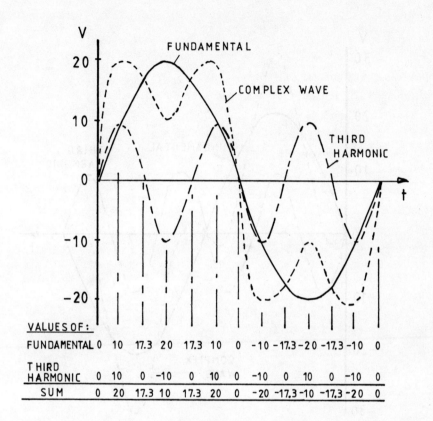

Fig. 3.3

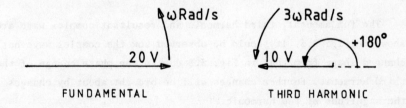

Fig. 3.4

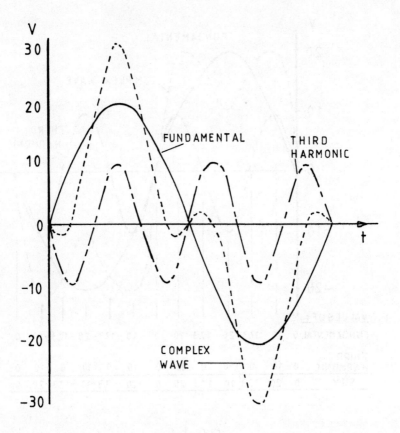

Fig. 3.5

The fundamental, third harmonic and resultant complex wave are shown in Fig. 3.5. It should be observed how the complex wave has changed shape from that in Fig. 3.3 due to the phase change of the third harmonic. Further changes will be brought about by changes in the magnitude of the harmonic.

In both Fig. 3.3 and Fig. 3.5 the negative half cycle is a mirror image about the time axis of the positive half cycle. This is true for all complex waves comprising a fundamental plus odd harmonics only. Looking back to Fig. 3.1(a), it will be observed that this condition is fulfilled so that the wave contains only odd harmonics.

Now consider the complex wave described by

COMPLEX WAVEFORMS

$v = 100 \sin \omega t + 35 \sin (2\omega t - 90°)$ V.

The phasors at t = 0 are shown in Fig. 3.6. It should be observed that at this instant the fundamental is at zero about to go positive while the second harmonic is at a maximum negative.

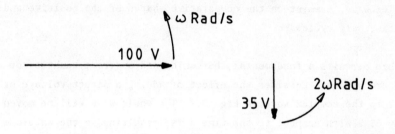

Fig. 3.6

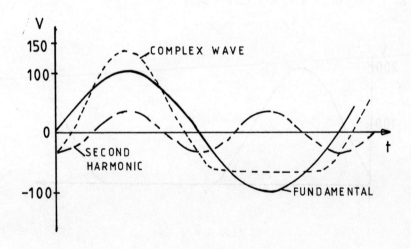

Fig. 3.7

Figure 3.7 shows the resulting complex wave and it can be seen that the negative half cycle is not a mirror image of the positive half cycle about the time axis. This is true for all complex waves comprising a fundamental plus even harmonics only.

<u>Self-assessment example 3.1</u>   A current wave is described by the

equation

$i = 10 \sin \omega t + 4 \sin 2\omega t$ A.

Draw the waveform of (a) the fundamental for 1½ complete cycles, (b) the second harmonic for the same time interval, (c) the resultant complex wave. Comment on the comparative shapes of the positive and negative half cycles.

In some circuits a fundamental, harmonics and a direct quantity or bias are present. Consider the effect of adding a direct voltage of +65 V to the complex wave in Fig. 3.7. The whole wave will be moved up by 65 V with respect to the time axis, resulting in the waveform shown in Fig. 3.8. This is the voltage waveform one would expect from a half-wave rectifier. The equation of this wave is
$v = 65 + 100 \sin \omega t + 35 \sin (2\omega t - 90°)$ V.

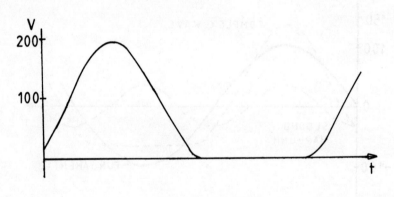

Fig. 3.8

Waves may be given a positive or a negative bias which shifts the waves up or down with respect to the time axis.

3.2 PRODUCTION OF HARMONICS

Harmonics are generated in devices which have a non-linear response to their inputs. Principally these are

(a) circuits containing iron

COMPLEX WAVEFORMS

(b) circuits containing transistors, rectifiers or valves.

Alternators generate voltages which contain odd harmonics due to inequalities in the flux distribution within the machine.

3.2.1 Free magnetisation of ferrous material

When a sinusoidal alternating voltage is applied to a coil wound on a ferrous core, a magnetising current flows which creates a flux in the core. Suppose the coil has N turns and a cross-sectional area of A m^2. From Faraday

$$e \propto \frac{Nd\Phi}{dt} \quad \text{and in S.I. units} \quad e = \frac{Nd\Phi}{dt} \text{ volts.}$$

Now $\Phi = BA$ webers, where B = flux density in tesla (Wb/m^2), hence

$$e = \frac{NAdB}{dt}.$$

Integrating both sides of the equation with respect to t gives

$$\int e \, dt = NAB.$$

By Lenz's law, the induced e.m.f. (e volts) opposes the applied voltage (v volts), and neglecting the resistance of the coil, e = v.

If the applied voltage is of sinusoidal form then e will be of the same form when $e = E_m \sin \omega t$ volts. Hence

$$\int E_m \sin \omega t \, dt = NAB.$$

Performing the integration yields

$$-\frac{E_m}{\omega} \cos \omega t = NAB.$$

Transposing gives

$$B = -\frac{E_m}{\omega NA} \cos \omega t \text{ tesla.}$$

Hence the waveform of B is a cosine wave while that of both the applied voltage and the induced e.m.f. is a sine wave. The waveforms of voltage and flux are the same shape but displaced by 90°.

An applied voltage of sinusoidal form results in the flux in the core also varying sinusoidally.

Now consider applying a sinusoidal voltage to a coil wound on a core of magnetic material with a hysteresis loop, as shown in Fig. 3.9.

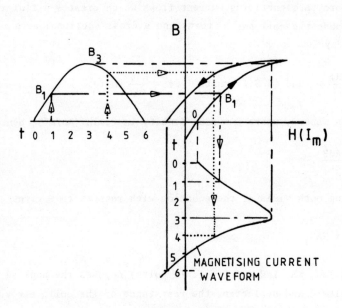

Fig. 3.9

In Fig. 3.9 there are four quadrants. In the top, left-hand quadrant a sinusoidal wave representing flux density is drawn. In the top, right-hand quadrant is the top part of a hysteresis loop for the core material. The horizontal axis (H) has dimensions amperes/metre so that it may be directly scaled as magnetising current (I_m) knowing the effective length of the core.

In the bottom, right-hand quadrant is developed the magnetising current waveform in the following manner.

At $t(0)$, $B = 0$. In order for the flux density to be zero, H must have some positive value (the coercive force) to overcome the effect from the negative half cycle (point 0 on the hysteresis loop). This

COMPLEX WAVEFORMS

is projected down into the fourth quadrant of the diagram to meet horizontal t(0).

At t(1), the flux density has increased to value (B_1). This is projected on to the rising side of the hysteresis loop and then down to meet a horizontal drawn through t(1) in the fourth quadrant. The process is repeated at regular intervals until we have seven values of I_m corresponding to the seven values of flux density, starting at zero and returning to zero on the top side of the hysteresis loop. At t(6), negative coercive force is required so that I_m has become negative for the last point plotted.

The resulting magnetising current waveform is distinctly non-sinusoidal. Repeating the process for the negative half cycle of flux density results in an identical wave shape so that it may be deduced that the magnetising current contains a fundamental plus odd harmonics only.

3.2.2 Forced magnetisation of ferrous material

If the resistance of a circuit is high compared with its inductance, and the applied voltage is of sinusoidal form, the effect of the resistance will be to cause the current to be of sinusoidal form. In three-phase, star-connected circuits where the supply has only three wires (no neutral), as, for example, with some transformers and induction motors, the line currents will be of sinusoidal form. Since it is necessary to have a current containing harmonics to produce a sinusoidal flux, it follows that if the current is of sinusoidal form the flux wave must be distorted. Further, if the flux wave is distorted then any voltages induced by this flux must also be distorted. In Fig. 3.10 it is assumed that the current in the fourth quadrant is sinusoidal. Projecting values from this curve on to the hysteresis loop results in a flux wave which is distinctly flat topped.

Now since

$$e = NA\frac{dB}{dt} \text{ volts}$$

the rate of change of flux with respect to time must be considered.

A tangent drawn to the flux wave at t(0) has a large positive slope so that $\frac{dB}{dt}$ is large and hence the induced e.m.f. will be large.

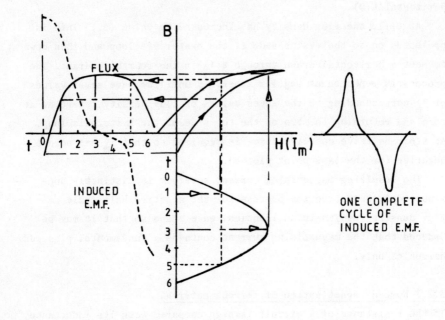

Fig. 3.10

At t(3), there is virtually no change in flux with respect to time so the induced e.m.f. is near zero. At t(5), $\frac{dB}{dt}$ is again large and with negative slope which implies that the induced e.m.f. is large and negative. A complete cycle of induced e.m.f. is also shown in Fig. 3.10; it is distinctly non-sinusoidal and contains a substantial amount of third harmonic.

3.2.3 Transistor and valve amplifiers

All amplifying devices are non-linear to some extent in their response to an input. This means that with a sine-wave input the positive half cycle is amplified by a different amount to the negative half cycle so that the two output half cycles have different amplitudes. Being different in shape, even harmonic distortion is indicated. It is, in part, the presence of second harmonic distortion which informs the listener that it is an amplifier output that is being listened to and not a live performance. The better the equipment the less is this distortion and the nearer the approach to the

COMPLEX WAVEFORMS

live performance.

3.3 THE r.m.s. VALUE OF A COMPLEX WAVE

Consider a voltage wave described by the equation:

$$v = V_0 + V_{m(1)} \sin(\omega t + \phi_1) + V_{m(2)} \sin(2\omega t + \phi_2) +$$

$$V_{m(3)} \sin(3\omega t + \phi_3) + \ldots + V_{m(n)} \sin(n\omega t + \phi_n) \text{ volts}$$

The r.m.s. value of a wave = square root of the mean value of v^2 over a complete cycle.

The instantaneous value of the above wave is equal to the sum of the instantaneous values of the individual waves

$$v = V_0 + v_1 + v_2 + v_3 + \ldots + v_n.$$

Hence

r.m.s. value of wave = $\sqrt{\text{mean value of } (V_0 + v_1 + v_2 + v_3 + \ldots + v_n)^2}$

over a complete cycle.
Multiplying out the squared term gives

$$(V_0 + v_1 + v_2 + v_3 + \ldots + v_n)^2 = V_0^2$$

$$+ V_0 v_1 + V_0 v_2 + V_0 v_3 + \ldots + V_0 v_n$$

$$+ v_1^2 + v_1 v_2 + v_1 v_3 + \ldots + v_1 v_n$$

$$+ v_2^2 + v_2 v_3 + \ldots + v_2 v_n$$

$$\ldots + v_n^2$$

There are four distinct types of term present
(a) V_0^2, (b) $V_0 v_1, V_0 v_2, \ldots, V_0 v_n$, (c) $v_1^2, v_2^2, \ldots, v_n^2$,
(d) $v_1 v_2, v_2 v_3, v_3 v_n$.

By the superposition principle each such term may be considered separately and the total effect summed at the end.

Examining each type of term in turn

(a) V_0^2. V_0 is a direct bias and so is constant in magnitude. The average value of V_0^2 over any period is V_0^2.

(b) $V_0 v_1$, $V_0 v_n$. These terms are all a constant V_0 multiplied by a voltage which is sinusoidally varying. For example, $v_n = 50 \sin(n\omega t + \phi_n)$ and $V_0 = 100$ V. Then $V_0 v_n = 100 \times 50 \sin(n\omega t + \phi_n)$ V. The average value of a sine wave over any number of complete cycles is zero since there are equal areas above and below the time axis. The effect of V_0 is merely to increase the amplitude of the wave in both positive and negative half cycles so that the average value is still zero.

(c) v_1^2, v_n^2. Each of these terms is the square of a sinusoidally varying voltage. Considering the general case for v_n^2, the average value of v_n^2 over a complete cycle is the average value of $v_{m(n)}^2 \sin^2(n\omega t + \phi_n)$ over that period. Now $\cos(A + B) = \cos A \cos B - \sin A \sin B$, so that if $A = B = (n\omega t + \phi_n)$

$$\cos 2(n\omega t + \phi_n) = \cos^2(n\omega t + \phi_n) - \sin^2(n\omega t + \phi_n). \qquad (3.1)$$

Also $\cos(A - B) = \cos A \cos B + \sin A \sin B$, so for $A = B = (n\omega t + \phi_n)$

$$\cos 0 = \cos^2(n\omega t + \phi_n) + \sin^2(n\omega t + \phi_n) \qquad (\cos 0 = 1)$$

Transposing gives

$$\cos^2(n\omega t + \phi_n) = 1 - \sin^2(n\omega t + \phi_n) \qquad (3.2)$$

Substituting 3.2 into 3.1 gives

$$\cos 2(n\omega t + \phi_n) = 1 - 2\sin^2(n\omega t + \phi_n)$$
$$\sin^2(n\omega t + \phi_n) = \tfrac{1}{2}(1 - \cos 2(n\omega t + \phi_n))$$

So that

COMPLEX WAVEFORMS

$$V_{m(n)}^2 \sin^2(n\omega t + \phi_n) = \tfrac{1}{2}V_{m(n)}^2(1 - \cos 2(n\omega t + \phi_n))$$

$$= \tfrac{1}{2}V_{m(n)}^2 - \tfrac{1}{2}V_{m(n)}^2 \cos 2(n\omega t + \phi_n).$$

This expression comprises a constant term and a cosine term. It is of the same form as that shown in Fig. 1.33(a). The mean value of the cosine term over a complete cycle is zero. The mean value of the expression is therefore $\tfrac{1}{2}V_{m(n)}^2$. Now

$$\frac{V_{m(n)}^2}{2} = \frac{V_{m(n)}}{\sqrt{2}} \times \frac{V_{m(n)}}{\sqrt{2}} = V_{(n)}^2$$

where V_n = r.m.s. value of wave.

For particular waves

The average value of $v_{(1)}^2$ over a complete cycle = $\dfrac{V_{m(1)}^2}{2} = V_{(1)}^2$

and of $v_{(2)}^2$ over a complete cycle = $\dfrac{V_{m(2)}^2}{2} = V_{(2)}^2$, etc.

(d) $v_1 v_2$, $v_2 v_n$. Each of these terms is the product of two sinusoidally varying voltages with different frequencies. Generally

$$V_{m(p)} \sin(p\omega t + \phi_p) \times V_{m(q)} \sin(q\omega t + \phi_q) \qquad \text{where } p \neq q$$

$$\cos(A + B) = \cos A \cos B - \sin A \sin B \tag{3.3}$$

$$\cos(A - B) = \cos A \cos B + \sin A \sin B \tag{3.4}$$

Subtracting 3.3 from 3.4 gives

$$\cos(A - B) - \cos(A + B) = 2 \sin A \sin B$$

and transposing gives

$$\sin A \sin B = \tfrac{1}{2}(\cos(A - B) - \cos(A + B)). \tag{3.5}$$

Letting $A = (p\omega t + \phi_p)$ and $B = (q\omega t + \phi_q)$ in equation 3.5

ELECTRICAL PRINCIPLES FOR HIGHER TEC

$$V_{m(p)} \sin(p\omega t + \phi_p) \times V_{m(q)} \sin(q\omega t + \phi_q)$$
$$= \tfrac{1}{2}(\cos((p\omega t + \phi_p) - (q\omega t + \phi_q)) - \cos((p\omega t + \phi_p) + (q\omega t + \phi_q)))$$
$$= \tfrac{1}{2}(\cos((p-q)\omega t + (\phi_p - \phi_q)) - \cos((p+q)\omega t + (\phi_p + \phi_q))).$$

Both terms represent sinusoidally varying waves, each having a different frequency. Once again, the average value of a sine wave over any number of complete cycles is zero so that the average value of the complete expression is zero.

The mean value of the product of two sine waves of different frequencies is zero.

To sum up the preceding results

The average value of v^2 is the sum of all the non-zero terms: V_0^2, $\dfrac{V_{m(1)}^2}{2}$, $V_{m(2)}^2$, ..., $V_{m(n)}^2$.

The r.m.s. value is the square root of this sum. Hence

$$V = \sqrt{\left(V_0^2 + \frac{V_{m(1)}^2}{2} + \frac{V_{m(2)}^2}{2} + \ldots + \frac{V_{m(n)}^2}{2}\right)} \qquad (3.6)$$

Similarly for current

$$I = \sqrt{\left(I_0^2 + \frac{I_{m(1)}^2}{2} + \frac{I_{m(2)}^2}{2} + \ldots + \frac{I_{m(n)}^2}{2}\right)} \qquad (3.7)$$

Now, since $\dfrac{V_m}{\sqrt{2}} = V$, where V is the r.m.s. value of voltage, $\dfrac{V_m^2}{2} = V^2$. Equation 3.6 may therefore be expressed as

$$V = \sqrt{\left(V_0^2 + V_1^2 + V_2^2 + \ldots + V_n^2\right)}$$

using r.m.s. values throughout.
Similarly equation 3.7 may be re-written

$$I = \sqrt{\left(I_0^2 + I_1^2 + I_2^2 + \ldots + I_n^2\right)}.$$

<u>Worked example 3.2</u>   Determine the r.m.s. value of a complex voltage wave described by the equation

$$v = 10 + 50 \sin(\omega t + 30°) + 10 \sin(3\omega t + 15°) + 5 \sin(5\omega t - 90°) \text{ volts.}$$

COMPLEX WAVEFORMS

Using equation 3.6

$$V = \sqrt{\left(10^2 + \frac{50^2}{2} + \frac{10^2}{2} + \frac{5^2}{2}\right)} = \sqrt{(100 + 1250 + 50 + 12.5)}$$

$$= 37.58 \text{ V}.$$

<u>Self-assessment example 3.3</u>   Determine the r.m.s. value of a complex current wave described by the equation

$$i = -5 + 20 \sin(\omega t - 10°) + 4 \sin(2\omega t + 30°) + 0.6 \sin(4\omega t + 25°) \text{ A}.$$

<u>Self-assessment example 3.4</u>   A complex voltage wave comprising a fundamental and third harmonic only has a r.m.s. value of 39.53 V. The r.m.s. value of the fundamental is $50/\sqrt{2}$ volts. Determine the r.m.s. value of the third harmonic.

3.4   POWER CONVEYED BY COMPLEX WAVES

Consider a wave described by the equation

$$v = V_0 + V_{m(1)} \sin(\omega t + \phi_a) + V_{m(2)} \sin(2\omega t + \phi_b) + \ldots + V_{m(n)} \sin(n\omega t + \phi_x) \text{ volts}.$$

It is applied to a circuit and a current flows given by the equation

$$i = I_0 + I_{m(1)} \sin(\omega t + \phi_a - \phi_1) + I_{m(2)} \sin(2\omega t + \phi_b - \phi_2) + \ldots + I_{m(n)} \sin(n\omega t + \phi_x - \phi_n) \text{ amperes}$$

where $\phi_1, \phi_2, \ldots, \phi_n$ are phase angles determined by the circuit components.

The instantaneous power in the circuit is vi watts.

The average power over a complete cycle is the average value of vi over this period.

As in section 3.3, the instantaneous value of the voltage wave may be written as the sum of the instantaneous values of the constituent voltages

$$v = V_0 + v_1 + v_2 + \ldots + v_n \text{ V}.$$

Similarly the instantaneous value of the current is

$$i = I_0 + i_1 + i_2 + \ldots + i_n \text{ A}.$$

Hence

$$v \times i = V_0 I_0$$
$$+ V_0 i_1 + V_0 i_2 + \ldots + V_0 i_n$$
$$+ v_1 i_1 + v_1 i_2 + \ldots + v_1 i_n$$
$$+ v_2 i_2 + \ldots + v_2 i_n, \text{ etc.}$$

There are four different types of terms here, as in section 3.3, from which the r.m.s. value has to be determined
(a) $V_0 I_0$, (b) $V_0 i_1$, $V_0 i_2$, ..., $V_0 i_n$, (c) $v_1 i_1$, $v_2 i_2$, $v_n i_n$, (d) $v_1 i_2$, $v_2 i_3$, $v_n i_m$ where $n \neq m$.
Dealing with each of these types in turn.
(a) $V_0 I_0$. Since these are direct components, the power contributed to the circuit is $V_0 I_0$ W.
(b) $V_0 i_1$, $V_0 i_n$. These terms are all the product of a constant term multiplied by a sinusoidally varying quantity. The average value over a complete cycle is zero as already reasoned in section 3.3 (part b).
(c) $v_1 i_1$, $v_n i_n$. The average value of power in the circuit is the average value of $V_{m(n)} \sin(n\omega t + \phi_x) \times I_{m(n)} \sin(n\omega t + \phi_x - \phi_n)$ over a complete cycle.
Using equation 3.5: $\sin A \sin B = \tfrac{1}{2}(\cos(A - B) - \cos(A + B))$.
Let $A = (n\omega t + \phi_x)$ and $B = (n\omega t + \phi_x - \phi_n)$, which gives

$$V_{m(n)} \sin(n\omega t + \phi_x) \times I_{m(n)} \sin(n\omega t + \phi_x - \phi_n)$$
$$= \frac{V_{m(n)} I_{m(n)}}{2}(\cos((n\omega t + \phi_x) - (n\omega t + \phi_x - \phi_n)) - \cos((n\omega t + \phi_x) + (n\omega t + \phi_x - \phi_n)))$$
$$= \frac{V_{m(n)} I_{m(n)}}{2}(\cos \phi_n - \cos(2(n\omega t + \phi_x) - \phi_n)) \text{ watts}.$$

ϕ_n is determined at the particular frequency by the circuit components and is therefore a constant at that frequency.

Looking back to Fig. 1.33 it will be seen that this expression for power reduces to

$$P = \frac{V_{m(n)} I_{m(n)}}{2} \cos \phi_n \text{ watts}$$

(again the average value of the time-varying cosine term is zero). Using r.m.s. values of voltage and current, $P = V_n I_n \cos \phi_n$ watts.

(d) $v_1 i_2$, all terms like $v_p i_q$, where $p \neq q$.

The average value of $v_p i_q$ is the average value over a complete cycle of

$$V_{m(p)} \sin (p\omega t + \phi_p) \times I_{m(q)} \sin (q\omega t + \phi_p - \phi_q).$$

This is the product of two sine terms with different frequencies which has already been examined in section 3.3 (part d), the average value of which was found to be zero.

Summing the non-zero terms, the power supplied by complex voltages and currents is therefore given by

$$P = V_0 I_0 + V_{(1)} I_{(1)} \cos \phi_1 + V_{(2)} I_{(2)} \cos \phi_2 + \ldots + V_{(n)} I_{(n)} \cos \phi_n \text{ watts} \qquad (3.8)$$

Thus, the total circuit power = sum of the powers due to the individual harmonics.

3.5 POWER FACTOR

The power factor of an a.c. circuit $= \dfrac{\text{circuit power in watts}}{\text{circuit volt-amperes}}$

$$= \frac{V_0 I_0 + \Sigma V_n I_n \cos \phi_n}{VI}.$$

(equation 1.7)

Notice that whereas power factor is equal to cosine ϕ when a sinusoidal voltage and current are involved, this may not be used where complex waveforms are present since each harmonic may involve a different phase angle and, in addition, there may be direct power present.

Where the circuit series resistance is known the power may be found by using the formula

$$P = I^2 R \text{ watts.}$$

3.6 PERCENTAGE HARMONIC CONTENT

The size of the nth harmonic is sometimes expressed by comparing it with the size of the fundamental.

$\dfrac{V_{m(n)}}{V_{m(1)}} \times 100\%$ is defined as the percentage nth harmonic content of the wave.

For example, for $V_{m(2)} = 20$ V, and $V_{m(1)} = 80$ V

$$\frac{20}{80} \times 100 = 25\%.$$

The complex wave contains 25% second harmonic.

Since power is only contributed to a circuit by a current and voltage of the same frequency, a dynamometer wattmeter may be employed as a simple complex wave analyser at power frequencies giving information as to the number of harmonics present and their magnitudes but not their phase with respect to the fundamental.

The complex wave to be analysed is fed to the current coil of the wattmeter while the voltage coil is fed at constant voltage from an oscillator which delivers a sine-wave output at varying frequency. By altering the oscillator frequency over a range from the fundamental value upward, the particular frequencies at which the wattmeter is deflected are noted. Whenever there is a deflection, the current in the coil must have a component which is of the same frequency as that presently applied to the voltage coil. All frequencies present in the current wave are therefore detected as the oscillator frequency is varied. The magnitude of the wattmeter deflection at a particular harmonic frequency is a measure of the relative size of that harmonic.

$$\frac{\text{Wattmeter deflection for the nth harmonic}}{\text{wattmeter deflection at fundamental frequency}} \times 100\% =$$

percentage nth harmonic content.

COMPLEX WAVEFORMS

<u>Worked example 3.5</u>   A voltage given by

$$v = 100 \sin \omega t + 50 \sin (3\omega t - 30°) + 10 \sin (5\omega t + 120°) \text{ V}$$

is applied to a circuit and the resultant current is found to be

$$i = 7.07 \sin (\omega t - 45°) + 1.58 \sin (3\omega t - 101.5°) + 0.196 \sin (5\omega t + 41.31°) \text{ A}.$$

Determine (a) total power supplied, (b) the overall power factor.

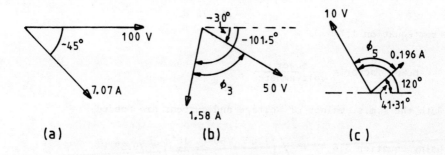

Fig. 3.11

From equation 3.8 it should be observed that only voltages and currents of the same frequency contribute power to a circuit. In this case there is no direct component.

Consider the fundamental voltage and current (Fig. 3.11(a))

$V_{m(1)} = 100$ V $I_{m(1)} = 7.07$ A   The angle between them is $45°$.

$$\text{Power} = \frac{V_m I_m}{2} \cos \phi \text{ W}$$

$$= \frac{100 \times 7.07}{2} \cos 45° = 250 \text{ W}$$

(or convert both quantities to r.m.s. values and use VI cos ϕ).

Consider the third harmonic voltage and current (Fig. 3.11(b))

$V_{m(3)} = 50$ V $I_{m(3)} = 1.58$ A

The angle between them is $\phi_3 = (101.5 - 30)°$.

Power = $\dfrac{50 \times 1.58}{2}$ cos $71.5°$ = 12.53 W.

Finally, consider the fifth harmonic voltage and current (Fig. 3.11(c))

$V_{m(5)}$ = 10 V $I_{m(5)}$ = 0.196 A   The angle between them is
$\phi_5 = (120 - 41.31)°$.

Power = $\dfrac{10 \times 0.196}{2}$ cos $78.69°$ = 0.192 W.

Answer (a) Total power supplied = Σ VI cos ϕ = 250 + 12.53 + 0.192 = 262.72 W.

From equation 1.7

$$\text{power factor} = \dfrac{\text{power}}{\text{volt-amperes}}.$$

Both the r.m.s. values of voltage and current are needed

Using equation 3.6, V = $\sqrt{\left(\dfrac{100^2}{2} + \dfrac{50^2}{2} + \dfrac{10^2}{2}\right)}$ = 79.37 V

and equation 3.7, I = $\sqrt{\left(\dfrac{7.07^2}{2} + \dfrac{1.58^2}{2} + \dfrac{0.196^2}{2}\right)}$ = 5.12 A.

Answer (b) Power factor = $\dfrac{262.72}{79.37 \times 5.12}$ = 0.646.

<u>Self-assessment example 3.6</u>   A voltage expressed by the equation

$$v = 10 + 50 \sin \omega t + 20 \sin (2\omega t + 20°) \text{ V}$$

is applied to an inductive circuit when the resulting current was found to be of the form

$$i = 10 + 4.47 \sin (\omega t - 26.6°) + 1.414 \sin (2\omega t - 25°) \text{ A}.$$

Determine (a) the r.m.s. value of the voltage, (b) the r.m.s. value of the current, (c) the circuit power, (d) the overall power factor, (e) the resistance of the circuit.

COMPLEX WAVEFORMS

3.7 HARMONICS IN SINGLE-PHASE CIRCUITS

Consider a voltage given by the equation

$$v = V_{m(1)} \sin \omega t + V_{m(2)} \sin 2\omega t + \ldots + V_{m(n)} \sin n\omega t \text{ volts.}$$

The effect will be considered of applying such a wave to linear circuit elements which have (a) resistance, (b) inductance, and (c) capacitance.

3.7.1 Pure resistance

The value of current in a resistor is expressed by Ohm's law ($I = V/R$ amperes).

The current is in phase with the voltage, and the resistance value is independent of frequency. Hence the current will be expressed by the equation

$$i = \frac{V_{m(1)}}{R} \sin \omega t + \frac{V_{m(2)}}{R} \sin 2\omega t + \ldots + \frac{V_{m(n)}}{R} \sin n\omega t \text{ amperes.}$$

The current and voltage waveforms are therefore identical in shape and the percentage nth harmonic current is the same as the percentage nth harmonic voltage (section 3.6).

3.7.2 Pure inductance

The reactance of an inductor with inductance L henry is given by $X_L = \omega L$ Ω, where ω rad/s is the angular velocity of the fundamental. For the nth harmonic the reactance will be $n\omega L$ Ω.

The inductive reactance is therefore proportional to the angular velocity and to frequency. The current in a pure inductor lags the driving voltage by $90°$ and this will be true for each constituent voltage of the complex wave. The general expression for the current wave in an inductor will therefore be

$$i = \frac{V_{m(1)}}{\omega L} \sin (\omega t - 90°) + \frac{V_{m(2)}}{2\omega L} \sin (2\omega t - 90°) + \ldots + \frac{V_{m(n)}}{n\omega L} \sin (n\omega t - 90°) \text{ A.}$$

The nth harmonic current expressed as a percentage of the fundamental current is

$$\left[\frac{\frac{V_{m(n)}}{n\omega L}}{\frac{V_{m(1)}}{\omega L}}\right] \times 100 = \left[\frac{V_{m(n)}}{V_{m(1)}} \frac{\omega L}{n\omega L}\right] \times 100 = \frac{1}{n} \times \text{percentage nth harmonic content of voltage wave.}$$

This reduction in magnitude of successive harmonics in addition to the 90° phase shift causes the current wave shape to differ considerably from that of the voltage wave.

3.7.3 Pure capacitance

The capacitive reactance of a capacitor with capacitance C farads is given by $X_C = (1/\omega C)$ Ω, where ω is the angular velocity of the fundamental. For the nth harmonic, $X_C = (1/n\omega C)$ Ω.

The current in a pure capacitor leads the voltage by 90°.

The general expression for the current wave will therefore be

$$i = \frac{V_{m(1)}}{\frac{1}{\omega C}} \sin(\omega t + 90°) + \frac{V_{m(2)}}{\frac{1}{2\omega C}} \sin(2\omega t + 90°) + \ldots +$$

$$\frac{V_{m(n)}}{\frac{1}{n\omega C}} \sin(n\omega t + 90°) \text{ A}$$

$$= V_{m(1)} \omega C \sin(\omega t + 90°) + V_{m(2)} 2\omega C \sin(2\omega t + 90°) + \ldots +$$

$$V_{m(n)} n\omega C \sin(n\omega t + 90°) \text{ A}.$$

The nth harmonic current expressed as a percentage of the fundamental current is

$$\left[\frac{V_{m(n)}}{V_{m(1)}} \frac{n\omega C}{\omega C}\right] \times 100 = n \times \text{nth percentage harmonic content of voltage wave.}$$

Once again the current wave shape will be different from that of the voltage wave due this time to a magnification by a factor n of the voltage content and to the phase advance of 90°.

Having seen how individual linear circuit elements affect the harmonic content of the current wave, series combinations of

COMPLEX WAVEFORMS

resistance and inductance and resistance and capacitance will now be examined.

3.7.4 Resistance and inductance in series

Worked example 3.7   A coil with resistance 10 Ω and inductance 10 mH is connected to a power source which provides a voltage given by the equation

$$v = 200 \sin \omega t + 50 \sin (3\omega t + 30°) + 10 \sin (5\omega t - 90°) \text{ V}$$

(ω = 400 rad/s).

Derive an expression for the instantaneous value of the circuit current.

Each of the voltages has to be considered separately since the impedance of the coil will be different at each of the frequencies present. Consider the fundamental

$$Z_1 = 10 + j400 \times 10 \times 10^{-3} = (10 + j4) = 10.77\underline{/21.8°} \text{ } \Omega.$$

$$I_{m(1)} = \frac{V_{m(1)}}{Z_1} = \frac{200}{10.77\underline{/21.8°}} = 18.57\underline{/-21.8°} \text{ A}.$$

The fundamental current has a peak value of 18.57 A and it lags the fundamental voltage by 21.8°. Hence

$$i_1 = 18.57 \sin (\omega t - 21.8°) \text{ A}.$$

Consider the third harmonic

$$Z_3 = 10 + j(3 \times 400) \times 10 \times 10^{-3} = 10 + j12 = 15.62\underline{/50.19°} \text{ } \Omega.$$

$$I_{m(3)} = \frac{V_{m(3)}}{Z_3} = \frac{50}{15.62\underline{/50.19°}} = 3.2\underline{/-50.19°} \text{ A}.$$

The third harmonic current has a peak value of 3.2 A and it lags the third harmonic voltage by 50.19°. Hence

$$i_3 = 3.2 \sin (3\omega t + 30° - 50.19°) = 3.2 \sin (3\omega t - 20.19°) \text{ A}.$$

An alternative method, if preferred, is to introduce the 30° lead of the voltage wave as follows

$$I_{m(3)} = \frac{V_{m(3)}}{Z_3} = \frac{50\underline{/+30°}}{15.62\underline{/50.19°}} = 3.2\underline{/-20.19°} \text{ A}.$$

Finally consider the fifth harmonic

$$Z_5 = 10 + j(5 \times 400) \times 10 \times 10^{-3} = 10 + j20 = 22.36\underline{/63.43°} \ \Omega.$$

$$I_{m(5)} = \frac{10}{22.36\underline{/63.43°}} = 0.447\underline{/-63.43°} \text{ A}.$$

$I_{m(5)}$ lags the fifth harmonic voltage by 63.43°. Hence

$$i_5 = 0.447 \sin (5\omega t - 90 - 63.43) = 0.447 \sin (5\omega t - 153.43°) \text{ A}.$$

The required expression for the current is

$$i = 18.57 \sin (\omega t - 21.8°) + 3.2 \sin (3\omega t - 20.19°) + 0.447 \sin (5\omega t - 153.43°) \text{ A}.$$

The phasor diagrams are shown in Fig. 3.12.

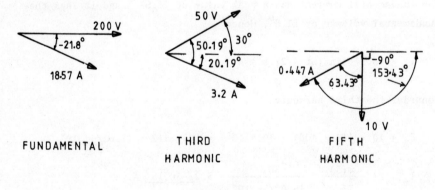

Fig. 3.12

Self-assessment example 3.8   A voltage described by the equation

$$v = 250 \sin \omega t + 50 \sin (3\omega t + 60°) + 20 \sin (5\omega t + 150°) \text{ V}$$

COMPLEX WAVEFORMS

is applied to a series circuit comprising a 15 Ω pure resistor and an inductor which has a resistance of 5 Ω and an inductance of 0.05 H. (ω = 314 rad/s.)

Derive an expression for the instantaneous current in the inductor and hence determine (a) the r.m.s. value of the current, (b) the power lost in the inductor.

3.7.5 Resistance and capacitance in series

<u>Worked example 3.9</u>   A voltage described by the equation

$$v = 100 \sin \omega t + 10 \sin (2\omega t - 180°) + 1 \sin (4\omega t + 30°) \text{ V}$$

is applied to a circuit comprising a 50 Ω resistor in series with a 10 μF capacitor. Determine the r.m.s. value of the circuit current for ω = 628 rad/s.

As with the inductive circuit in worked example 3.7, it is necessary to consider each constituent voltage separately.

(a) Fundamental

$$Z_1 = 50 + \frac{1}{j628 \times 10 \times 10^{-6}} = 50 - j159.2 = 166.9 \underline{/-72.57°} \text{ Ω}.$$

$$I_{m(1)} = \frac{V_{m(1)}}{Z_1} = \frac{100}{166.9 \underline{/-72.57°}} = 0.6 \underline{/72.57°} \text{ A}.$$

$$i_1 = 0.6 \sin (\omega t + 72.57°) \text{ A}.$$

(b) Second harmonic

$$Z_2 = 50 + \frac{1}{j(2 \times 628) \times 10 \times 10^{-6}} = 50 - j79.62 = 94 \underline{/-57.87°} \text{ Ω}.$$

$$I_{m(2)} = \frac{V_{m(2)}}{Z_2} = \frac{10}{94 \underline{/-57.87°}} = 0.106 \underline{/57.87°} \text{ A}.$$

$$i_2 = 0.106 \sin (2\omega t - 180° + 57.87°) = 0.106 \sin (2\omega t - 122.13°) \text{ A}.$$

(c) Fourth harmonic

$$Z_3 = 50 - j39.81 = 63.9\underline{/-38.53^\circ}\ \Omega.$$

$$I_{m(4)} = \frac{1}{63.9\underline{/-38.53^\circ}} = 0.0156\underline{/38.53^\circ}\ A.$$

$$i_4 = 0.0156\sin(4\omega t + 30 + 38.53) = 0.0156\sin(4\omega t + 68.53^\circ)\ A.$$

From equation 3.7

$$I = \sqrt{\left(\frac{0.6^2}{2} + \frac{0.106^2}{2} + \frac{0.0156^2}{2}\right)}$$

$$= 0.431\ A.$$

Self-assessment example 3.10   A circuit comprising an 18 Ω resistor in series with a capacitor of value 0.884 μF is connected to a source of e.m.f. when the current in the circuit is described by the equation

$$i = 0.982\sin(\omega t + 45^\circ) + 0.527\sin(3\omega t + 108.43^\circ) + 0.1089\sin(5\omega t - 38.69^\circ)\ A$$

($\omega = 2\pi \times 10{,}000$ rad/s).

Derive an expression for the instantaneous voltage applied to the circuit and hence determine (a) the r.m.s. value of the circuit voltage, (b) the power developed in the resistor, (c) the circuit power factor.

3.8 SELECTIVE RESONANCE

The impedance of a circuit comprising resistance, inductance and capacitance in series is given by

$$Z = R + j\left(\omega L - \frac{1}{\omega C}\right)\ \Omega.$$

At one particular frequency, $\omega L = \frac{1}{\omega C}$, when $Z = R\ \Omega$.

The circuit appears to be purely resistive, the impedance is a

COMPLEX WAVEFORMS

minimum and the circuit current will therefore be at a maximum value. The circuit is said to exhibit series resonance.

In the case of a parallel combination of a resistive inductor and a capacitor, at a particular frequency the circuit appears to be purely resistive. The value of this apparent resistance is known as the dynamic impedance. The dynamic impedance is the largest impedance that the particular circuit can possess. The circuit current is at a minimum and the circuit is said to exhibit parallel resonance.

If a supply voltage containing harmonics is applied to either circuit, resonance may occur, not at the fundamental frequency but at that of one of the harmonics. Such an occurrence is termed selective or harmonic resonance.

In the series circuit a large current at the particular harmonic frequency will flow giving rise to large voltages across the reactive components, i.e. voltage magnification. In the parallel circuit, current magnification at the harmonic frequency takes place, the currents in the reactive components at that frequency being much larger than that taken from the supply.

Selective resonance is one reason why it is undesirable to have non-sinusoidal supply voltages except for special purposes. Alternating current generators frequently generate waveforms with appreciable odd harmonic content. This is eliminated in the generator transformer, sometimes wound with a delta-connected tertiary, or by the use of tuned saturable reactors.

<u>Worked example 3.11</u>   A 50 Hz, three-phase alternator has an internal impedance of $(1 + j10)$ Ω per phase at the fundamental frequency. It is connected to a cable, each core of which has an effective total capacitance to earth of 1.884 µF. Each phase of the alternator delivers a voltage described by the equation

$$v = 26944 \sin 314t + 50 \sin (13 \times 314)t \text{ volts.}$$

Calculate the value of the maximum voltage impressed on each core of the cable to earth while no load is connected. (The resistance and inductance of the cable may be ignored.)

Consider one phase of the symmetrical system as shown in Fig. 3.13.

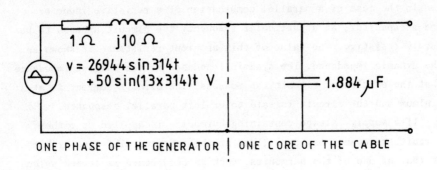

Fig. 3.13

At the fundamental frequency
Resistance = 1 Ω. Inductive reactance = +j10 Ω.

Capacitive reactance of the cable, $X_C = \dfrac{1}{314 \times 1.884 \times 10^{-6}}$

$$= 1690.4 \quad (-j1690.4 \text{ Ω}).$$

With no load on the cable the effective impedance is that of a series circuit comprising R, L and C.

$$Z = R + j(X_L - X_C) = 1 + j(10 - 1690.4) = -j1680.4 \text{ Ω (ignoring R)}.$$

Peak value of fundamental current $I_{m(1)} = \dfrac{V_{m(1)}}{Z_1} = \dfrac{26944}{-j1680.4} = j16.03$ A.

Peak voltage impressed on the cable = peak voltage across the capacitor

$$= I_{m(1)} \times -jX_C$$
$$= j16.03 \times (-j1690.4) = 27104 \text{ V} \quad \text{(in phase with the supply voltage)}.$$

At the 13th harmonic frequency
Resistance = 1 Ω. Inductive reactance = 13 × 10 = 130 Ω.

COMPLEX WAVEFORMS

Capacitive reactance of the cable = $\dfrac{1}{13 \times 314 \times 1.884 \times 10^{-6}} = 130$

$(-j130\ \Omega)$.

Circuit impedance = $1 + j(130 - 130) = 1\ \Omega$.

Peak value of 13th harmonic current = $\dfrac{V_{m(13)}}{Z_{13}} = \dfrac{50}{1} = 50$ A.

Peak 13th harmonic voltage = $50 \times -j130$
$= -j6500$ V $(6500\underline{/-90^\circ}$ V$)$.

Total voltage on the cable = $27104 \sin 314t + 6500 \sin (13 \times 314t - 90^\circ)$ V.

Each complete cycle of the 13th harmonic takes the same time as $360/13 = 27.69^\circ$ of the fundamental. In Fig. 3.14 the horizontal scale is marked in degrees of the fundamental. The 13th harmonic starts at a maximum negative (-90° at harmonic frequency or $90/13 = 6.92^\circ$ on the fundamental scale). During the first 90° of the fundamental the 13th harmonic completes 3¼ cycles. Adding the fundamental and harmonic waves together gives the complex wave which is observed to have a maximum value at 96.9°. Here, the fundamental has a value $27104 \sin 96.9^\circ = 26907$ V. Adding the peak of the harmonic, 6500 V, gives a value of 33407 V.

This is the required answer. It is not necessary to plot the negative half cycle of the fundamental voltage since an odd harmonic, which is known to produce identical positive and negative half cycles, is being added.

<u>Self-assessment example 3.12</u>   A voltage described by the equation

$v = 2000 \sin \omega t + 400 \sin 3\omega t + 100 \sin 5\omega t$ V

is applied to a series circuit comprising a resistance of 10 Ω, a capacitance of 30 µF, and an inductance of 0.04114 H. ($\omega = 300$ rad/s.)
  Determine (a) the r.m.s. value of current in the circuit, (b) the peak value of third harmonic voltage across the capacitor, (c) the r.m.s. value of the supply voltage, (d) the circuit power, (e) the power factor.

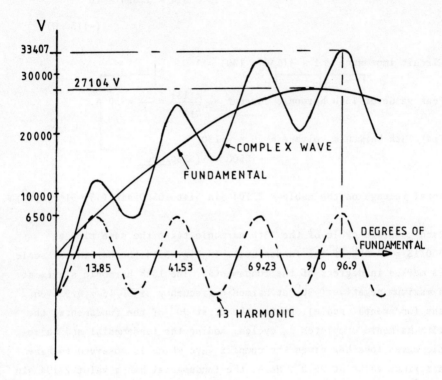

Fig. 3.14

Where circuits are made up of parallel branches and are fed from a voltage which is complex, the circuit impedance to each particular frequency contained in the complex wave must be determined separately. The circuit current and hence, where required, the branch currents must also be calculated separately for each frequency.

Worked example 3.13 The circuit shown in Fig. 3.15 is supplied with a complex voltage described by the equation $v = 50 \sin \omega t + 5 \sin 3\omega t$ V. Determine (a) an expression for the total circuit current, (b) an expression for the current in the 20 Ω resistor.

At the fundamental frequency, $X_C = \dfrac{1}{500 \times 100 \times 10^{-6}} = 20\ \Omega$.

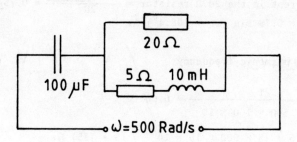

Fig. 3.15

$Z_{coil} = 5 + j500 \times 10 \times 10^{-3} = (5 + j5)\ \Omega.$

Equivalent impedance of the parallel branches $= \dfrac{20(5 + j5)}{20 + 5 + j5}$

$\hspace{6cm} = (4.62 + j3.08)\ \Omega.$

Adding the reactance of the capacitor

$\hspace{1cm}$ total impedance $= 4.62 + j3.08 - j20$
$\hspace{3.5cm} = (4.62 - j16.92)\ \Omega.$

Total circuit current at the fundamental frequency
$= \dfrac{50}{(4.62 - j16.92)}\ A.$

Rationalising and converting to polar form yields

$I = 2.85\underline{/74.72°}\ A.$
$i_1 = 2.85\ \sin\ (\omega t + 74.72°)\ A.$

Potential difference across the parallel arms $= I_1 \times$ equivalent impedance $= 2.85\underline{/74.72°} \times (4.62 + j3.08) = 2.85\underline{/74.72°} \times 5.55\underline{/33.69°} = 15.82\underline{/108.4°}\ V.$

Therefore

peak current in the 20 Ω resistor = $\dfrac{15.82 \underline{/108.4°}}{20}$ = $0.79 \underline{/108.4°}$ A.

$i_{(20\,\Omega)} = 0.79 \sin(\omega t + 108.4°)$ A.

At the third harmonic frequency

$$X_C = \dfrac{1}{3 \times 500 \times 100 \times 10^{-6}} = 6.667 \; \Omega.$$

$$Z_{coil} = 5 + j3 \times 500 \times 10 \times 10^{-3} = (5 + j15) \; \Omega.$$

Impedance of the parallel branches = $\dfrac{20(5 + j15)}{20 + 5 + j15}$

= $(8.25 + j7.05) \; \Omega.$

Total impedance = $8.25 + j7.05 - j6.667 = (8.25 + j0.386) \; \Omega.$

Total circuit current at third harmonic frequency = $\dfrac{5}{(8.25 + j0.386)}$

= $0.605 \underline{/-2.71°}$ A.

$i_3 = 0.605 \sin(3\omega t - 2.71°)$ A.

Potential difference across the parallel arms = I_3 × equivalent impedance = $0.605 \underline{/-2.71°} \times (8.25 + j7.05) = 0.605 \underline{/-2.71°} \times 10.85 \underline{/40.51°} = 6.57 \underline{/37.8°}$ V.

Hence

peak current in the 20 Ω resistor = $\dfrac{6.57 \underline{/37.8°}}{20}$ = $0.328 \underline{/37.8°}$ A.

$i_{(20\,\Omega)} = 0.328 \sin(3\omega t + 37.8°)$ A.

(a) The total current is expressed as

$i_{(T)} = 2.85 \sin(\omega t + 74.2°) + 0.605 \sin(3\omega t - 2.71°)$ A.

(b) The current in the 20 Ω resistor is expressed as

COMPLEX WAVEFORMS

$$i_{(20\ \Omega)} = 0.79 \sin(\omega t + 108.4°) + 0.382 \sin(3\omega t + 37.8°) \text{ A}.$$

3.9 ERRORS IN THE MEASUREMENT OF IMPEDANCE CREATED BY THE PRESENCE OF HARMONICS

Considerable error may occur when measuring the impedance of a circuit containing reactive components if any harmonic content of the supply voltage is not allowed for. For example, if the reactance of a coil is to be determined, using instruments which indicate r.m.s. values, the impedance is given by

$$Z = \frac{V}{I}\ \Omega.$$

Also

$$Z = \sqrt{(R^2 + X_L^2)}\ \Omega.$$

For small resistance values the impedance is very nearly the same as the reactance. Now if sinusoidal waveforms are assumed, only one frequency is present, and the apparent inductance of the coil is found from

$$X_L = 2\pi f L\ \Omega.$$

Hence

$$L = \frac{X_L}{2\pi f}\ \text{H}.$$

However let us consider the true effect when a complex voltage is applied to a coil with inductance L henry and a resistance small enough to be ignored.

Let the voltage wave be expressed as

$$v = V_{m(1)} \sin \omega t + V_{m(2)} \sin 2\omega t + \ldots + V_{m(n)} \sin n\omega t \text{ volts}.$$

At the fundamental frequency: $i_1 = \frac{V_{m(1)}}{\omega L} \sin(\omega t - 90°)$ A.

At the second harmonic frequency: $i_2 = \frac{V_{m(2)}}{2\omega L} \sin(2\omega t - 90°)$ A and so on, up to the nth harmonic: $i_n = \dfrac{V_{m(n)}}{n\omega L} \sin(n\omega t - 90°)$ A.

The r.m.s. value of the current (equation 3.7) gives

$$I = \sqrt{\left[\tfrac{1}{2}\left(\dfrac{V_{m(1)}}{\omega L}\right)^2 + \tfrac{1}{2}\left(\dfrac{V_{m(2)}}{2\omega L}\right)^2 + \ldots + \tfrac{1}{2}\left(\dfrac{V_{m(n)}}{n\omega L}\right)^2\right]}$$

$$= \dfrac{1}{\sqrt{2}\,\omega L}\sqrt{\left[V_{m(1)}^2 + \left(\dfrac{V_{m(2)}}{2}\right)^2 + \ldots + \left(\dfrac{V_{m(n)}}{n}\right)^2\right]} \text{ A.}$$

The r.m.s. value of the voltage (equation 3.6) gives

$$V = \sqrt{\left[\dfrac{(V_{m(1)})^2}{2} + \dfrac{(V_{m(2)})^2}{2} + \ldots + \dfrac{(V_{m(n)})^2}{2}\right]}$$

$$= \dfrac{1}{\sqrt{2}}\sqrt{\left(V_{m(1)}^2 + V_{m(2)}^2 + \ldots + V_{m(n)}^2\right)}.$$

The impedance of the circuit as $\dfrac{V}{I}$

$$= \dfrac{\dfrac{1}{\sqrt{2}}\sqrt{\left(V_{m(1)}^2 + V_{m(2)}^2 + \ldots + V_{m(n)}^2\right)}}{\dfrac{1}{\sqrt{2}\,\omega L}\sqrt{\left[V_{m(1)}^2 + \left(\dfrac{V_{m(2)}}{2}\right)^2 + \ldots + \left(\dfrac{V_{m(n)}}{n}\right)^2\right]}}$$

$$= \omega L \sqrt{\left[\dfrac{V_{m(1)}^2 + V_{m(2)}^2 + \ldots + V_{m(n)}^2}{V_{m(1)}^2 + \dfrac{(V_{m(2)})^2}{4} + \ldots + \dfrac{(V_{m(n)})^2}{n^2}}\right]} \qquad (3.9)$$

Now ωL is the reactance of the coil at the fundamental frequency. When harmonics are present the value of the reactance calculated as V/I is too large by the factor

$$\sqrt{\left[\dfrac{V_{m(1)}^2 + V_{m(2)}^2 + \ldots + V_{m(n)}^2}{V_{m(1)}^2 + \dfrac{(V_{m(2)})^2}{4} + \ldots + \dfrac{(V_{m(n)})^2}{n^2}}\right]}$$

COMPLEX WAVEFORMS

Where a pure capacitor is concerned a similar proof yields the result

$$\frac{V}{I} = \frac{1}{\omega C} \sqrt{\frac{V_{m(1)}^2 + V_{m(2)}^2 + \cdots + V_{m(n)}^2}{V_{m(1)}^2 + 4(V_{m(2)})^2 + \cdots + n^2(V_{m(n)})^2}} \qquad (3.10)$$

The calculated value of capacitive reactance (V/I) is less than the true value since in this case the value of the square root term is always less than unity.

For circuits containing resistance and a reactive component an error will still be present but it will be reduced by the presence of the resistance.

<u>Worked example 3.14</u> A coil with negligible resistance is supplied with an e.m.f. of the form

$v = 28.24 \sin \omega t + 33.88 \sin 3\omega t + 35.3 \sin 5\omega t$ volts.
($\omega = 2\pi \times 50$ rad/s)

The r.m.s. value of the current in the coil is measured using a moving-iron instrument and is found to be 15.64 A.

Determine the value of the coil inductance and the percentage error when calculating it as V/I.

Equation 3.6

$$V = \sqrt{\left(\frac{28.24^2}{2} + \frac{33.88^2}{2} + \frac{35.3^2}{2}\right)} = 39.95 \text{ V}.$$

$$\frac{V}{I} = \frac{39.95}{15.64} = 2.55 \text{ }\Omega.$$

Apparent inductance $= \dfrac{2.55}{2\pi \times 50} = 8.12$ mH.

Using equation 3.9

$$\frac{V}{I} = \omega L \sqrt{\frac{28.24^2 + 33.88^2 + 35.3^2}{28.24^2 + \frac{33.88^2}{9} + \frac{35.3^2}{25}}} = \sqrt{\left(\frac{797.5 + 1147.8 + 1246.1}{797.5 + 127.54 + 49.48}\right)}$$

$2.55 = \omega L\sqrt{(3.27)} = 1.809 \omega L.$ ($\frac{V}{I} = 2.55$ Ω, from above)

Therefore $L = \dfrac{2.55}{2\pi \times 50 \times 1.809} = 4.49$ mH.

Error = 8.12 - 4.49 = 3.63 mH.

As a percentage of the true value this is

$\dfrac{3.63}{4.49} \times 100\% = 80.85\%$ error.

<u>Self-assessment example 3.15</u>   A pure capacitor with capacitance 10 µF is fed with a complex e.m.f. The supply current is given by the equation

$$i = 3 \sin(\omega t + 90°) + 4.3 \sin(3\omega t + 150°) \text{ A}$$

($\omega = 2\pi \times 5000$ rad/s).
Determine (a) the equation for the voltage wave, (b) the apparent value of the capacitance calculated using the r.m.s. values of voltage and current ($V/I = X_C = 1/\omega C$ Ω).

<u>Further problems</u>
3.16. What are the fundamental and harmonic frequencies of a voltage wave described by the equation

$$v = 100 \sin 314.2t + 50 \sin(942.5t + 60°) + 10 \sin(1571t - 45°) \text{ V?}$$

3.17. Determine the r.m.s. value of a complex voltage whose instantaneous value is given by the equation

$$v = 100 \sin(\omega t + 30°) + 20 \sin(3\omega t - 90°) + 10 \sin(5\omega t + 15°) \text{ V}.$$

3.18. (a) Explain what 'harmonics' of a non-sinusoidal, periodic alternating wave are. (b) A series circuit, comprising a resistor with a resistance of 10 Ω, an inductor with value 0.0375 H and negligible resistance and a capacitor, is supplied from a power source with an output voltage described by the equation

COMPLEX WAVEFORMS

$$v = 2000 \sin \omega t + 600 \sin 3\omega t + 400 \sin 5\omega t \text{ V}$$

($\omega = 2\pi \times 50$ rad/s).

Determine (i) the required value of capacitance to give resonance at the third harmonic frequency, then with this value of capacitance in circuit (ii) the r.m.s. value of current, (iii) the circuit power, (iv) the peak voltage at 3rd and 5th harmonic frequencies developed across the capacitor.

3.19. A complex voltage waveform is described by the equation:

$$v = 200 \sin \omega t + 40 \sin 3\omega t \text{ V}$$

($\omega = 100\pi$ rad/s). It is applied to a coil of inductance 200 mH and resistance 50 Ω.

Calculate (a) the r.m.s. value of the voltage, (b) the r.m.s. value of the current, (c) the power, (d) the power factor.

3.20. (a) Explain what is meant by the term 'selective resonance'.
(b) a coil with resistance 10 Ω and inductance 50 mH is connected in parallel with a capacitor across a supply voltage given by the equation

$$v = 0.5 \sin \omega t + 0.1 \sin (3\omega t + 30°) + 0.05 \sin (5\omega t + 90°) \text{ V}$$

($\omega = 2\pi \times 7500$ rad/s).

Calculate the value of capacitance necessary to give resonance at the fifth harmonic frequency.

With this value of capacitance in circuit, deduce an expression for the instantaneous current in the capacitor.

3.21. Sketch the shapes of the following complex waves over a complete cycle of the fundamental (a) $10 + 30 \sin \omega t + 10 \sin (2\omega t - 90°)$, (b) $5 \sin \omega t + 2 \sin (3\omega t + 180°)$, (c) $2 + 4 \sin \omega t + 1.5 \sin (3\omega t + 90°)$.

3.22. The generator in Fig. 3.16 produces a complex e.m.f. with a fundamental having an r.m.s. value of 20 V and a third harmonic of 10 V r.m.s. The capacitor has a reactance of 6 Ω at the fundamental frequency. Use Thevenin's theorem to find the current in the 3 Ω

resistor due to the fundamental and third harmonic voltages separately and hence the r.m.s. value of the current due to these currents combined.

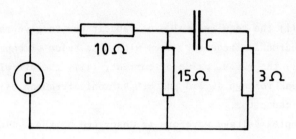

Fig. 3.16

3.23. (a) Explain briefly with the aid of a diagram why the magnetising current of a single-phase transformer contains harmonics. (b) A current containing a fundamental, third and fifth harmonics of peak amplitudes 5 A, 2.45 A and 1 A respectively flows in an inductive circuit. The power input to the circuit is 48 W. The inductance of the coil is 0.01 H.

Determine (i) the r.m.s. value of potential difference developed across the coil, (ii) the circuit power factor. (The fundamental frequency is 100 Hz.)

3.24. A coil with inductance L henry and resistance R Ω is supplied with a complex e.m.f. given by the equation

$$v = 100 \sin \omega t + V_{m(2)} \sin (2\omega t + 60°) + V_{m(4)} \sin (4\omega t - 10°) \text{ V}.$$

The resulting current in the coil is described

$$i = 2.685 \sin (\omega t - 57.51°) + 0.758 \sin (2\omega t - 12.34°) + 0.039 \sin (4\omega t - 90.95°) \text{ A}$$

($\omega = 200\pi$ rad/s).

Calculate (a) the impedance of the circuit at the fundamental frequency and hence the values of R and L, (b) the value of $V_{m(2)}$, (c) the value of $V_{m(4)}$, (d) the circuit power, (e) the power factor.

3.25. A voltage given by the equation

$$v = 50 + 120 \sin(\omega t + 30°) + 25 \sin(3\omega t - 45°) \text{ V}$$

is applied to the circuit in Fig. 3.17 ($\omega = 2\pi \times 100$ rad/s). Derive an expression for the current in the branch AB.

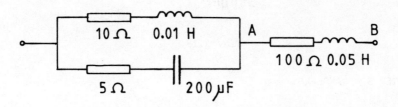

Fig. 3.17

3.26. A voltage wave containing a fundamental and 20% third harmonic is applied to (a) an inductor with negligible resistance, (b) a pure capacitor.
Compare the apparent reactance, calculated as V/I, with the true value in each case.

3.27. A current containing a fundamental, 2nd and 4th harmonics of peak amplitudes 1.0 A, 0.5 A and 0.25 A respectively flows in a coil with negligible resistance. The r.m.s. value of voltage developed across the coil is 15 V. Given that $\omega = 2\pi \times 100$ rad/s, determine the inductance of the coil.

3.28. A voltage described by the equation

$$v = 25 + 75 \sin 314t + 10 \sin(3 \times 314)t$$

is applied to the circuit shown in Fig. 3.18. Determine (a) the r.m.s. value of current in each branch, (b) the circuit power.

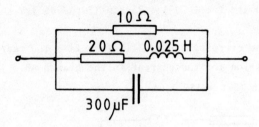

Fig. 3.18

CHAPTER 3   ANSWERS

3.1

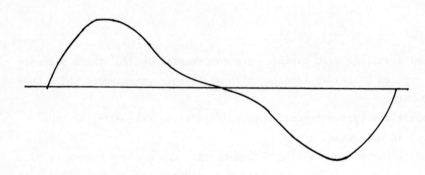

Fig. 3.19

3.3. 15.27 A

3.4. $\dfrac{25}{\sqrt{2}}$ V

3.6. (a) 39.37 V   (b) 10.54 A   (c) 209.92 W   (d) 0.506
     (e) 10 Ω

3.8. $i = 9.84 \sin(\omega t - 38.1°) + 0.98 \sin(3\omega t - 7°) + 0.247 \sin(5\omega t + 74.3°)$; $I = 6.99$ A; power = 244.6 W

3.10. $v = 25 \sin \omega t + 10 \sin(3\omega t + 90°) + 2 \sin(5\omega t - 50°)$ V
      (a) $V = 19.1$ V   (b) 11.28 W   (c) 0.746

3.12. (a) 31.72 A   (b) 1481 V   (c) 1445 V   (d) 10069 W
      (e) 0.218

3.15. $v = 9.55 \sin \omega t + 4.56 \sin(3\omega t + 60°)$ V; $V = 7.48$ V; $I = 3.707$ A; $X_C = 2.018$ Ω; apparent $C = 15.77$ μF

COMPLEX WAVEFORMS

3.16. 50 Hz; 150 Hz; 250 Hz

3.17. 72.46 V

3.18. (b) (i) 30 μF　　(ii) 45.53 A　　(iii) 20.73 kW　　(iv) 10.26 A: X_C = 21.22 Ω, 217.7 V; 60 A: X_C = 35.4 Ω, 2122 V

3.19. (a) 144.2 V　　(b) 1.77 A　　(c) 156 W　　(d) 0.613

3.20. (b) 360.25 pF; i = 8.488 sin (ωt + 90°) + 5.093 sin (3ωt + 120°) + 4.244 sin (5ωt + 180°) μA

3.22. Fundamental: (0.922 + j0.616) A; third: (0.635 + j0.141) A; r.m.s. = 1.29 A

3.23. (b) R = 3 Ω　　(i) 46.83 V r.m.s.　　(ii) 0.256

3.24. (a) R = 20 Ω; X_L = 31.41 Ω; L = 50 mH　　(b) 50 V　　(c) 5 V (d) 77.85 W　　(e) 0.499

3.25. i = 4.55 + 1.083 sin (ωt + 14.84°) + 0.178 sin (3ωt - 86.44°) A

3.26. (a) 1.0175 × true value　　(b) 0.874 × true value

3.27. 19.5 mH

3.28. (a) Top: 5.905 A; middle: 2.776 A; bottom: 5.38 A　　(b) 503 W

4 A C Bridges and Potentiometers

4.1 STANDARDS

A piece of equipment which is known as a standard has a precisely known value for one of its properties. The standard kilogram or the standard metre are examples, the particular mass and length respectively being held in a secure place. By comparison with these standards secondary standards are created which are used by equipment manufacturers to calibrate their various products. Secondary standards must have long-term stability and have a high degree of accuracy of adjustment for the original setting up.

Measurements of electrical quantities in terms of the standard mass and length together with time and the permeability of free space are called absolute measurements, and absolute standards exist for the measurement of mutual inductance, resistance and current. These standards are few in number and are held by such bodies as the National Physical Laboratory (Britain) and the National Bureau of Standards (USA).

The remaining electrical quantities are determined by comparison with the absolute standards. For example, e.m.f. and p.d. can be measured in terms of the potential difference developed across an absolutely measured resistance carrying an absolutely measured current. Capacitance and self-inductance may be determined using absolutely measured mutual inductance and resistance in a bridge circuit.

In this chapter some of the simpler bridges will be examined which are used to determine values of unknown components in terms of inductance and capacitance and then to measure frequency using a bridge made up of components, the values of which are all known.

A C BRIDGES AND POTENTIOMETERS

4.2   GENERAL ARRANGEMENT OF MEASURING BRIDGES

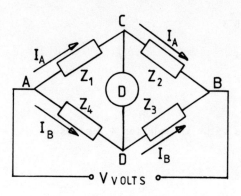

Fig. 4.1

An alternating supply is fed to points A and B of the bridge shown in Fig. 4.1. A detector is connected between points C and D (see section 4.3). Z_1, Z_2, Z_3 and Z_4 are impedances, some, or all, of which are complex and hence will be expressed generally as $Z = (R \pm jX)$ Ω. When zero current flows in the detector the bridge is said to be balanced. Measurements of unknown components are made by putting such a component into one of the arms, and with three known arms at least one of which is variable, the bridge is brought to balance. In this book, in each case, Z_2 will be considered to be the unknown.

For zero detector current the potential difference between A and C must be the same as that from A to D when the points C and D will have the same potential with respect to either of the supply terminals. Similarly the potential difference C to B must match that from D to B.

Let $Z_1 + Z_2 = Z_A$ and $Z_4 + Z_3 = Z_B$. Then at balance, with no detector current,

$$I_A = \frac{V}{Z_A} \quad \text{and} \quad I_B = \frac{V}{Z_B}.$$

Potential difference from A to C = $I_A Z_1$ and from A to D, p.d. = $I_B Z_4$. Therefore, at balance

$$I_A Z_1 = I_B Z_4 \tag{4.1}$$

$$\text{and } I_A Z_2 = I_B Z_3. \tag{4.2}$$

From equation 4.1

$$I_A = \frac{I_B Z_4}{Z_1}.$$

Substituting this value in equation 4.2 gives

$$\frac{I_B Z_4 Z_2}{Z_1} = I_B Z_3.$$

Cancelling I_B on both sides of the equation and rearranging gives

$$\frac{Z_4}{Z_1} = \frac{Z_3}{Z_2} \tag{4.3}$$

$$\text{or } Z_2 Z_4 = Z_1 Z_3. \tag{4.4}$$

Expressing the results in words, balance is achieved when:

(a) the ratios of adjacent arms are equal (equation 4.3)
(b) the products of opposite arms are equal (equation 4.4).

Before analysing some of the available bridges, the types of detector in common use will be looked at.

4.3 DETECTORS

To detect a null in a bridge operated at mains frequency and up to a few kilohertz, a vibration galvanometer may be used. This comprises a moving-coil movement suspended between the poles of a permanent magnet. The suspension is a strip of phosphor-bronze which may be tensioned mechanically so that the mechanical resonant frequency of the movement is equal to that of the supply being used to drive the bridge. There are no hair springs as in the case of the d.c. moving-coil meter. When any current of the correct frequency passes, the coil will vibrate being deflected by the instantaneous value of the current. The coil motion is detected by shining a beam of light on to a small mirror fixed to the suspension, vibration causing the reflected light beam to appear as a wide band on a screen. Zero current or a null is indicated by no movement of the mirror and a pin-sharp image of a line painted on the mirror becoming apparent on the screen.

There are also a number of electronic detectors available which use a tuned circuit to detect current at the right frequency and then, via a high-gain amplifier and rectifier, give a read-out on a conventional moving-coil meter.

Earphones may be used for frequencies between a few hundred and about 12 kilohertz above which the sensitivity of the ear decreases.

A cathode-ray oscilloscope is useful as a detector where harmonics are present because it shows clearly when the fundamental disappears leaving only the harmonic. Measuring the inductance of an iron-cored coil can be difficult since harmonics in the coil current will be present and a bridge containing reactive components can only be balanced at one frequency at a time. Attempting to balance at the fundamental frequency leaves the harmonic currents in the detector. Attempting to balance at the harmonic frequency leaves the fundamental currents in the detector.

4.4 TYPES OF BRIDGE

In section 4.2 it was shown that bridges become balanced when

(a) the ratios of adjacent arms are equal
(b) the products of opposite arms are equal.

A very large number of combinations of components in bridges is possible, but for bridges to come quickly to a point of balance the requirement is that either the adjacent arms are pure components, both resistors, both air-spaced capacitors or one of each, when the bridge is known as a ratio bridge; or that a pair of opposite arms are pure components, when the bridge is known as a product bridge.

A number of bridges will now be examined in detail.

4.4.1 The Maxwell comparison bridge

This bridge is used to measure the resistance and inductance of coils with fairly high Q-factors in terms of known values of resistance and inductance. R_x and L_x are the resistance and inductance respectively to be determined. R_s and L_s are the corresponding values for a standard coil together with some additional resistance if required. It may be necessary for this extra resistance to be added to the arm containing the unknown inductance when its Q-factor is higher than that of the standard coil. The bridge is shown in Fig. 4.2.

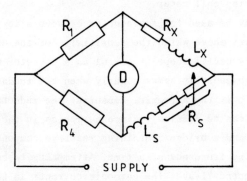

Fig. 4.2

Analysis

This is a ratio bridge, R_1 and R_4 being pure components situated in adjacent arms. From equation 4.3 it is known that the bridge balances when the ratios of adjacent arms are equal, i.e. when

$$\frac{R_1}{R_4} = \frac{(R_x + j\omega L_x)}{(R_s + j\omega L_s)}.$$

Multiplying both sides of the equation by $(R_s + j\omega L_s)$ gives

$$\frac{R_1}{R_4}(R_s + j\omega L_s) = (R_x + j\omega L_x).$$

For balance, both sides of the equation must be identical and with complex numbers this is realised when both real and j terms of the expression equate. Hence

Equating real parts: $R_x = \dfrac{R_1}{R_4} \times R_s \; \Omega.$

Equating j parts: $\omega L_x = \dfrac{R_1}{R_4} \times \omega L_s$ or $L_x = \dfrac{R_1}{R_4} L_s$ H.

4.4.2 The Owen bridge

This is a ratio bridge, the pure components R_1 and C_4 being situated in adjacent arms. It is used mainly for the measurement of fairly large values of inductance. By the addition of a direct power supply to the normal alternating one, the incremental inductance and permeability of an iron-cored coil may be determined. The incremental values are those applicable to the coil when operating in a partly magnetically-saturated state. Direct current is fed through a coil of very high inductance L_m while the alternating supply is taken through a capacitor. The capacitor blocks direct current and the high inductance blocks alternating current. In the bridge itself C_4 and C_3 block the direct current from the bottom arms so that it all flows in the coil under test without upsetting the detector.

Balance is obtained in the normal manner by the adjustment of C_3 and R_3. For measurement of L_x with no magnetic saturation, the d.c. supply is omitted. The Owen bridge together with the two power supplies is shown in Fig. 4.3.

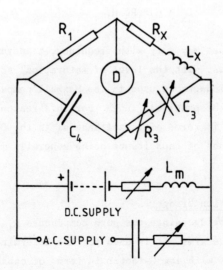

Fig. 4.3

Analysis
Using equation 4.3

$$\frac{R_1}{\frac{1}{j\omega C_4}} = \frac{R_x + j\omega L_x}{R_3 + \frac{1}{j\omega C_3}}.$$

Bringing the bottom line on the right-hand side to a common denominator $j\omega C_3$

$$j\omega C_4 R_1 = \frac{R_x + j\omega L_x}{\frac{j\omega C_3 R_3 + 1}{j\omega C_3}} = \frac{j\omega C_3 (R_x + j\omega L_x)}{j\omega C_3 R_3 + 1}.$$

Multiplying both sides of the equation by $\frac{j\omega C_3 R_3 + 1}{j\omega C_3}$ yields

$$(j\omega C_3 R_3 + 1)\frac{C_4 R_1}{C_3} = (R_x + j\omega L_x).$$

Equating real parts: $R_x = \frac{C_4 R_1}{C_3}\ \Omega.$

Equating j parts: $\omega L_x = \frac{C_4 R_1}{C_3} \times \omega C_3 R_3$

$$L_x = C_4 R_1 R_3\ H.$$

Inductance values may vary with frequency, temperature and for other than air cores, with the level of saturation, so that the standard value of inductance used in the Maxwell comparison bridge (Fig. 4.2) may not be precisely known. For this reason a determination of inductance in terms of capacitance as in the Owen bridge is more accurate, values of capacitance being generally more definitely known.

4.4.3 The Maxwell-Wien bridge
This is a product bridge since the pure components R_1 and R_3 are situated in opposite arms. It is used for the determination of the inductance of coils with low Q-factor in terms of capacitance and resistance. The arrangement is shown in Fig. 4.4.

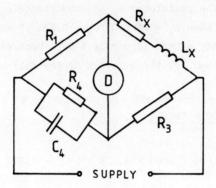

Fig. 4.4

Analysis
Using equation 4.4

$$R_1 R_3 = (R_x + j\omega L_x) \times Z_4$$

where Z_4 is the equivalent impedance of R_4 in parallel with C_4. Therefore

$$\frac{R_1 R_3}{Z_4} = (R_x + j\omega L_x)$$

Now since $1/Z_4 = Y_4$, the admittance of the arm

$$R_1 R_3 Y_4 = (R_x + j\omega L_x) \quad \text{and} \quad Y_4 = \frac{1}{R_4} + \frac{1}{\frac{1}{j\omega C_4}} = \frac{1}{R_4} + j\omega C_4.$$

Hence

$$R_1 R_3 \left(\frac{1}{R_4} + j\omega C_4\right) = (R_x + j\omega L_x).$$

Equating real parts: $R_x = \dfrac{R_1 R_3}{R_4} \; \Omega.$

Equating j parts: $\omega L_x = R_1 R_3 \, \omega C_4$

$$L_x = R_1 R_3 C_4 \; H.$$

ELECTRICAL PRINCIPLES FOR HIGHER TEC

Worked example 4.1   A Maxwell-Wien bridge as shown in Fig. 4.4 is used to determine the resistance R_x and inductance L_x of a coil. The bridge is balanced when R_4 = 9896 Ω, C_4 = 0.238 μF and R_1 = R_3 = 1000 Ω. The frequency of the supply is 1 kHz. Determine the values of (a) R_x, (b) L_x, and (c) the Q-factor of the coil at the test frequency.

(a) $R_x = \dfrac{R_1 R_3}{R_4} = \dfrac{1000 \times 1000}{9896} = 101.05$ Ω.

(b) $L_x = R_1 R_3 C_4 = 1000 \times 1000 \times 0.238 \times 10^{-6} = 0.238$ H.

(c) $Q = \dfrac{\omega L}{R} = \dfrac{2\pi \times 1000 \times 0.238}{101.05} = 14.97$.

4.4.4 The Hay bridge

Replacing the R_4, C_4 parallel combination with a series one in the Maxwell-Wien bridge transforms it into what is known as the Hay bridge. It is again a product bridge and it is used to measure the inductance and resistance of coils with large Q-factors. The solutions for R_x and L_x contain the angular velocity, ω, of the supply so it is essential that this be known precisely. When balancing the bridge, C_4 is usually kept constant while successive changes to R_1, R_4 and R_3 are made. The Hay bridge is shown in Fig. 4.5.

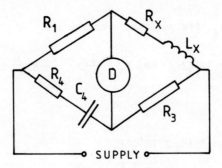

Fig. 4.5

Analysis
Using equation 4.4

A C BRIDGES AND POTENTIOMETERS

$$R_1R_3 = (R_x + j\omega L_x)\left(R_4 + \frac{1}{j\omega C_4}\right).$$

Bring right-hand term to common denominator $(j\omega C_4)$

$$R_1R_3 = (R_x + j\omega L_x)\frac{(j\omega C_4 R_4 + 1)}{j\omega C_4}.$$

Cross-multiply to leave $(R_x + j\omega L_x)$ on the right-hand side

$$\frac{j\omega C_4 R_1 R_3}{1 + j\omega C_4 R_4} = (R_x + j\omega L_x).$$

Rationalise the left-hand side

$$\frac{j\omega C_4 R_1 R_3 (1 - j\omega C_4 R_4)}{1 + \omega^2 C_4^2 R_4^2} = (R_x + j\omega L_x).$$

Multiplying out the top bracket

$$\frac{\omega^2 C_4^2 R_1 R_3 R_4 + j\omega C_4 R_1 R_3}{1 + \omega^2 C_4^2 R_4^2} = R_x + j\omega L_x.$$

Equating real terms: $R_x = \dfrac{\omega^2 C_4^2 R_1 R_3 R_4}{1 + \omega^2 C_4^2 R_4^2}\ \Omega$.

Equating j terms: $L_x = \dfrac{C_4 R_1 R_3}{1 + \omega^2 C_4^2 R_4^2}\ H$.

<u>Worked example 4.2</u>   The Hay bridge shown in Fig. 4.5 is used to measure the resistance and inductance of a coil (R_x and L_x respectively). Under balanced conditions R_3 = 750 Ω, R_1 = 2410 Ω, R_4 = 64 Ω, C_4 = 0.35 μF. Determine the resistance, R_x, and the inductance, L_x, of the coil under test given that the frequency is 500 Hz.

$$L_x = \frac{R_1 R_3 C_4}{1 + \omega^2 C_4^2 R_4^2} = \frac{750 \times 2410 \times 0.35 \times 10^{-6}}{1 + (2\pi \times 500)^2 (0.35 \times 10^{-6})^2 (64)^2}$$

$$= \frac{0.6326}{1.0049}$$

$$= 0.63\ H.$$

Comparing the equations for R_x and L_x it may be observed that the denominator is the same in each case. In the numerator the expression for R_x is $(\omega^2 C_4 R_4)$ times that for L_x. Hence

$$R_x = \omega^2 C_4 R_4 \times 0.63$$

$$= (2\pi \times 500^2 \times 0.35 \times 10^{-6} \times 64 \times 0.63)$$

$$= 139 \ \Omega.$$

All the values can of course be substituted in the full expression for R_x if preferred.

4.4.5 The Schering bridge

The Schering bridge is a product bridge since the pure components C_1 and R_3 are situated in opposite arms. It is used for the measurement of the effective series resistance and capacitance of an imperfect capacitor (R_x and C_x respectively) in terms of the capacitance of air-spaced standard capacitors C_1 and C_4, the latter being variable, and two resistance standards R_3 and R_4. (See Chapter 5 for dielectric losses in capacitors and their series and parallel representations.) The Schering bridge is shown in Fig. 4.6.

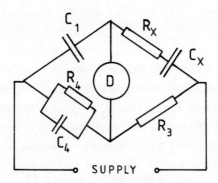

Fig. 4.6

Analysis
Using equation 4.4

A C BRIDGES AND POTENTIOMETERS

$$\frac{1}{j\omega C_1} \times R_3 = \left(R_x + \frac{1}{j\omega C_x}\right) \times Z_4$$

where Z_4 is the equivalent impedance of R_4 and C_4 in parallel. Dividing both sides by Z_4 gives

$$\frac{R_3}{j\omega C_1 Z_4} = \left(R_x + \frac{1}{j\omega C_x}\right).$$

Writing Y_4 for $\frac{1}{Z_4}$ (see section 4.4.3) gives

$$\frac{R_3}{j\omega C_1}\left(\frac{1}{R_4} + j\omega C_4\right) = \left(R_x + \frac{1}{j\omega C_x}\right).$$

Multiplying out the bracket

$$\frac{R_3}{j\omega C_1 R_4} + \frac{R_3 C_4}{C_1} = \left(R_x + \frac{1}{j\omega C_x}\right).$$

Equating real parts: $R_x = \dfrac{R_3 C_4}{C_1}$ Ω.

Equating j parts: $\dfrac{1}{\omega C_x} = \dfrac{R_3}{\omega C_1 R_4}$ then cancel ω on both sides and invert

$$C_x = \frac{C_1 R_4}{R_3} \text{ F.}$$

<u>Worked example 4.3</u> The conditions of balance of a Schering bridge (Fig. 4.6) set up to measure the capacitance and effective series resistance of a capacitor with a solid dielectric are C_1 = 0.1 μF, R_3 = 163 Ω, R_4 = 500 Ω, C_4 = 0.0033 μF, f = 1 kHz. Determine the values of (a) R_x, (b) C_x, (c) the phase angle of the unknown capacitor, (d) the power loss in the unknown capacitor when connected across a 450 V supply assuming the power factor does not change.

(a) $R_x = \dfrac{R_3 C_4}{C_1} = \dfrac{163 \times 0.0033 \times 10^{-6}}{0.1 \times 10^{-6}} = 5.38$ Ω.

(b) $C_x = \dfrac{C_1 R_4}{R_3} = \dfrac{0.1 \times 10^{-6} \times 500}{163} = 0.307 \times 10^{-6}$ F.

(c) For a resistor and pure capacitor in series, the phasor diagram is drawn in Fig. 4.7. In a pure capacitor, I leads V by $90°$. Because of the resistance the angle is less than $90°$ by δ (delta) which is known as the loss angle.

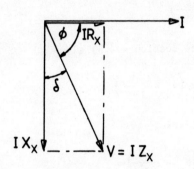

Fig. 4.7

$$\tan \phi = \frac{IX_x}{IR_x} = \frac{\frac{1}{\omega C_x}}{R_x} = \frac{1}{\omega C_x R_x} = \frac{1}{2\pi \times 1000 \times 0.307 \times 10^{-6} \times 5.38}$$

$$= 96.46$$

Therefore

$$\phi = 89.4°.$$

Alternatively the component values from the original bridge could be used when

$$\frac{1}{\omega C_x R_x} = \frac{1}{\frac{\omega C_1 R_4}{R_3} \times \frac{R_3 C_4}{C_1}} = \frac{1}{\omega C_4 R_4}.$$

Hence $\tan \phi = \dfrac{1}{2\pi \times 1000 \times 500 \times 0.0033 \times 10^{-6}} = 96.46$ as before.

(d) Power = VI cos ϕ watts.

Now $I = \dfrac{V}{Z_x} = \dfrac{V}{\sqrt{\left[R_x^2 + \left(\dfrac{1}{\omega C_x}\right)^2\right]}} = \dfrac{V}{518.5}$ A.

A C BRIDGES AND POTENTIOMETERS

Hence, power = $450 \times \frac{450}{518.5} \cos 89.4° = 4.06$ W.

Self-assessment example 4.4   The arms of a Maxwell-Wien bridge have the following values at balance
Arm AB = 28 Ω pure resistance
Arm BC = 500 Ω pure resistor in parallel with a capacitor with capacitance 5.3 μF
Arm CD = 50 Ω pure resistance
Arm DA is an air-cored coil with resistance R_x and inductance L_x considered as a series equivalent.
The supply is connected to A and C and the detector to B and D.

Working from first principles develop the balance equations for the bridge and hence determine (a) R_x, (b) L_x, and (c) the Q-factor for the coil at a frequency of 200 Hz.

Self-assessment example 4.5   A Schering bridge which is being used to measure the series resistance and capacitance of a capacitor has the following impedances in its arms when at balance using a 50 Hz supply
Arm AB = 0.15 μF (standard capacitor)
Arm BC = unknown capacitor considered as R_x and C_x in series
Arm CD = pure resistance of 58 Ω
Arm DA = pure resistance with value 1056 Ω in parallel with a variable standard capacitor set to 65.9 nF.
The supply is connected to A and C and the detector to B and D.

Determine the values of (a) R_x, (b) C_x, and (c) the phase angle of the capacitor ϕ.

4.5   THE WIEN BRIDGE

The Wien bridge is a ratio bridge but is treated separately here since it is not generally used to measure components but is used either to measure frequency in terms of known components or to provide the necessary positive feedback to maintain oscillation in the Wien bridge oscillator. The bridge is shown in Fig. 4.8.

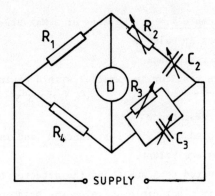

Fig. 4.8

Analysis
Using equation 4.3

$$\frac{R_1}{R_4} = \frac{R_2 + \frac{1}{j\omega C_2}}{Z_3}$$

where Z_3 is the equivalent impedance of R_3 and C_3 in parallel.

$$\frac{1}{Z_3} = Y_3 = \left(\frac{1}{R_3} + j\omega C_3\right) \quad \text{(see also, sections 4.4.3 and 4.4.5)}.$$

Hence $\dfrac{R_1}{R_4} = \left(R_2 + \dfrac{1}{j\omega C_2}\right)\left(\dfrac{1}{R_3} + j\omega C_3\right).$

Bringing both brackets to common denominators, $j\omega C_2$ and R_3 respectively gives

$$\frac{R_1}{R_4} = \frac{(j\omega C_2 R_2 + 1)}{j\omega C_2} \times \frac{(1 + j\omega C_3 R_3)}{R_3}.$$

Multiplying both sides by $j\omega C_2 R_3$

$$\frac{j\omega C_2 R_3 R_1}{R_4} = (1 + j\omega C_2 R_2)(1 + j\omega C_3 R_3).$$

Multiplying out the brackets

A C BRIDGES AND POTENTIOMETERS

$$\frac{j\omega C_2 R_3 R_1}{R_4} = 1 + j\omega C_3 R_3 + j\omega C_2 R_2 - \omega^2 C_2 R_2 C_3 R_3.$$

Equating real terms: $0 = 1 - \omega^2 C_2 R_2 C_3 R_3$

hence $1 = \omega^2 C_2 R_2 C_3 R_3$ and $\omega = \sqrt{\left(\dfrac{1}{C_2 R_2 C_3 R_3}\right)}.$

To simplify the expression, capacitors C_2 and C_3 are ganged so that whatever adjustments are made they remain equal in value

$$C_2 = C_3 = C.$$

Similarly resistors R_2 and R_3 are ganged so that $R_2 = R_3 = R$. Then

$$\omega = \sqrt{\left(\frac{1}{CRCR}\right)} = \frac{1}{CR}.$$

Now equating the j terms

$$\frac{\omega C_2 R_3 R_1}{R_4} = \omega C_3 R_3 + \omega C_2 R_2.$$

Again $C_2 = C_3 = C$ and $R_2 = R_3 = R$. Hence

$$\frac{CRR_1}{R_4} = CR + CR \quad \text{so that} \quad \frac{R_1}{R_4} = 2.$$

To use the Wien bridge either to measure frequency or as a frequency stabilising network, $R_1/R_4 = 2$ and to achieve this R_1 is made equal to twice R_4.

The resistances R_2 and R_3 are changed in steps to alter the range of frequency while C_2 and C_3 are continuously variable air-spaced capacitors to give fine adjustment of frequency within a given range.

<u>Worked example 4.6</u> The arms of a Wien bridge associated with a Wien bridge oscillator have the following values
Arm AB = pure resistor, value 2000 Ω
Arm BC = pure resistor, value 65 kΩ in series with a 500 pF pure capacitor
Arm CD = pure resistor, value 65 kΩ in parallel with a 500 pF pure

capacitor

Arm DA = pure resistor, value 1000 Ω.

Calculate (a) the operating frequency of the oscillator, (b) the necessary values of the capacitors in arms BC and CD in order that the operating frequency shall be 6.5 kHz, (c) with the capacitors set as in (a) above, the values of the resistances in arms BC and CD in order to double the operating frequency.

(a) Operating frequency = $\frac{1}{CR}$ = $\frac{1}{500 \times 10^{-12} \times 65000}$ = 30769.2 rad/s

$$f = 4897 \text{ Hz}.$$

(b) 6.5 kHz = $2\pi \times 6500$ = 40840.7 rad/s.

If the new capacitor values are C_x then

$$40840.7 = \frac{1}{C_x \times 65000}$$

$$C_x = 376.7 \text{ pF}.$$

(c) Double the angular velocity in (a) is 61538.4 rad/s. Therefore

$$61538.4 = \frac{1}{500 \times 10^{-12} \times R_x}$$

$$R_x = 32500 \text{ Ω } (= R_2 = R_3).$$

<u>Self-assessment example 4.7</u>   In the Wien bridge shown in Fig. 4.8, $R_1 = 2R_4$, $R_2 = R_3 = 15000$ Ω, and capacitors C_2 and C_3 are ganged so that at all times their values are equal.

(a) Determine the range over which C_2 and C_3 must be variable in order that the bridge shall operate between 5 kHz and 50 kHz.

(b) Over what range of frequencies would the bridge operate for the range of capacitance in (a), if R_2 and R_3 were changed to 1500 Ω?

4.6 THE ANDERSON BRIDGE

With the Anderson bridge very precise measurement of inductance in terms of a standard capacitor is possible. It is included here partly

A C BRIDGES AND POTENTIOMETERS

for that reason and partly because it relies on previous network analysis, the delta-star transformation, for a solution.

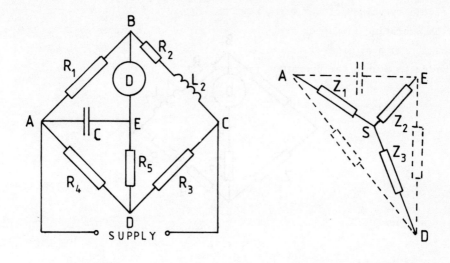

Fig. 4.9   Fig. 4.10

The Anderson bridge is shown in Fig. 4.9. To balance the bridge a battery is connected to the supply terminals and balance obtained by manipulating R_3 and the ratio arms R_1 and R_4. An a.c. supply is substituted for the battery and final balance is achieved by altering R_5 only.

Considering the delta AED in the original bridge, the values of Z_1, Z_2 and Z_3 will be determined to form an equivalent star as shown in Fig. 4.10 (see section 2.4.2 for the transformation).

The sum of the delta impedances $= R_4 + R_5 + \dfrac{1}{j\omega C}$

$$= \dfrac{j\omega C R_4 + j\omega C R_5 + 1}{j\omega C}$$

Hence: $Z_1 = \dfrac{R_4 \times \dfrac{1}{j\omega C}}{\dfrac{j\omega C R_4 + j\omega C R_5 + 1}{j\omega C}} = \dfrac{R_4}{1 + j\omega C(R_4 + R_5)}$.

Similarly, $Z_2 = \dfrac{R_5}{1 + j\omega C(R_4 + R_5)}$

$$Z_3 = \frac{R_4 R_5}{\frac{j\omega CR_4 + j\omega CR_5 + 1}{j\omega C}} = \frac{j\omega C R_4 R_5}{1 + j\omega C(R_4 + R_5)} .$$

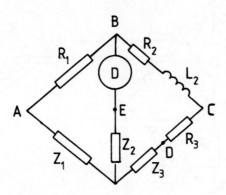

Fig. 4.11

The bridge may be re-drawn as in Fig. 4.11. Z_2 will be seen to be in series with the detector and its only significance is that it will slightly de-sensitise it.

For balance: $\dfrac{R_1}{Z_1} = \dfrac{R_2 + j\omega L_2}{Z_3 + R_3}$.

Firstly, to evaluate $(Z_3 + R_3)$

$$Z_3 + R_3 = \frac{j\omega C R_4 R_5}{1 + j\omega C(R_4 + R_5)} + R_3 .$$

Bringing R_3 over the common denominator

$$Z_3 + R_3 = \frac{j\omega C R_4 R_5 + R_3(1 + j\omega C(R_4 + R_5))}{1 + j\omega C(R_4 + R_5)} .$$

In addition, using the value for Z_1, the balance equation becomes

$$\frac{R_1(1 + j\omega C(R_4 + R_5))}{R_4} = \frac{(R_2 + j\omega L_2)(1 + j\omega C(R_4 + R_5))}{j\omega C R_4 R_5 + R_3(1 + j\omega C(R_4 + R_5))} .$$

Cancel $(1 + j\omega C(R_4 + R_5))$ on both sides and cross-multiply

$$\frac{R_1}{R_4}(j\omega CR_4 R_5 + R_3(1 + j\omega C(R_4 + R_5))) = R_2 + j\omega L_2.$$

Equating real parts: $\frac{R_1}{R_4}R_3 = R_2$.

Equating j parts: $\frac{R_1}{R_4}(\omega CR_4 R_5 + \omega CR_3(R_4 + R_5)) = \omega L_2$

$$L_2 = \frac{R_1}{R_4}C(R_4 R_5 + R_3(R_4 + R_5)).$$

<u>Worked example 4.8</u> The arms of an Anderson bridge as shown in Fig. 4.9 have the following values when at balance with an unknown inductor R_2 and L_2 in arm BC. R_1 = pure resistor, value 300 Ω, R_3 = R_4 = pure resistor, value 500 Ω, C = 0.3 μF and R_5 = pure resistor, value 150 Ω.

Determine the values of R_2 and L_2.

$$R_2 = \frac{R_1}{R_4}R_3 = \frac{300}{500} \times 500 = 300 \text{ Ω}.$$

$$L_2 = \frac{R_1}{R_4}C(R_4 R_5 + R_3(R_4 + R_5))$$

$$= \frac{300 \times 0.3 \times 10^{-6}}{500}(500 \times 150 + 500(500 + 150))$$

$$= 0.072 \text{ H}.$$

4.7 POTENTIOMETERS

4.7.1 d.c. type

The potentiometer is a device for measuring potential difference by comparison with a standard e.m.f. derived from a source such as the Weston cadmium cell which has an e.m.f. of 1.01859 V at 20° C. The simplest form of potentiometer is shown in Fig. 4.12.

With the selector switch selected to A, the standard cell is in circuit. Under the influence of a supply battery a current flows in a perfectly uniform slide wire which is often 1 m in length.

By sliding the slider along the wire a position is found for which the galvanometer current is zero (point P). This means that the potential difference between X and P due to the current in the

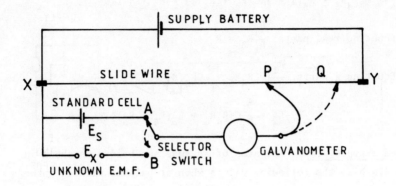

Fig. 4.12

wire is exactly matched by the e.m.f. of the standard cell. Since the wire is uniform, the voltage drop along the wire may be expressed as (E_s/XP) volts per unit length.

Changing the selector switch to B brings the unknown e.m.f. into circuit and provided that its value is less than that of the main supply battery a new null point may be found on the wire at Q. Then

$$E_x = \frac{E_s}{XP} \times XQ \text{ volts.}$$

4.7.2 The a.c. polar potentiometer (Drysdale)

The polar potentiometer has a single slide wire fed from a phase-shifting transformer which provides a constant r.m.s. value of output voltage which is variable in phase over the range $0°$–$360°$. The phase-shifting transformer is similar to an induction motor except that the rotor is not allowed to rotate freely. The stator carries two windings which are displaced by $90°$. One winding is connected directly to a single-phase supply while the other is fed through a capacitor. In this manner the two winding currents are made to be $90°$ out of phase electrically and a rotating magnetic field is created which links with a single coil R on the rotor. By turning the rotor through $360°$ the phase of the induced voltage may be varied by this amount since the instant at which the maximum flux sweeps past the rotor coil will be earlier or later according to its physical position.

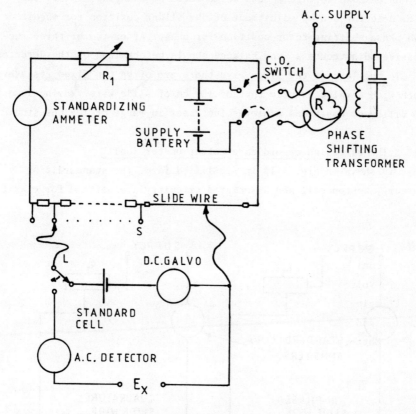

Fig. 4.13

Initially the potentiometer is standardised as in the d.c. case using a Weston cell, the d.c. galvanometer and the supply battery. The standard cell is connected to the slide wire at S by moving the wandering lead L. If the slider is situated on the slide wire at 101.86 cm from the left-hand end and rheostat R_1 varied to give the required slide wire current to produce a null, then the slide wire is standardised at 0.01 V/cm since there is a p.d. of 1.0186 V on 101.86 cm. The reading on the standardising ammeter is noted.

The power supply is changed from the battery to that from the phase-shifting transformer by operating the change-over switch. By readjusting R_1 to give precisely the same current as for d.c. standardising, the same fall of potential is established in the slide wire on a.c.

An unknown voltage E_x can be measured by changing over to the a.c. detector and by adjustment of the slider position for magnitude and phase-shifting rotor position for phase. After two or three successive adjustments a fine balance should be obtained on the detector.

To the left of the slide wire there are often ten fixed resistors each with a value exactly that of 100 cm of slide wire. Moving lead L to different positions achieves increases in range of 1 V per stud.

4.7.3 The cartesian co-ordinate potentiometer (Gall)

This is shown in Fig. 4.14 in simplified form. The standardizing battery, Weston cell and associated circuitry are omitted for clarity.

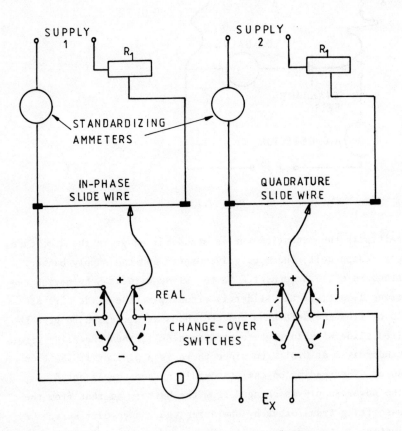

Fig. 4.14

The cartesian co-ordinate potentiometer has two wires each of which is standardised as described for the polar potentiometer. The currents in the two wires are out of phase by precisely 90°.

Measurement of an unknown voltage is achieved by initially setting both change-over switches at random, say both positive. A null on the detector D is sought by moving the sliders on each wire. If by moving one slider towards the left-hand (zero) end it seems that balance could be achieved by going beyond the zero, the unknown voltage is applied to the slide wire in question with reversed polarity by operating the change-over switch. Balance may then be achieved by moving the slider back up the wire. Once the correct switch positions have been ascertained, successively smaller movements of each slider in turn will obtain a null on the detector, the sensitivity of which is increased in steps as balance is approached until only a single millimetre of slider movement will give a noticeable deflection.

The in-phase wire gives ± real quantities
The quadrature wire gives ± j quantities.

4.7.4 Uses of the potentiometer
(a) Measurement of p.d.
Potentiometers can handle only low voltages, the polar type up to possibly 10 V and the cartesian type 1 V when in the forms described in sections 4.7.2 and 4.7.3. To measure higher voltages it is necessary to employ a voltage-reducing device or 'volt-ratio box'. This feeds a proportion of the unknown voltage into the potentiometer and is often in the form of a resistance voltage divider (see Fig. 4.15).
(b) Measurement of current
A standard resistor is series connected into the circuit in which the current is to be measured. Such a resistor generally replaces the conventional ammeter. A value of resistance is chosen such that when the current flows in it the potential difference across it is in the region of 1 V. By Ohm's law

$$\text{Circuit current} = \frac{\text{measured potential difference}}{\text{value of the standard resistor in ohms}}.$$

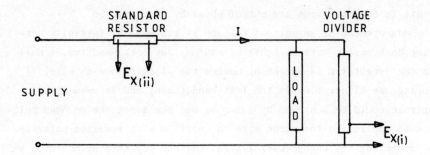

Fig. 4.15

Worked example 4.9

(a) The voltage across a coil was measured by means of a cartesian co-ordinate potentiometer and a volt-ratio box. The following readings were taken with a null on the detector: volt-ratio box ratio 20 : 1. In-phase wire +20 cm, quadrature wire +50 cm. Potentiometer standardised at 0.01 V/cm.

(b) A standard resistor with value 1 Ω was connected in series with the coil and the direct readings on the potentiometer used to measure the p.d. across this resistor were: in-phase wire +65 cm, quadrature wire -6 cm.

Determine (i) the resistance of the coil, (ii) the inductive reactance of the coil, (iii) the power developed in the coil.

(a) Voltage
Real (in-phase): + 20 cm at 0.01 V/cm = 20 × 0.01 = +0.2 V.
This was derived from a 20 : 1 volt-ratio box.
Input voltage = 20 × (+0.2) = +4 V.
Quadrature: + 50 cm at 0.01 V/cm on the quadrature wire = +50 × j0.01 = + j0.5 V.
Input voltage to the volt-ratio box = 20 × (+ j0.5) = +j10 V.
The coil potential difference = (4 + j10) V.

(b) Current
Real (in-phase): 65 × 0.01 = 0.65 V.
Quadrature: -6 × 0.01 = -j0.06 V.

A C BRIDGES AND POTENTIOMETERS

These were derived from a 1 Ω standard resistor so that current is numerically equal to voltage.

$$I = (0.65 - j0.06) \text{ A.}$$

$$Z_{coil} = \frac{V}{I} = \frac{(4 + j10)}{(0.65 - j0.06)} = \frac{10.77\underline{/68.2°}}{0.653\underline{/-5.27°}} = 16.49\underline{/73.47°}$$

$$= 4.69 + j15.8 \text{ Ω.}$$

Answers
(i) $R = 4.69$ Ω.
(ii) $X = 15.8$ Ω.
(iii) Power $= I^2 R = 0.653^2 \times 4.69 = 2$ W.

At this stage it is worthwhile observing that the phasors representing voltage and current may appear anywhere on the Argand diagram, their position being fixed by the supply voltage to the potentiometer. The potentiometer measures values with respect to the reference which is the supply to the device.

In Fig. 4.16 $|V_1| = |V_2|$ and $|I_1| = |I_2|$. These values were derived by using two potentiometers, one fed from the red phase of a three-phase supply and the other from the blue phase. The effect is to move V_2 on by 240° from the position of V_1 but the component is the same, the power factor is the same and in each case, power = VI cos 60° W.

Self-assessment example 4.10   A polar potentiometer is used to determine the power and power factor in a circuit by measuring the supply voltage and current in polar form. The voltage is measured using a volt-ratio box with ratio 150 : 1 and the current by using a 0.1 Ω standard resistor. The readings are

Output from volt-ratio box = $0.75\underline{/159°}$ V
Potential difference across the 0.1 Ω standard resistor = $0.35\underline{/105°}$ V.

Determine (a) the circuit power, (b) the power factor.

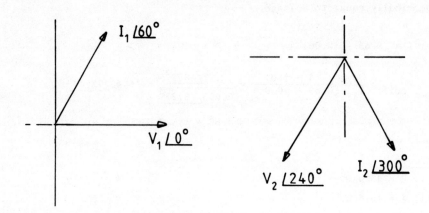

Fig. 4.16

Further problems

4.11. The arms of a balanced a.c. bridge are as follows
AB: an unknown inductive resistor
BC: a non-reactive resistor of 1000 Ω
AD: a non-reactive resistor of 500 Ω
DC: a non-reactive resistor of 1250 Ω in parallel with a capacitor with value 0.3 μF.
The detector is connected to B and C and the a.c. supply to A and C. Determine the resistance and inductance of the unknown resistor.

4.12. The inductance of a coil is measured at 1.5 kHz by the following bridge
arm AB: the coil under test
arm BC: a non-inductive resistance of 100 Ω
arm CD: a standard capacitor with value 0.9 μF
arm DA at balance has a resistance of 1070 Ω in series with a capacitance with value 0.11 μF.
The detector is connected from B to D.
Deduce from first principles the balance equations for the bridge and hence find the impedance of the test coil.

4.13. A Schering bridge is used to test a sample of cellulose acetate film. The bridge consists of
arm AB: the sample under test

arm BC: 100π pF capacitor
arm CD: $1000/\pi$ Ω resistor in parallel with a variable capacitor
arm DA: variable resistor.
A 500 Hz supply is connected to points B and D and a vibration galvanometer between points A and C. Derive the balance equations for the bridge and hence obtain expressions for the effective series resistance and capacitance of the specimen. To obtain balance the variable capacitor must be set to 0.023 μF and the variable resistor to 2500 Ω.

Calculate the capacitance and power factor of the sample.

4.14. A bridge used to determine the incremental inductance of an iron-cored coil has the following values in its arms at balance
arm AB: a non-inductive resistance of 500 Ω
arm BC: the iron-cored coil with resistance R_x and inductance L_x
arm CD: a loss-less capacitor of value 4 μF in series with a non-inductive resistor with value 300 Ω.
arm DA: a loss-less capacitor with value 2 μF.
The supplies are connected to A and C and the detector to D and B. Derive the balance equations for R_x and L_x and determine the values of these for the conditions given.

4.15. A Schering bridge used to measure the capacitance and loss angle of a capacitor with paper dielectric was balanced at 500 Hz under the following conditions
arm AB is a standard 0.1 μF air-spaced capacitor
arm BC is the capacitor under test
arm CD is a 265 Ω non-inductive resistor
arm DA is a 650 Ω non-inductive resistor shunted by a capacitor with value 0.0056 μF.
Determine the power loss in the capacitor when operating at 650 V, 500 Hz assuming that the power factor does not change.

4.16. For the bridge shown in Fig. 4.17 deduce the ranges required for C_3 and P in order that a range of coils can be measured between 0.1 H and 1 H with resistances between 2 Ω and 100 Ω.

4.17. Derive the balance equations for L and R_1 in the bridge shown in Fig. 4.18. The bridge is balanced. Calculate the values of L and R_1 given that R_2 = 550 Ω, R_3 = 450 Ω, R_4 = 5000 Ω, C = 1.06 μF and f = 4 kHz. Determine also the Q-factor of the coil.

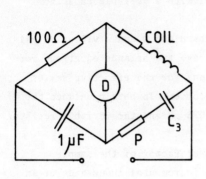
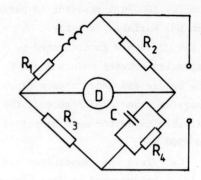

Fig. 4.17   Fig. 4.18

4.18. Describe, with the aid of a simple circuit diagram, one form of a.c. potentiometer explaining its action. State how it may be used to measure current.

An iron-cored coil in series with a 10 Ω standard resistor is connected to an a.c. supply. The potential difference across the resistor and coil are measured using a cartesian co-ordinate potentiometer and are found to be

$$V_{resistor} = (0.576 + j0.688) \text{ V}$$

$$V_{coil} = (-0.378 + j0.534) \text{ V}.$$

The coil has a resistance of 0.8 Ω. Calculate the value of the iron loss in the core of the coil.

4.19. A 50 Hz supply was connected to a choke in series with a standard 1 Ω resistor. The voltage across the choke was found to be (13 + j37) V and across the resistor (0.8 - j0.1) V. Calculate the inductance and effective resistance of the choke.

4.20. An a.c. polar potentiometer used to measure the value of an unknown impedance Z, gave the following readings
23.4$/27.9°$ V when connected across Z
1.3$/-17.1°$ V when connected across a 0.5 Ω standard resistor in series with Z.
Determine (a) the resistance, (b) the reactance of the unknown

A C BRIDGES AND POTENTIOMETERS

impedance stating whether it is inductive or capacitive.

4.21. An iron-cored coil and a standard 1 Ω resistor are connected in series and the potential difference across each is measured using a cartesian co-ordinate potentiometer. The readings are $(3.6 + j1.2)$ V for the coil and $(2.4 - j1.8)$ V for the resistor. Calculate (a) the power developed by the coil, (b) the value of the magnetising current.

4.22. In the measurement of power by means of a cartesian co-ordinate potentiometer, the potential difference readings across a load were taken through a potential dividing network (volt-ratio box) of 100 : 1. The current was measured using a 0.5 Ω standard resistor connected in series with the load.

Voltage output from the volt-ratio box = $(1.14 + j0.6)$ V
Voltage across the standard resistor = $(0.65 - j0.5)$ V.
Determine (a) the phase angle of the load, (b) the power developed by the load, (c) the resistance of the load.

4.23. A wattmeter was calibrated using an a.c. potentiometer. The following readings were taken

voltage output from a 10 : 1 potential divider connected across the wattmeter voltage coil = $(3 + j5.2)$ V

potential difference across a 0.5 Ω standard resistor connected in series with the wattmeter current coil = $(1.78 + j1)$ V.

The wattmeter indication was 200 W.

Determine the value of the true circuit power and the percentage error of the wattmeter at this load.

CHAPTER 4   ANSWERS

4.4. R_x = 2.8 Ω; L_x = 7.42 mH; Q = 3.33
4.5. (a) 25.48 Ω   (b) 2.73 μF   (c) ϕ = 88.75°
4.7. (a) 2.122 nF to 212.2 pF   (b) 50 kHz to 500 kHz
4.10. (a) 231.4 W   (b) 0.588
4.11. 400 Ω; 0.15 H
4.12. R_x = 817 Ω; L_x = 0.096 H; Z_{coil} = 1222 Ω
4.13. C_x = 40 pF; R_x = 183000 Ω; ϕ = $(90 - 1.317)°$; power factor = 0.023
4.14. R_x = 250 Ω; L_x = 0.3 H
4.15. 3.72 W
4.16. C_3: 50 μF to 1 μF; R_3: 1000 Ω to 10000 Ω

4.17. $R_1 = 49.5 \, \Omega$; $L = 0.262 \, H$; $Q = 133$
4.18. Total loss = 0.015 W; copper loss = 0.00644 W; core loss (by difference) = 0.00856 W
4.19. 0.15 H; 10.3 Ω
4.20. $R = 6.36 \, \Omega$; $X = 6.36 \, \Omega$ inductive
4.21. 6.47 W; 2.47 A
4.22. (a) 65.3° (b) 88.2 W (c) 32.8 Ω
4.23. 210.83 W; 5.13% error

5 Electrostatics

5.1 COULOMB'S LAW

The basic model of the atom comprises a positively charged nucleus and orbiting electrons which in total carry the same charge as the nucleus but of opposite polarity. Benjamin Franklin discovered that when certain insulating substances are rubbed vigorously with fabric they become electrically charged. Glass rubbed with silk becomes positively charged since the silk removes electrons from the glass leaving some atoms with insufficient electrons to neutralise the positive charge on the nucleus. Ebonite rubbed with fur becomes negatively charged since the fur donates electrons to the ebonite.

It must be said that the terms 'positive' and 'negative' were chosen quite arbitrarily since the reasons for the phenomena were not understood at the time. All our conventions with respect to currents, forces, charges and the like have been established on the basis of that original choice.

It was found that two charged glass rods repelled one another and two charged ebonite rods repelled one another. However a charged glass rod attracted a charged ebonite rod. This implies that like charges of either polarity repel one another while unlike charges attract one another.

Coulomb experimented with such charges and concluded that the force between two charges q_1 and q_2 could be expressed in the form:

$$F \propto \frac{q_1 q_2}{d^2} \quad \text{or introducing a constant k,} \quad F = \frac{k q_1 q_2}{d^2} \qquad (5.1)$$

where F = force and d = distance between the charges, both in a particular system of units.

5.2 LINES OF ELECTRIC FORCE

A charge of q_1 coulombs at a point will have an effect all round it. Another charge q_2 of like polarity brought near to q_1 from any direction will be repelled along a line drawn through the two charges. From equation 5.1 it will be deduced that the effect will be identical for any particular distance (d) between the charges whatever the direction of approach of q_2.

Suppose charge q_1 is positive and is situated at the centre of an imaginary sphere with radius r metres. For charge $+q_2$ at any point on the surface of that sphere the values of the forces will all be the same. The directions of the forces are all normal (at right angles) to the surface of the sphere. For a straight conductor carrying charge $+q_1$ coulombs per metre length, a charge $+q_2$ brought near to it from any direction will be repelled along a radius drawn normal to the surface of the conductor. Again for any particular distance (r) from the conductor, all the forces will be equal in magnitude. The force directions will be reversed for q_1 negative. A plan showing the directions of forces acting on a positive charge $(+q_2)$ is known as an electrostatic field plot and the line along which the charge moves is a line of electric force. Figures 5.1 and 5.2 show the two cases discussed.

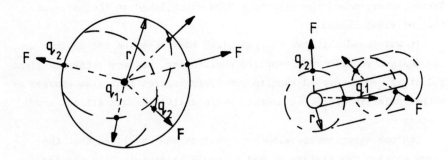

Fig. 5.1　　　　　　　　　　　Fig. 5.2

For a parallel-plate capacitor the field plot will be as shown

ELECTROSTATICS

in Fig. 5.3.

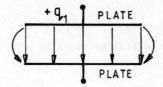

Fig. 5.3

5.3 CHARGE OR FLUX DENSITY

The convention used comes from an electrostatic theorem due to Gauss. This states that the total outward flux normal to the enclosing surface is numerically equal to the charge enclosed.

$$D = \frac{q}{A} \; C/m^2 \tag{5.2}$$

where D = charge or flux density, q = charge (in coulombs) enclosed, A = area (in m^2) across which the flux crosses normally. As an example, consider Fig. 5.1. The charge enclosed is q_1 and the area across which the lines of force and hence electrostatic flux pass is the surface area of the sphere, i.e. $4\pi r^2 \; m^2$. Hence at any radius r m,

$$D = \frac{q_1}{4\pi r^2} \; C/m^2 \tag{5.3}$$

Similarly for the straight conductor in Fig. 5.2, if a 1 m length of conductor is considered upon which the charge is q_1 coulombs, at any radius r m from the conductor, the surface area of the enclosing cylinder is $2\pi r \; m^2$. Hence

$$D = \frac{q_1}{2\pi r} \; C/m^2 \; . \tag{5.4}$$

Finally, for the parallel-plate capacitor in Fig. 5.3, neglecting the curved lines or fringing at the edges of the plates, all lines cross the space between the plates, i.e. the dielectric, normally so that for the dielectric area A m^2,

$$D = \frac{q_1}{A} \text{ C/m}^2 \tag{5.5}$$

5.4 ELECTRIC FORCE OR POTENTIAL GRADIENT E

The electric force or potential gradient in a material is expressed in volts per metre.

$$E = \frac{\text{potential difference in volts}}{\text{distance in metres}} . \tag{5.6}$$

This is analogous to magnetising force (A/m) in electromagnetism.

For the conductor shown in Fig. 5.4, the potential difference between the ends of the conductor is IR volts. This is on one metre length so that E = IR V/m. In a dielectric material the resistance will be much higher and values of potential gradient will be extremely high in consequence. As an example, for the parallel-plate capacitor in Fig. 5.4(b), d is 2 mm and the applied voltage is 4000 V, so that

$$E = \frac{4000}{2} = 2000 \text{ V/mm} \; (= 2 \times 10^6 \text{ V/m}).$$

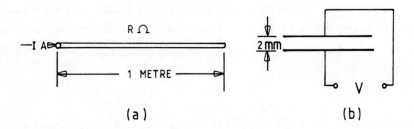

Fig. 5.4

5.5 FORCE ON A CHARGE IN AN ELECTROSTATIC FIELD

The energy (work done) in an electric circuit is given by

$$W = VIt = Vq \text{ joules}. \tag{5.7}$$

where W = work done, I = current in amperes, t = time in seconds and V = potential difference.

ELECTROSTATICS

Charge in coulombs, q = It.

Transposing equation 5.7 gives

$$\frac{W}{q} = V. \qquad (5.8)$$

In words, equation 5.8 states 'potential difference is equal to work done per coulomb of charge moved in a circuit'.

This is a useful concept when it comes to the absolute potential of a point in space. The absolute potential may be expressed in terms of work done in bringing unit charge (1 coulomb) from an infinite distance at which there is no force felt, up to the point at which the potential is to be determined.

Absolute potential at a point = energy required to bring the unit positive charge to this point from one at which no force is felt.

Transposing equation 5.6 gives

p.d. = E × distance.

From equation 5.8

$$\text{p.d.} = \frac{\text{work done}}{\text{charge}}$$

Therefore

$$E \times \text{distance} = \frac{\text{work done}}{\text{charge}} = \frac{\text{force (N)} \times \text{distance}}{\text{charge}}.$$

Cancelling distance on both sides of the equation gives

$$E = \frac{\text{force (N)}}{\text{charge}}. \qquad (5.9)$$

Transposing: Eq = force in newtons. (5.10)

Making the charge 1 coulomb, equation 5.10 becomes

E = force in newtons per coulomb of charge. (5.11)

5.6 PERMITTIVITY

For any arrangement of conductors and dielectric the relationship between electric flux density (D) and potential gradient (E) is expressed as

$$\frac{D}{E} = \varepsilon$$

where ε is the permittivity of the dielectric.

For vacuum and air the permittivity of free space ε_o (= 8.85 × 10^{-12} farad/m) is used. For other materials a relative number must be included. This expresses how much better than vacuum the particular material is and is called the relative permittivity and is denoted by ε_r.

$$\frac{D}{E} = \varepsilon_o \varepsilon_r \quad \text{or} \quad E = \frac{D}{\varepsilon_o \varepsilon_r} \qquad (5.12)$$

5.7 ELECTRIC POTENTIAL IN TERMS OF E

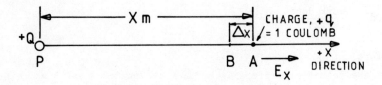

Fig. 5.5

In Fig. 5.5, a charge +Q is located at P. A charge +q is presently situated at A and may be moved along the x-axis.

Between the two charges there will be a force of repulsion since they are of like polarity. At point A an electric field of strength E_x is present due to charge +Q at P.

ELECTROSTATICS

To move the unit charge +q a distance Δx to point B will require work to be done.

Work done = force × distance = $F \Delta x$ Nm. (5.13)

But from equation 5.11, E = force in newtons per coulomb of charge. Hence, E = F and work done = $E \Delta x$ Nm (from equation 5.13).

From equation 5.8, the p.d. between two points is expressed in newton metres of work done per coulomb of charge displaced. Therefore

p.d. between A and B = work done on 1 coulomb = $E \Delta x$.

Coulomb's law states (equation 5.1) that force is inversely proportional to the square of the distance between the charges, so that as x decreases in Fig. 5.5, i.e. as the unit charge is moved in the -x direction, more work has to be done due to the increasing force. This means that the potential is increasing as the charge is moved to the left, i.e. in the -x direction.

To take this into account when required the following is usually written

p.d.$_{(AB)}$ = $-E \Delta x$.

In the limit as $\Delta x \to 0$

$$V_{BA} = - \int_a^b E \, dx \text{ volts.} \qquad (5.14)$$

V_{BA} is the potential difference between two points A and B which are situated at distances a and b respectively from the charge +Q. The potential difference between the points is a scalar quantity which depends only on the distance of each point from the charge and not on the distance between the two points. Thus any route may be taken between A and B as it gives the same result. From this can be deduced that moving a charge round any closed loop in a static electric field from A and back to A requires no energy expenditure.

5.8 RELATIONSHIP BETWEEN ELECTRIC LINES OF FORCE AND EQUI-POTENTIAL SURFACES

Consider a point charge of q coulombs. At any radius r m from the charge,

$$\text{flux density, } D = \frac{q}{4\pi r^2} \text{ C/m}^2. \qquad \text{(equation 5.3)}$$

$$\text{Also} \quad E = \frac{D}{\varepsilon} \text{ V/m}. \qquad \text{(equation 5.12)}$$

Substituting equation 5.3 in equation 5.12 gives

$$E = \frac{q}{4\pi r^2 \varepsilon}. \qquad (5.15)$$

Equation 5.14 gives

$$V_{BA} = -\int_a^b E \, dx.$$

Therefore, substituting equation 5.15 in equation 5.14 gives

$$V_{BA} = -\frac{q}{\varepsilon} \int_a^b \frac{dr}{4\pi r^2} = \frac{q}{4\pi\varepsilon}\left[\frac{1}{r}\right]_a^b$$

$$= \frac{q}{4\pi\varepsilon}\left[\frac{1}{b} - \frac{1}{a}\right] \text{ volts.} \qquad (5.16)$$

If a is an extremely large radius at which no effect is felt, the absolute potential of point B at radius b becomes

$$V_B = \frac{q}{4\pi\varepsilon}\left[\frac{1}{b}\right] \text{ volts.}$$

Since q, 4π and ε are constants it can be seen that all points at radius b have the same potential so that a sphere with radius b metres is an equi-potential surface. The equi-potential surface is at right angles to all the lines of force emanating from the charge (see Fig. 5.1). Equi-potential points may be joined by conductors without any effect since if both ends of the conductor are at the same potential no current will flow.

Any sphere with a point charge at its centre is an equi-potential surface so it would be possible to fabricate thin metal spheres and

ELECTROSTATICS

dispose them around such a charge without affecting the electrostatic field.

For a straight conductor

$$D = \frac{q}{2\pi r} \text{ C/m}^2. \qquad \text{(equation 5.4)}$$

Hence

$$E = \frac{q}{2\pi r \varepsilon} \text{ (from equation 5.12)}$$

$$V_{BA} = -\frac{q}{2\pi\varepsilon} \int_a^b \frac{1}{r} dr = \frac{q}{2\pi\varepsilon} \log_e \frac{a}{b} \text{ volts.}$$

Fixing either a or b as a datum point allows the potential difference between this and any other point to be determined. For example, suppose b is the radius r of the conductor itself, then the potential difference between the surface of the conductor and any point outside the conductor at a distance a normal to its surface is given by

$$V = \frac{q}{2\pi\varepsilon} \log_e \frac{a}{r}.$$

All such points lie on the surface of a cylinder with radius a surrounding the conductor.

Any concentric cylinder surrounding a conductor forms an equipotential surface and cylindrical metal sheaths can be disposed within a solid dielectric with no effect on the field. In fact some bus-bars for power application and transformer bushings contain aluminium foil cylinders to smooth out any irregularity in the field which might be present due to non-uniformity within the dielectric, especially in the surface layers where paper, air and porcelain may be present in varying thicknesses.

<u>Worked example 5.1</u> In Fig. 5.6, conductors A and B are situated in air and carry charges of 10×10^{-9} C/m length with opposite polarity. The conductors are 0.7 m apart.

Calculate (a) the field strength at conductor B due to the charge carried on conductor A, (b) the force developed between the conductors.

(a) Using equation 5.4, the flux density D due to charge on conductor A at any radius r m is given by

$$D = \frac{q}{2\pi r} = \frac{10 \times 10^{-9}}{2\pi \times 0.7}.$$

From equation 5.12,

$$E = \frac{D}{\varepsilon_o}$$

$$= \frac{10 \times 10^{-9}}{2\pi \times 0.7 \times 8.85 \times 10^{-12}} = 256.9 \text{ V/m}.$$

(b) Force = Eq (equation 5.10). Conductor B, which carries a charge of 10×10^{-9} C, is situated in an electrostatic field of strength 256.9 V/m set up by the charge on conductor A. Therefore

$$\text{force} = 256.9 \times 10 \times 10^{-9} = 2.569 \text{ μN/m length}$$

(this is a force of attraction, since unlike charges attract one another).

<u>Self-assessment example 5.2</u> Figure 5.6 shows the electrostatic field plot between a pair of long parallel conductors carrying opposite charges. Using the fact that equi-potential lines are at right angles to force lines, deduce the shape of the equi-potential surfaces between the conductors.

5.9 CAPACITANCE OF A PARALLEL-PLATE CAPACITOR

The capacitance of a capacitor is measured in farads. By definition, a capacitor has a capacitance of 1 farad if, when a potential difference of 1 volt is applied between the plates, 1 coulomb of charge is stored. In symbols

$$q = CV.$$

where C = capacitance in farads.
For two parallel plates as shown in Fig. 5.3

ELECTROSTATICS

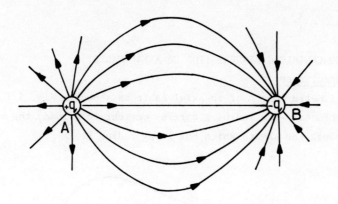

Fig. 5.6

$$E = \frac{D}{\varepsilon}.$$ (equation 5.12)

When the voltage is applied to the capacitor a charge q coulombs is displaced, being taken from one plate and stored on the other.

The charge density, $D = \frac{q}{A}$ C/m^2. (equation 5.5)

Substituting equation 5.5 into equation 5.12 gives

$$E = \frac{q}{A\varepsilon}$$

but $V = -\int_a^b E \, dx$ (equation 5.14)

b and a are general limits giving the potential difference between two points A and B. In the parallel-plate capacitor the potential difference exists between two plates, d metres apart so that the limits become 0 and d.

$$V = -\int_0^d \left(\frac{q}{A\varepsilon}\right) dx = -\left[\frac{qx}{A\varepsilon}\right]_0^d = -\frac{qd}{A\varepsilon} \text{ volts.} \quad (5.17)$$

In this case the negative sign has no significance since either polarity can be obtained by reversing the connections to the plates.

Substituting q = CV in equation 5.17 gives

$$V = \frac{CVd}{A\varepsilon}$$

Hence, $C = \frac{A}{d}\varepsilon$ F (5.18)

5.10 CONCENTRIC CYLINDERS (THE CO-AXIAL CABLE)

5.10.1 Capacitance

Consider 1 metre length of co-axial cable as shown in Fig. 5.7. The centre conductor with radius r carries current to a load; the return being through the sheath which has inner radius R.

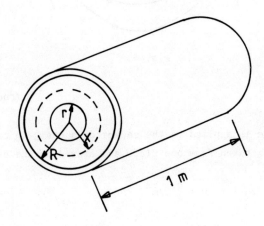

Fig. 5.7

The dielectric has absolute permittivity ε.
The surface area of a cylinder within the dielectric with radius x is $2\pi x$ m^2.

The charge density at this radius $= \frac{q}{2\pi x}$ C/m^2. (equation 5.4)

From equation 5.12, $E = \frac{D}{\varepsilon} = \frac{q}{2\pi x \varepsilon}$ V/m. (5.19)

The potential difference between core and sheath, $V = -\int_a^b E\, dx$.

(equation 5.14)

Now, if the potential of the core is higher than that of the sheath, the potential is highest at radius r and equation 5.14 becomes

ELECTROSTATICS

$$V = -\int_R^r \left(\frac{q}{2\pi\varepsilon}\frac{1}{x}\,dx\right)$$

Substituting equation 5.19 in 5.14

$$V = -\frac{q}{2\pi\varepsilon}[\log_e x]_R^r = -\frac{q}{2\pi\varepsilon}\log_e \frac{r}{R}. \tag{5.20}$$

Substituting $q = CV$ in equation 5.20 gives

$$V = -\frac{CV}{2\pi\varepsilon}\log_e \frac{r}{R} \text{ volts}.$$

Hence, $C = \dfrac{-2\pi\varepsilon}{\log_e \frac{r}{R}}$ farads per metre length of the cable.

This may be rearranged to eliminate the negative sign

$$C = \frac{2\pi\varepsilon}{\log_e \frac{R}{r}} \text{ F/m}. \tag{5.21}$$

5.10.2 Stresses in the dielectric

Transposing equation 5.19 gives

$$q = 2\pi x\varepsilon\ E\ \ C.$$

Substituting in equation 5.20

$$V = \frac{-2\pi x\varepsilon\ E}{2\pi\varepsilon}\log_e \frac{r}{R} = xE \log_e \frac{R}{r}.$$

Transposing: $E = \dfrac{V}{x \log_e \frac{R}{r}}$ V/m. $\tag{5.22}$

Thus, the potential gradient in a co-axial cable has a maximum value when x has a minimum value. The smallest possible value for x within the dielectric is when $x = r$, the inner conductor radius. Thus

$$E_m = \frac{V}{r \log_e \frac{R}{r}} \text{ V/m}. \tag{5.23}$$

Worked example 5.3   A concentric cable has inner sheath radius 5 cm and core radius 2 cm. The core potential is 53.7 kV r.m.s. with respect to the sheath. The relative permittivity of the dielectric is 2.5.

Calculate the capacitance of the cable per kilometre length.

Draw a graph showing the variation of potential gradient through the dielectric between core and sheath.

Substituting the values in equation 5.21 gives

$$C = \frac{2\pi \times 8.85 \times 10^{-12} \times 2.5}{\log_e \frac{5}{2}} \qquad (\varepsilon = \varepsilon_o \varepsilon_r)$$

$$= 151.7 \text{ pF/m} = 151.7 \text{ nF/km} \qquad (\times 1000)$$

The potential gradient throughout the dielectric is determined from equation 5.22.

53.7 kV r.m.s. = 53700 × √2 = 75943.3 V peak.

Substitute in equation 5.22

$$E = \frac{75943.3}{x \log_e \frac{5}{2}} = \frac{75943.3}{x(0.916)} = \frac{82881.2}{x}.$$

For different values of x the corresponding values of E are calculated

x (cm)	2	3	4	5
E (V/cm)	41440	27627	20720	16576

These are the stresses at the various radii which are present at the peak of the voltage wave and are therefore the values which the dielectric must withstand if it is not to fail in service (Fig. 5.8).

<u>Self-assessment example 5.4</u> Investigate the effect on the values of potential gradient throughout the dielectric in the concentric cable in worked example 5.3 if the core has a radius of 1.839 cm, all other parameters remaining unchanged.

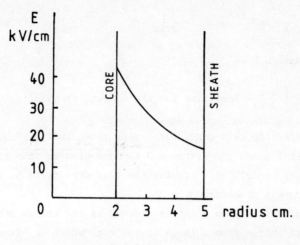

Fig. 5.8

It may be proved mathematically that the minimum value of stress at the core for a fixed sheath radius and working voltage occurs when R/r = natural e (2.718). In self-assessment example 5.3, R/r = 5/1.839 = e and the stress has been reduced from that found in worked example 5.3.

5.11   CAPACITANCE OF A TWIN LINE NOT IN CLOSE PROXIMITY TO EARTH

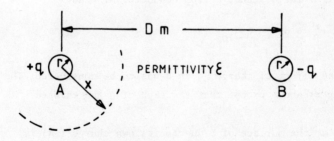

Fig. 5.9

Consider firstly the general case pertaining to a 1 metre length of a single conductor carrying a charge of q coulombs per metre length. At any radius x metres outside the conductor the surface area of the enclosing cylinder is $2\pi x$ m^2. Therefore

$$D = \frac{q}{2\pi x} \text{ C/m}^2 \quad \text{and} \quad E = \frac{q}{2\pi x \varepsilon} \text{ V/m} \qquad \text{(equation 5.19)}$$

Using equation 5.14

$$V = - \int \frac{q \, dx}{2\pi \varepsilon x} \quad \text{(omitting for the moment, the limits)}. \qquad (5.24)$$

Now consider the twin conductor system shown in Fig. 5.9. The length of line is 1 m and the distance D is much greater than the radii of the conductors (r). This enables the flux density to be assumed constant across a diameter.

The potential at a point is defined as the energy required to bring unit charge from a large distance at which no effect is felt up to that point (see section 5.5). Let this large distance be R m. Hence the limits in equation 5.24 become R, the large distance, and r, the distance from the centre of the conductor to its surface. The potential at the surface of A due to its own charge +q is given by

$$V_A = - \int_R^r \frac{q \, dx}{2\pi \varepsilon x} = \frac{q}{2\pi \varepsilon} \log_e \frac{R}{r} \text{ V. (Similar method to equation 5.20.)}$$
$$(5.25)$$

The potential at the surface of conductor B due to charge +q on A is equal to the work done in bringing unit charge from R up to its surface, i.e. the distance D from conductor A. Hence

$$V_B = - \int_R^D \frac{q \, dx}{2\pi \varepsilon x} = \frac{q}{2\pi \varepsilon} \log_e \frac{R}{D} \text{ V}. \qquad (5.26)$$

Now the effect of charge -q on B has to be considered. The method is exactly as above except that the sign of q has changed.

Potential at the surface of B due to its own charge (-q) = $\frac{-q}{2\pi \varepsilon} \log_e \frac{R}{r}$ V. \hfill (5.27)

Potential at the surface of A due to charge (-q) on B = $\frac{-q}{2\pi \varepsilon} \log_e \frac{R}{D}$ V. \hfill (5.28)

ELECTROSTATICS

Total potential at A = $\frac{q}{2\pi\varepsilon}\left(\log_e \frac{R}{r} - \log_e \frac{R}{D}\right)$ (equation 5.25 + equation 5.28).

Subtracting logarithms is to perform the division function, hence

total potential at A = $\frac{q}{2\pi\varepsilon} \log_e\left(\frac{R/r}{R/D}\right) = \frac{q}{2\pi\varepsilon} \log_e \frac{D}{r}$ volts.

Total potential at B = $\frac{-q}{2\pi\varepsilon}\left(\log_e \frac{R}{r} - \log_e \frac{R}{D}\right)$ (equation 5.27 + equation 5.26)

= $\frac{-q}{2\pi\varepsilon} \log_e \frac{D}{r}$ volts.

Total potential at A - total potential at B = p.d. V_{AB}

$V_{AB} = \frac{q}{2\pi\varepsilon} \log_e \frac{D}{r} - \left(\frac{-q}{2\pi\varepsilon} \log_e \frac{D}{r}\right) = \frac{2q}{2\pi\varepsilon} \log_e \frac{D}{r} = \frac{q}{\pi\varepsilon} \log_e \frac{D}{r}$ V.

But q = CV, hence

$V = \frac{CV}{\pi\varepsilon} \log_e \frac{D}{r}$.

Cancelling V on both sides and transposing

$C = \frac{\pi\varepsilon}{\log_e \frac{D}{r}}$ farads per metre length.

<u>Worked example 5.5</u> A long straight cylindrical wire of small diameter carrying a charge of q C/m length is suspended at a large distance from earth. Determine (a) the electric field strength at a distance x m from the conductor, (b) the charge per unit length if the potential difference between two points 1.5 cm and 8 cm from the wire is 125 V.

(a) $D = \frac{q}{2\pi x}$ C/m^2 considering 1 m length of the wire. (equation 5.5)

$E = \frac{D}{\varepsilon_o} = \frac{q}{2\pi x \varepsilon_o}$ V/m (equation 5.12)

(b) Using equation 5.14

$$V_{AB} = -\int_a^b E\,dx = -\int_8^{1.5} \frac{q}{2\pi\varepsilon_0}\frac{1}{x}\,dx = \frac{q}{2\pi\varepsilon_0}\log_e \frac{8}{1.5}.$$

But V_{AB} is given as 125 V, therefore

$$125 = \frac{q}{2\pi\varepsilon_0}\log_e \frac{8}{1.5}.$$

Transposing gives

$$q = \frac{2\pi\varepsilon_0 \times 125}{\log_e \frac{8}{1.5}} = \frac{2\pi \times 8.85 \times 10^{-12} \times 125}{1.674}$$

$$= 4.15 \times 10^{-9} \text{ C/m}.$$

Worked example 5.6 A parallel-plate capacitor has the following details:
area of plates 0.16 m^2
distance between the plates 0.8 mm
working voltage 3000 V (r.m.s.)
relative permittivity of the dielectric 3.3.

Determine (a) the peak electric stress in the dielectric assuming the applied voltage is of sine form, (b) the capacitance of the capacitor.

(a) 3000 V r.m.s. = 3000 × √2 = 4242.6 V peak.

$$\text{Peak stress} = \frac{\text{peak voltage}}{\text{distance between the plates}} = \frac{4242.6}{0.8} = 5303.3 \text{ V/mm}.$$

(b) $C = \dfrac{\varepsilon_0 \varepsilon_r A}{d}$ F (equation 5.18)

$$= \frac{8.85 \times 10^{-12} \times 3.3 \times 0.16}{0.8 \times 10^{-3}} = 5840 \text{ pF}.$$

Worked example 5.7 A concentric cable is required to have a capacitance of 160 pF/m length. The core diameter, which is determined by its current carrying capacity, is 1 cm. The relative permittivity of the dielectric is 4.

Determine the necessary inside diameter of the sheath.

ELECTROSTATICS

The capacitance of concentric cable per metre length =

$$\frac{2\pi\varepsilon_o \varepsilon_r}{\log_e \frac{R}{r}} \text{ F} \qquad \text{(equation 5.21)}.$$

Hence

$$160 \times 10^{-12} = \frac{2\pi \times 8.85 \times 10^{-12} \times 4}{\log_e \frac{R}{r}}.$$

Transposing

$$\log_e \frac{R}{r} = \frac{2\pi \times 8.85 \times 10^{-12} \times 4}{160 \times 10^{-12}} = 1.39$$

Therefore

$$e^{1.39} = \frac{R}{r} = 4.015$$

Now $\frac{D}{d} = \frac{R}{r}$, where D = inner diameter of sheath, d = core diameter.
Hence

$$\frac{D}{d} = \frac{D}{1} = 4.015$$

Therefore D = 4.015 cm.

Self-assessment example 5.8 A single-phase circuit comprises two parallel conductors 0.6 cm in diameter spaced 1 m apart in air.
 Determine for 1 km of this circuit (a) the capacitance, (b) the capacitive reactance, (c) the charging current if the potential difference between the conductors is 19052 V at 50 Hz. (Line charging current = normal capacitive current = V/X_C A.)

Worked example 5.9 A concentric cable is to be operated with an r.m.s. voltage to ground of 159000 V. The maximum potential gradient within the cable is not to exceed 60 kV/cm. Determine (a) the radius of the core and the inner radius of the sheath for ideal operation, (b) the stress on the dielectric at the inner surface of the sheath.

(a) $V_m = \sqrt{2} \times 159000 = 224860$ V.

To give minimum stress at the core (ideal operation) $\frac{R}{r} = e$.

Equation 5.23 gives

$$E_m = \frac{V}{r \log_e \frac{R}{r}}$$

and for ideal operation this becomes

$$E_m = \frac{V}{r \log_e e} = \frac{V}{r} .$$

Therefore $60000 = \frac{224860}{r}$ or $r = \frac{224860}{60000} = 3.75$ cm.

Outer sheath inner radius = $e \times 3.75 = 10.2$ cm.

(b) At radius x, stress = $\dfrac{V}{x \log_e \frac{R}{r}}$ (equation 5.22).

Therefore, at 10.2 cm, $E = \frac{224860}{10.2} = 22045$ V/cm.

5.12 ENERGY STORED IN AN ELECTROSTATIC FIELD

Consider a capacitive system initially charged such that the potential difference between the electrodes is v volts.

A small quantity of charge Δq coulombs is supplied to the system which causes the voltage to rise to $(v + \Delta v)$ volts.

The mean value of voltage during the change = $\left(v + \dfrac{\Delta v}{2}\right)$ volts.

The energy supplied to an electric circuit is given as $W = Vq$ J (equation 5.7), hence the change in energy, ΔW, due to arrival of charge Δq will be

$$\Delta W = \left(v + \frac{\Delta v}{2}\right)\Delta q = v \Delta q + \frac{\Delta v \, \Delta q}{2} \text{ J}.$$

If the changes are made small, as Δq approaches zero, Δv approaches zero so that the product $\Delta v \, \Delta q$ is small enough to be ignored. In the limit, $\Delta v \to 0$, $\Delta q \to 0$ and $dW = v \, dq$.

The total energy supplied as the charge is increased from zero to a final value Q is obtained by integration

$$\int dW = \int_0^Q v\, dq.$$

But since $q = Cv$, $v = \frac{q}{C}$, which gives

$$\int dW = \int_0^Q \frac{q\,dq}{C}.$$

Total energy, $W = \left[\frac{q^2}{2C}\right]_0^Q = \frac{Q^2}{2C}$.

Again, $Q = CV$, where V = final potential difference. Substituting gives

$$W = \frac{CV^2}{2}\ J. \qquad (5.29)$$

Now consider a very small element within the dielectric of the capacitive system. Whatever the arrangement of electrodes, i.e. concentric cable, parallel conductors, etc., with a small element the force lines may be considered parallel and the equi-potential lines may be considered parallel. In Fig. 5.10, AD is parallel to CB and DC is parallel to AB.

$DA = CB = \Delta x.$ $DC = AB = \Delta y.$

A 1 m length of such an element is now considered. Surface DAPS is crossed by the electrostatic force lines and is an equi-potential surface. CBQR is also an equipotential surface (see section 5.8). These surfaces can be made conducting with no change in the field and when in this form could form the plates of a parallel-plate capacitor.

Capacitance, $C = \varepsilon_o \varepsilon_r \frac{A}{d}$ (equation 5.18)

$$= \varepsilon_o \varepsilon_r \frac{\Delta x \times 1}{\Delta y}\ F. \qquad (5.30)$$

The potential difference across Δy is V volts, therefore $E = \frac{V}{\Delta y}$ (equation 5.6).

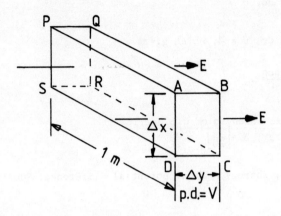

Fig. 5.10

Transposing gives

$$V = E \, \Delta y. \tag{5.31}$$
$$\text{Energy stored} = \tfrac{1}{2}CV^2 \tag{equation 5.29}.$$

Substituting 5.30 and 5.31 into 5.29 gives

$$\text{energy stored} = \tfrac{1}{2}\varepsilon_o \varepsilon_r \frac{\Delta x \times 1}{\Delta y} (E \, \Delta y)^2$$

$$= \tfrac{1}{2}\varepsilon_o \varepsilon_r \, \Delta x \, E^2 \, \Delta y \, \text{J}.$$

Now, for 1 m length the volume of the dielectric = $1 \times \Delta x \times \Delta y \, \text{m}^3$.

Therefore, the energy stored per unit volume = $\tfrac{1}{2}\varepsilon_o \varepsilon_r E^2 \, \text{J}.$ (5.32)

Substituting for E from equation 5.12 into equation 5.32 gives

$$\text{energy stored per unit volume} = \tfrac{1}{2}\varepsilon_o \varepsilon_r \left(\frac{D}{\varepsilon_o \varepsilon_r}\right)^2$$

$$= \frac{D^2}{2\varepsilon_o \varepsilon_r} \, \text{J/m}^3 \tag{5.33}$$

ELECTROSTATICS

5.13 DIELECTRIC MATERIALS

A dielectric material is one which has a very high resistivity when compared with that of a conductor such as copper and aluminium, etc. Dielectrics are therefore used to maintain separation between conductors which are at different potentials, i.e. electric power lines from earth and from each other, capacitor plates, etc.

5.13.1 Electric strength of dielectrics

The breakdown voltage gradient (E) of an insulating material is dependent upon its thickness but not proportional to that thickness.

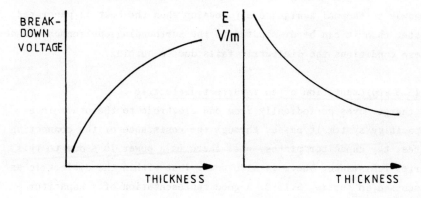

Fig. 5.11

Typical shapes for breakdown voltage versus thickness and potential gradient (E) versus thickness are shown in Fig. 5.11. The precise shapes of the curves depend on the shape of the test electrodes: flat, cylindrical, pointed, etc.; the material being tested: air, glass, plastics; and on the atmospheric pressure and humidity if the system is not sealed. Actual breakdown voltages quoted in reference works have to be treated with considerable care unless the precise conditions of the test are known.

5.13.2 Dielectric failure

Where a dielectric is subjected to alternating stresses, dielectric hysteresis has to be considered. This involves the outer electrons being attracted away from the nucleus in the direction of the electrostatic field. Provided that the force of attraction is not too great most electrons do not move very far through the material. However work is done in this re-orienting of the electrons during each reversal of stress and heat is produced. Some electrons do finally end up on the electrode system and constitute a small charge movement through the dielectric. Such a movement of charge is a leakage current in phase with the voltage and again heat is produced.

Dielectrics become conducting if they are supplied with sufficient energy either in the form of applied potential difference or directly as heat. The insulation resistance of most dielectrics falls rapidly as the temperature is increased so that leakage current increases thus generating further heat and a condition known as thermal runaway or thermal avalanche may develop when the heat is generated faster than it can be dissipated to the surrounding environment. Under these conditions the dielectric fails due to burning.

5.13.3 Representation of an imperfect dielectric

As charge moves periodically from one electrode to the other in a capacitive system it passes through the resistance of the connecting wires, the capacitor plates, etc. There is a power loss due to this series resistance. There is also power lost within the dielectric as discussed in section 5.13.2. A good representation of a capacitor would therefore be as shown in Fig. 5.12.

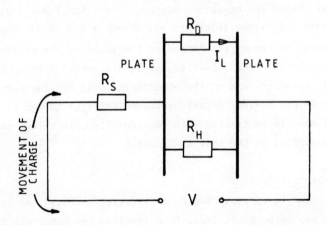

Fig. 5.12

ELECTROSTATICS

R_s = series resistance, R_D = resistance of the dielectric through which I_L, the leakage current, flows. R_H is a resistance which represents the hysteresis loss.

However it is impossible to establish the individual value of such resistors and measurements using equipment such as the Schering bridge (section 4.6.2) provide us with values of capacitance and effective resistance values for either series or parallel configurations.

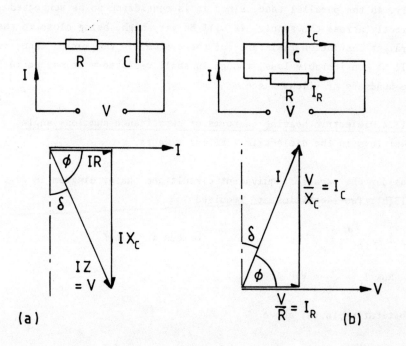

Fig. 5.13

The circuit and phasor diagrams for the series equivalent are shown in Fig. 5.13(a). The two voltage drops IR and IX_C sum phasorially to V, the supply voltage. The circuit phase angle is ϕ. Now if the resistance were zero, V would lag I by exactly 90°. The capacitor would then be perfect. The difference between the perfect and the practical case is the angle δ which is known as the loss angle. This will usually be a fraction of one degree and rarely more than two degrees.

185

Exactly the same result is achieved by considering the parallel equivalent shown in Fig. 5.13(b). The branch currents I_C and I_R sum to the total current which has the same value as in the series case. Notice that V/X_C is that value of current which would flow in a perfect capacitor. This is called the charging current in transmission line work.

Both values of capacitance C are in all practical cases identical, but the series and parallel resistance values will differ considerably. In the parallel case, since it is considered to be connected directly across the supply, it will be very high, being close to the straight insulation resistance of the system. In the series case, it will be considerably less, though in small capacitors it may be in the hundreds of kilohm range.

5.13.4 Dielectric heating in terms of capacitance and loss angle

Power loss in the dielectric = VI cos ϕ watts.

Consider the parallel equivalent circuit and phasor diagram in Fig. 5.13(b). Two identities are required

(a) $I = \dfrac{I_C}{\cos \delta}$ 	(b) $\cos \phi = \sin \delta$.

Now $I_C = \dfrac{V}{X_C} = V\omega C$ A.

Substituting in (a) above

$$I = \dfrac{V\omega C}{\cos \delta}.$$

Therefore, power loss = VI cos ϕ = $V \times \dfrac{V\omega C}{\cos \delta} \times \cos \phi$ W.

Using (b) above

$$\text{Power loss} = \dfrac{V^2 \omega C}{\cos \delta} \sin \delta$$

$$= V^2 \omega C \tan \delta \text{ W.} \quad \left(\dfrac{\sin \delta}{\cos \delta} = \tan \delta\right). \quad (5.34)$$

In a cable this power loss is a nuisance and provision has to be made to get rid of the heat produced. At extra-high voltages the V^2 term becomes very large and the dielectric loss can exceed the cable I^2R losses. In welding machines for joining plastics, the frequency is increased into the kilohertz or megahertz range and the dielectric loss is used to melt the internal surfaces to be joined so that they fuse together under pressure.

Self-assessment example 5.10   A concentric cable has core radius 2 cm and inside sheath radius 5.5 cm. The relative permittivity of the dielectric is 2.3. The loss angle, $\delta = 0.16°$, and the working voltage (V) = 194 kV at 50 Hz.
Calculate, for a 1 km length of cable, (a) the capacitance, (b) the charging current, (c) the power loss.

5.13.5 Properties of dielectrics

Dielectrics used in capacitor manufacture need to have high permittivities in order to give the greatest possible capacitance within the smallest physical size. Paper/paraffin wax insulated capacitors are used at all voltages up to 150 kV in circuits where the loss angle is not too significant. They are not polarity-sensitive and sizes from 250 pF to 10 µF are readily available. Electrolytic capacitors are used at voltages up to about 600 V where the applied potential is uni-directional. The dielectric is an extremely thin film of aluminium oxide which is chemically developed on aluminium foil. Electrical contact is made with the oxide through a conducting liquid, the electrolyte. The losses are greater than with the paper type but very large values of capacitance can be produced within a very small physical size.

Where the loss angle must be small, mica capacitors may be used. These comprise thin layers of mica and foil clamped together. These are not polarity-sensitive but are expensive. Sizes from 25 pF to 0.25 µF are available.

Capacitors using titanium oxide or polycarbonate have extremely high capacitance within a given physical size when used at low voltages, e.g. up to 10 V.

In power cable manufacture, high capacitance is a distinct disadvantage. This gives rise to system stability problems and a rising voltage on the system during periods of light loading. This is known as the Ferranti effect. In addition as has already been seen in section 5.13.4, cable heating is directly proportional to capacitance (equation 5.34). The capacitance of oil-impregnated paper insulated cables is relatively large (equation 5.21). This capacitance could be reduced if the relative permittivity could be made small, ideally unity. Several cable models have been evaluated and one which uses polythene tapes pressurised with sulphur hexafluoride gas at a pressure of approximately 14 bar gives a reduction in capacitance of 40%, of thermal resistivity 30% and dielectric losses of over 95% compared with a corresponding paper/oil insulated cable. The costs of the two types of cable are comparable.

Table of properties of selected dielectrics

Material	ε_r	Approx. breakdown strength (kV/mm)	Resistivity, Ωcm
Air	1.0006	3	
Transformer oil	2.3	13	7×10^{12}
Paraffin wax	2.2	7.5	0.9×10^{16}
Mica	7	180	2×10^{15}
Glass	6.5	5–7	2×10^{14}
Varnished paper	2	15	up to 10^6
Polythene	1.5	40	10^{17}
PTFE	2	15	10^{17}
Titanium oxide	70–80	Used in extremely thin layers in capacitors	—
Polycarbonate	2.8	— ditto —	$10^{17}+$

5.14 FIELD PLOTTING

Where electrode systems are asymmetric, mathematical analysis is difficult and equi-potential lines may be plotted using models of the systems involved. Such plots may be used to determine capacitance. One method is to use demineralised water as the dielectric. A model of the system in section is made from copper or brass and is supported

ELECTROSTATICS

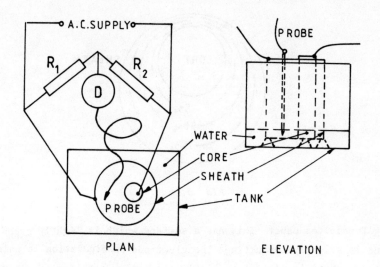

Fig. 5.14

on insulators in a tank so that it is partly immersed in the demineralised water. In Fig. 5.14, a section of cable is shown in which the core is not concentric with the sheath. A Wheatstone bridge is used to determine points of equal potential in the water. For example, with $R_1 = R_2$, the probe is moved about in the water until the detector indicates a zero. At this point in the water the potential difference between the core and the probe is equal to that between the probe and the sheath. Many such points can be found and an equipotential line drawn through them.

Other equi-potential lines are drawn by the same method using different ratios of R_1 and R_2. A typical result is shown in Fig. 5.15.

With $R_2 = 5000\ \Omega$ and $R_1 = 15000\ \Omega$, the probe will detect the 25% equi-potential line, i.e. for 100 V on the core and the sheath earthed, the p.d. between the core and any point on this line is 25 V.

For $R_2 = 15000\ \Omega$ and $R_1 = 5000\ \Omega$, the 75% equi-potential line may be drawn.

Since stress lines are at right angles to equi-potential lines, these may be drawn in by inspection. Similar results are achievable

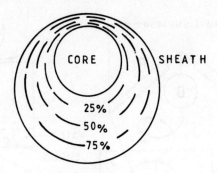

Fig. 5.15

using Teledeltos paper. This has a surface which is lightly graphited making it slightly conducting. The electrode configuration is painted on the graphite in plan using highly conducting metal paint. The procedure is as with the demineralised water, using a probe to mark the paper at the required null points, between which the equi-potential lines are drawn.

5.15 CAPACITANCE OF SYSTEMS DEDUCED FROM A FIELD PLOT
5.15.1 With stress and equi-potential lines making true squares

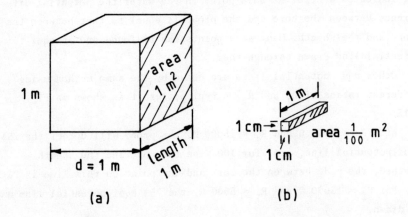

Fig. 5.16

ELECTROSTATICS

Consider the capacitance of a parallel-plate capacitor formed using a 1 m cube of dielectric as shown in Fig. 5.16(a)

$$C = \frac{\varepsilon_o \varepsilon_r A}{d} \text{ F} \qquad \text{(equation 5.18)}.$$

Now since $A = 1 \text{ m}^2$ and $d = 1 \text{ m}$

$$C = \varepsilon_o \varepsilon_r \text{ F}.$$

Reducing the area and distance between the plates in the same proportion, while leaving the length unchanged at 1 m, results in the same value of capacitance. The capacitor shown in Fig. 5.16(b) also has a capacitance of $\varepsilon_o \varepsilon_r$ farad. A further reduction to 1 mm height and 1 mm between the plates, provided that the length is still 1 m, leaves the capacitance unchanged.

One method of determination of the capacitance of a parallel-plate capacitor would be to consider the dielectric to be split up into such squares.

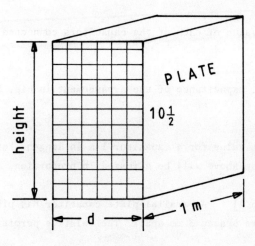

Fig. 5.17

In Fig. 5.17, the face of the dielectric is divided into squares. The distance (d) between the plates is divided into an exact number of parts. Let this number be n. In Fig. 5.17, n = 6.

Using the same scale, the height is divided into parts. There will usually be a part division involved. Let this number be m. In Fig. 5.17, m = 10½.

Each of the vertical lines is an equi-potential surface and so the overall capacitance of the capacitor would be unchanged if these were made conducting. There are m squares in parallel between each equi-potential surface and so there are m capacitors in parallel. Similarly there are n such groups in series.

Capacitors in parallel have their values added directly to give an equivalent capacitance. Hence m capacitors in parallel have capacitance $m\varepsilon_o\varepsilon_r$ farad.

Capacitors in series have their values added as reciprocals to give an equivalent capacitance. Therefore for n identical capacitors

$$\frac{1}{C_{eq}} = n\left(\frac{1}{C}\right).$$

Transposing

$$C_{eq} = \frac{C}{n}$$

where C is the value of each of the capacitors connected in series (= $m\varepsilon_o\varepsilon_r$ F).

Hence, the total capacitance of the arrangement in Fig. 5.17 =
$$\frac{m\varepsilon_o\varepsilon_r}{n} \text{ F} . \tag{5.35}$$

This will be the value for a capacitor 1 m in length. For other lengths the value above will be adjusted in proportion.

<u>Worked example 5.11</u>   A parallel-plate capacitor has plates 10 cm × 15 cm. They are spaced 3 mm apart. The relative permittivity of the dielectric is 4.

Determine the capacitance of the capacitor (a) by the use of squares, (b) directly from equation 5.18.

It is of no significance whether 10 cm is considered as the height or the length of the plates. In Fig. 5.18, 10 cm is arbitrarily chosen as the height.

ELECTROSTATICS

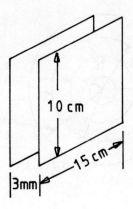

Fig. 5.18

(a) Any convenient size square may be used provided that the 3 mm distance between the plates is divided exactly. 3 mm squares, 1.5 mm squares or 1 mm squares, for example, could be used.
Using 1 mm squares, n = 3 and m = 100

For a 1 m length, $C = \dfrac{m\varepsilon_o \varepsilon_r}{n}$ F (equation 5.35)

$$= \frac{100 \times 8.85 \times 10^{-12} \times 4}{3} = 1.18 \times 10^{-9} \text{ F.}$$

For a 15 cm length, $C = \dfrac{15}{100} \times 1.18 \times 10^{-9} = 177$ pF.

(The reader might care to verify that the same result is obtained however many squares are used.)

(b) Equation 5.18 gives

$$C = \frac{\varepsilon_o \varepsilon_r A}{d} = 8.85 \times 10^{-12} \times 4 \times \frac{10 \times 15 \times 10^{-4}}{3 \times 10^{-3}}$$

$$= 177 \text{ pF.}$$

<u>5.15.2 With stress and equi-potential lines forming curvilinear squares</u>
Consider a 1 m length of co-axial cable, a cross-section of which

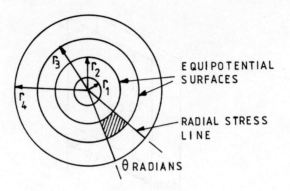

Fig. 5.19

is shown in Fig. 5.19. The core radius is r_1 and the inner sheath radius is r_4. By the correct choice of spacing between stress lines and between equi-potential surfaces in a field plot, the 'near-squares' formed (one 'square' hatched in Fig. 5.19), or curvilinear squares as they are called, can be made to have the same capacitance as a true square in section 5.15.1.

From equation 5.21, the capacitance between cylindrical equi-potential lines at radii r_p and r_q respectively is given by

$$C = \frac{2\pi\varepsilon_o \varepsilon_r}{\log_e \frac{r_q}{r_p}} \text{ farads per metre length.}$$

Hence for a segment θ radians the capacitance will be

$$C_\theta = \frac{\theta}{2\pi} \times \frac{2\pi\varepsilon_o \varepsilon_r}{\log_e \frac{r_q}{r_p}} \text{ F/m}$$

$$= \frac{\theta \varepsilon_o \varepsilon_r}{\log_e \frac{r_q}{r_p}} \text{ F/m}. \tag{5.36}$$

Hence if $\theta = \log_e \frac{r_q}{r_p}$ (5.37), equation 5.36 becomes $C = \varepsilon_o \varepsilon_r$ farad, which is the same as that for a perfect square.

ELECTROSTATICS

The dielectric may be subdivided by as many equi-potential surfaces as desired, three or four usually being selected to give reasonably-shaped curvilinear squares.

Now from equation 5.37

$$\theta = \log_e \frac{r_q}{r_p} \quad \text{hence,} \quad e^\theta = \frac{r_q}{r_p}. \tag{5.38}$$

For core radius r_1 and equi-potential surfaces at r_2 and r_3 with sheath inner radius r_4, each ratio of successive radii must satisfy the condition in turn. Thus

$$e^\theta = \frac{r_2}{r_1}, \quad e^\theta = \frac{r_3}{r_2} \quad \text{and} \quad e^\theta = \frac{r_4}{r_3}.$$

Multiplying together gives

$$e^\theta \times e^\theta \times e^\theta = \frac{r_2}{r_1} \times \frac{r_3}{r_2} \times \frac{r_4}{r_3}$$

$$e^{3\theta} = \frac{r_4}{r_1}. \tag{5.39}$$

This equation is used to determine the value of θ and hence the number of segments.

<u>Worked example 5.12</u> Draw a flux/equi-potential curvilinear square plot for a concentric cable with core radius 0.75 cm and inner sheath radius 4 cm. Determine from the plot, the capacitance of the cable per metre run if the relative permittivity of the dielectric is 3.5. Compare the result with that directly obtained from equation 5.21.

Using two equi-potential surfaces within the dielectric so forming three capacitors in series in each segment

$$e^{3\theta} = \frac{r_4}{r_1} \quad \text{(equation 5.39)}.$$

Substituting in values gives

$$e^{3\theta} = \frac{4}{0.75} = 5.33$$

Therefore

$$3\theta = \log_e 5.33 = 1.674$$

$$\theta = \frac{1.674}{3} = 0.558 \text{ radians } (31.9°).$$

Since 2π radians comprise a complete circle, with each segment having a value of 0.558 rad, there will be $2\pi/0.558 = 11.26$ segments in the plot.

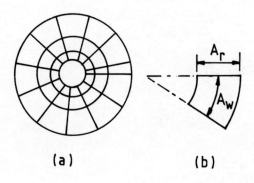

Fig. 5.20

The plot is shown in Fig. 5.20. If 1 metre length of the cable is considered, each segment has three capacitors in series between the core and sheath. There are 11.26 such groups of capacitors in parallel.

Each curvilinear square, 1 m in length, has a capacitance of $\varepsilon_o \varepsilon_r$ farad.

Therefore, total capacitance per metre length = $\varepsilon_o \varepsilon_r \times \frac{11.26}{3}$

$$= 8.85 \times 10^{-12} \times 3.5 \times \frac{11.26}{3}$$

$$= 116.26 \text{ pF.}$$

Using equation 5.21

ELECTROSTATICS

$$C = \frac{2\pi\varepsilon_o \varepsilon_r}{\log_e \frac{4}{0.75}} = 116.26 \text{ pF}.$$

It should be observed from Fig. 5.20 that in any size of curvilinear square the mean width of the segment A_w is equal to its length measured along a radius A_r. This fact will be helpful with asymmetric electrode plots.

<u>Self-assessment example 5.13</u>   Using three equi-potential surfaces within the dielectric (four series capacitors), use the principle of curvilinear squares to determine the capacitance of a concentric cable with core radius 2 cm and inner sheath radius 5.5 cm. The relative permittivity of the dielectric is 2.3.

Considerable use is made of this method to determine the capacitance of asymmetric systems. Consider again Fig. 5.15. By selecting (say) the 50% equi-potential line, the core and the sheath only, the dielectric space may be divided into curvilinear squares using dividers. The distance from the core to equi-potential line d_1 is stepped off midway between core and the equi-potential line circumferentially. Figure 5.21 should make this clear.

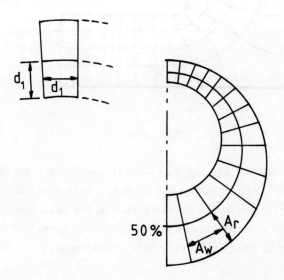

Fig. 5.21

Re-setting the dividers at each stage it is possible to create curvilinear squares progressing from the top to the bottom. In the half section shown in Fig. 5.21 there are 12.5 segments. The other half is symmetrical so there will be 25 sections in all. In each segment there are two capacitors in series between core and sheath.

Capacitance per metre length = $\varepsilon_o \varepsilon_r \times \dfrac{25}{2}$ F.

<u>Self-assessment example 5.14</u>   What is the capacitance per metre length of the L-shaped air-insulated bus-bar system shown in Fig. 5.22?

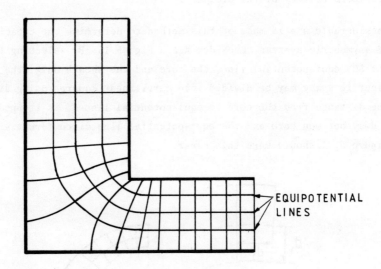

Fig. 5.22

(Notice that not all the curvilinear squares are perfectly regular especially those close to the corners. This makes the method a little imprecise but quite accurate enough for most applications. The alternative is a complicated calculation which itself is subject to error because of the simplifying assumptions made.)

<u>Further problems</u>
5.15. In Fig. 5.23, A and B are two long air-insulated conductors of

small diameter in a plane normal to that of the paper. Each carries a charge of 5×10^{-8} C/m length.

Determine (a) the electric field strength at point P due to the charge on conductor A, (b) the electric field strength at point P due to the charge on conductor B, (c) the resultant of (a) and (b), taking into account their directions and adding as vectors, and (d) the resultant force on a positive charge of 1 mC placed at point P.

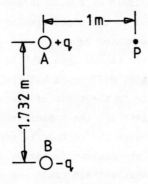

Fig. 5.23

5.16. Two charges q_1 and q_2 are separated in air by a distance of 1 mm. $q_1 = 2 \times 10^{-9}$ C, $q_2 = 1.6 \times 10^{-10}$ C.
Determine the value of the force between the two charges.

5.17. (a) Determine the potential difference between two points situated 10 cm and 20 cm respectively from a point charge of 3 μC in vacuum. (b) Suppose a charge were distributed uniformly on a long straight wire at a density of 3 μC/m length, what would be the potential difference between two points 10 cm and 20 cm from the wire?

5.18. At a point in air 1.4 m from an isolated charge-carrying conductor, an electrostatic energy density of 26.28×10^{-15} J/m^3 is present. What is the magnitude of the conductor charge in coulombs per metre run?

5.19. A three-phase circuit comprises three lead covered cables laid parallel in the ground. The core of each cable has a radius of 2.5 cm and an inner sheath radius of 6 cm. The relative permittivity of the dielectric is 3 and the loss angle δ is 0.2°. The circuit operates at

132 kV (line) and 50 Hz.

Determine, for a 10 km length of circuit, (a) the capacitance of each cable to earth, (b) the charging current, (c) the total dielectric power loss of the circuit. (For the charging current you will need the voltage to earth of each conductor.)

5.20. Using the table of properties of dielectrics in section 5.13.5, determine the area and spacing between two plates which form a standard capacitor of value 10 pF when they are immersed in transformer oil. The working voltage is 15 kV r.m.s. (assume sinusoidal form). Allow a safety factor of 3 on the plate spacing.

5.21. Two metal plates separated by a sheet of glass form a parallel-plate capacitor. Each plate has an area of 0.36×10^{-3} m^2. The glass is 3 mm thick. The potential difference between the plates is 1.5 kV d.c. Using the information in the table of properties of dielectrics in section 5.13.5, calculate (a) the capacitance of the capacitor, (b) the charge displaced when the voltage is applied, starting from the uncharged state, (c) the minimum factor of safety at which the capacitor is operating (what voltage could the glass support before breakdown?), (d) the value of capacitance if the glass plate were removed.

5.22. An isolated twin line comprises two air-insulated conductors each with radius 0.5 cm, spaced 1.5 m apart. The potential difference between the lines is 6350 V at 50 Hz.

Calculate, per kilometre length, (a) the line capacitance, (b) the r.m.s. value of charge carried by each wire, (c) the charging current.

5.23. A concentric cable has a capacitance of 2.5 µF and a dielectric power loss of 15.6 kW when operated at 66 kV and 50 Hz. Determine (a) the value of the loss angle, (b) the equivalent parallel resistance of the cable.

5.24. Determine the area of the plates of a parallel-plate capacitor and the spacing between them allowing a safety factor of 3, if the energy stored is to be 0.06 J at a steady voltage of 5000 V. The breakdown potential gradient of the dielectric is 35 kV/cm and it has a relative permittivity of 4.3.

5.25. A concentric cable has core radius 1.5 cm and sheath inner radius 3.5 cm. The permittivity of the dielectric is 2.7. Using the method of curvilinear squares, deduce the capacitance of the cable per metre length. (Use two equi-potential surfaces within the dielectric.)

5.26. A field plot between two metal plates is shown in Fig. 5.24. The relative permittivity of the dielectric is 2.5. What is the capacitance per metre length of the system?

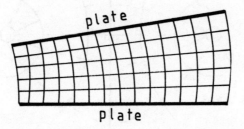

Fig. 5.24

5.27. A field plot between two copper bus-bars is shown in Fig. 5.25. They are air-insulated. Determine the capacitance of a 100 m length of the bus-bars using the method of curvilinear squares.

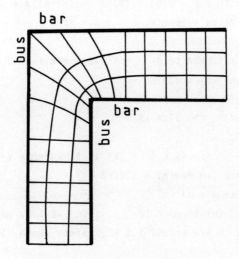

Fig. 5.25

CHAPTER 5 ANSWERS

5.2.

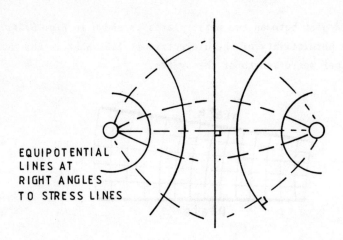

Fig. 5.26

5.4. 41296 V to 15188 V
5.8. (a) 4.786×10^{-9} F/km (b) $X_C = 665085$ (c) $I_c = 0.0286$ A
5.10. (a) 126.43×10^{-9} F (b) 7.705 A (c) 4174 W
5.13. $\theta = 14.49°$; 24.84 segments; C = 126.4 pF/m
5.14. 37.6 pF/m
5.15. (a) 899.2 V/m from A to P (b) 449.6 V/m from P to B
      (c) 935 V/m (d) 0.935 N
5.16. 2.88 mN
5.17. (a) 134.88 kV (b) 37.4 kV
5.18. 6×10^{-12} C/m
5.19. (a) 1.905 µF (b) 45.6 A (c) 12.13 kW/core (= 36.4 kW total)
      (Phase voltage (to earth) = $132/\sqrt{3}$ kV)
5.20. d = 4.9 mm; area = 24 cm^2
5.21. (a) 6.9 pF (b) 10.35×10^{-9} C (c) at 5 kV/mm, max. voltage would be 15 kV, actual 1.5 kV, safety factor 10
      (d) 1.06 pF
5.22. (a) 4.87 nF (b) 30.9 µC (c) 9.72 mA
5.23. (a) 0.26° (b) R = 280 kΩ
5.24. C = 4.8 nF; d = 0.428 cm; area = 0.54 m^2

ELECTROSTATICS

5.25. $\theta = 16.18°$; 22.25 segments; 177.2 pF/m
5.26. 61.95 pF
5.27. 4.72 nF

6 Electromagnetism

6.1 BASIC RELATIONSHIPS BETWEEN ELECTRIC CURRENT, MAGNETIC FIELDS AND FORCES

When a potential difference is created by an external source of e.m.f. between the ends of a conductor, an electric current is established. Such a current sets up a magnetic field within and surrounding the conductor.

When a magnetic field is established which links with a conductor, an e.m.f. is induced in that conductor during the period taken to establish the field. Lenz's law states that the induced e.m.f. is in such a direction as to oppose the establishment of the original current. During a field collapse, the induced e.m.f. will attempt to maintain the established current.

Associated with currents and magnetic fields, whether produced by those currents or provided externally, there are always mechanical forces.

In this section the principal relationships between the several quantities will be examined.

ELECTROMAGNETISM

6.1.1 The electric and magnetic circuits. Relationships between (a) current density, electric force and conductivity, (b) flux density, magnetising force and permeability

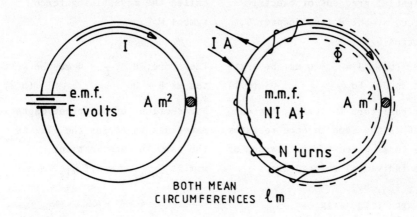

Fig. 6.1

Electric circuit	Magnetic circuit
Resistance, $R = \dfrac{\ell}{\sigma A}$	Reluctance $S = \dfrac{\ell}{\mu A}$
σ = conductivity of the material	μ = permeability of the material
e.m.f. = electromotive force (volts)	m.m.f. = magnetomotive force (N × I At)*
From Ohm's law: $R = \dfrac{e.m.f.}{I}$	$S = \dfrac{m.m.f.}{\Phi}$
transposing: e.m.f. = IR or $I = \dfrac{e.m.f.}{R}$	m.m.f. = ΦS or $\Phi = \dfrac{m.m.f.}{S}$
Resistance has units volts per ampere	Reluctance has units magnetomotive force per weber
I = JA (J = current density, A/m^2)	Φ = BA (B = flux density, Wb/m^2)
$I = \dfrac{e.m.f.}{R}$ can be re-written as $JA = \dfrac{E}{\ell/\sigma A}$ (cancel A both sides)	$\Phi = \dfrac{m.m.f.}{S}$ can be re-written as $BA = \dfrac{NI}{\ell/\mu A}$ (cancel A both sides)
$J = \dfrac{E}{\ell} \times \sigma$	$B = \dfrac{NI}{\ell} \times \mu$

Electric circuit	Magnetic circuit
$\frac{E}{\ell}$ volts per metre length is called potential gradient or electric force, symbol E (see chapter 5, section 5.4)	$\frac{NI}{\ell}$ amperes per metre length is called the magnetising force, symbol H
Therefore, $J = \frac{E}{\ell} \times \sigma$ may be written as $J = E\sigma$ (6.1)	Therefore, $B = \frac{NI}{\ell} \times \mu$ may be written as $B = H\mu$ (6.2)
For resistors in series, the current is the same in each resistor. The total circuit e.m.f. = sum of the individual voltage drops IR_1, IR_2, ... $E = IR_1 + IR_2 + IR_3 + ...$ $= I(R_1 + R_2 + R_3 + ...)$.	For sections of different magnetic materials in series the flux is the same in each section. $\text{m.m.f.}_{(total)} = \text{m.m.f.}_{(1)} + \text{m.m.f.}_{(2)} + ...$ $= \Phi S_1 + \Phi S_2 + \Phi S_3 + ...$ $= \Phi(S_1 + S_2 + S_3 + ...)$.

*A current I amperes flowing in N turns produces an m.m.f. of NI ampere-turns. Since turns have no dimensions the product N × I is expressed simply as amperes. Hence

$$H = \frac{NI}{\ell} \text{ A/m.}$$

The same effect as I amperes in N turns could be achieved by NI amperes in one turn.

For free space the permeability is μ_o and this has a value of $4\pi \times 10^{-7}$ H/m. For other materials $\mu = \mu_o \mu_r$, where μ_r is the relative permeability.

Materials are divided into two classifications (1) paramagnetic, (2) diamagnetic. Paramagnetic materials have relative permeabilities greater than unity and when in the form of a needle and suspended in a magnetic field will align themselves with the magnetic field. Diamagnetic materials have relative permeabilities less than unity and will not orientate themselves with the field. Iron, nickel and cobalt are examples of ferro-magnetic materials which are very strongly paramagnetic. Air is considered to have a relative permeability of

ELECTROMAGNETISM

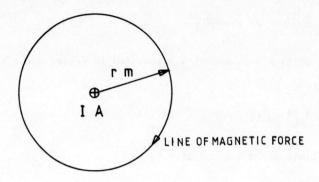

Fig. 6.2

unity and therefore to perform as free space.

Now a single current-carrying conductor will be considered. The current I amperes sets up a magnetic field one line of which, at radius r metres, is 2πr metres long.

M.m.f. = I amperes.

Magnetising force, $H = \frac{\text{amperes}}{\text{metre length}} = \frac{I}{2\pi r}$ A/m.

Now from equation 6.2, $B = H\mu$, hence, substituting for H,

$$B = \frac{I\mu}{2\pi r} \text{ tesla (Wb/m}^2\text{) at radius r metres.} \tag{6.3}$$

6.1.2 Induced e.m.f. and flux changes

When a magnetic field which links with a conductor changes in magnitude, during the period of the change an e.m.f. will be induced in the conductor.

From Faraday's original work the following relationships were obtained

induced e.m.f., $e \propto \Delta\Phi$

$e \propto N$ (N = number of turns linked by the flux)

$e \propto \frac{1}{\Delta t}$

Combining these quantities gives

$$e \propto \frac{N \Delta\Phi}{\Delta t} \quad \text{or} \quad e = \frac{kN \Delta\Phi}{\Delta t}.$$

In S.I. units k = 1, and with Φ measured in webers and t measured in seconds,

$$e = \frac{N \Delta\Phi}{\Delta t} \text{ volts}.$$

In the limit as $\Delta\Phi \to 0$ and $\Delta t \to 0$,

$$e = \frac{N \, d\Phi}{dt} \text{ volts}. \tag{6.4}$$

Considering a coil used either as a choke or as the primary to a transformer, and assuming that the flux varies sinusoidally, equation 6.4 becomes

$$e = \frac{N \, d\Phi_m \sin \omega t}{dt} = N \omega \Phi_m \cos \omega t.$$

The maximum value of e occurs when $\cos \omega t = 1$, when $E_m = N \omega \Phi_m$ ($\omega = 2\pi f$). Dividing both sides of the equation by $\sqrt{2}$ gives

$$\frac{E_m}{\sqrt{2}} = \frac{N \, 2\pi f \, \Phi_m}{\sqrt{2}}.$$

$\frac{2\pi}{\sqrt{2}} = 4.44$, $\Phi_m = B_m A$ (equation 6.1) and $\frac{E_m}{\sqrt{2}} = E$ (r.m.s.).

Therefore, $E = 4.44 B_m A f N$ volts. \qquad (6.5)

It should be noted that B_m is the peak value of flux density under the particular working conditions which will not necessarily be the maximum possible for the material employed.

6.1.3 Force, current and flux density

When an electric current passes through a magnetic field in which the lines of force are not parallel with the direction of the current, a force is exerted on that current. The current may be in a conductor or in the form of an electron beam in a vacuum.

ELECTROMAGNETISM

For a current in a conductor, with flux density B tesla, current I amperes and length of the conductor in the field ℓ metres, and where the direction of current and flux are at right angles, the value of the force F is given by

$$F = BI\ell \text{ newtons} \tag{6.6}$$

The force is at right angles to both current and flux.

Now consider a single conductor situated in air and carrying a current I_1 amperes. At radius r metres,

$$H = \frac{I_1}{2\pi r} \text{ A/m and } B = \frac{I_1 \mu_o}{2\pi r} \text{ tesla} \qquad \text{(equation 6.3, Fig. 6.2)}$$

Another conductor running parallel with the first and at a distance r metres from it will be situated in this field and will suffer a force $F = BI\ell$ newtons (equation 6.6). If the second conductor is carrying a current of I_2 amperes

$$F = \frac{I_1 \mu_o}{2\pi r} \times I_2 \times \ell \text{ newtons on length } \ell \text{ metres.}$$

Making $\ell = 1$ metre,

$$F = \frac{I_1 I_2 \mu_o}{2\pi r} \text{ N/m length.} \tag{6.7}$$

For I_1 and I_2 in the same direction, the force is one of attraction between the conductors and vice-versa.

For the particular case when $I_1 = I_2 = 1$ ampere and $r = 1$ metre

$$F = \frac{\mu_o}{2\pi} = \frac{4\pi \times 10^{-7}}{2\pi} = 2 \times 10^{-7} \text{ N/m length.}$$

This leads to the definition of the ampere which is:

Each of two infinitely long parallel conductors situated one metre apart in vacuum suffer a force of 2×10^{-7} newtons per metre length when the current in each conductor is one ampere.

6.2 THE DOMAIN THEORY OF MAGNETISATION OF FERROUS MATERIALS

The simplest suitable model of an atom consists of a positively charged nucleus around which electrons orbit. Since the movement of electrons constitutes an electric current, each orbiting electron sets up a magnetic field at right angles to the plane in which it moves. Now in an atom with many electrons, each in a different orbit, many fields in different directions are produced and the total effect is usually zero. However in certain materials the fields do not cancel and there is a resultant magnetic field in one particular direction. In ferro-magnetic materials groups of about 10^{15} atoms group themselves together in what are termed DOMAINS. Each domain is in effect a tiny magnet with a north pole on one side and a south pole on the other. In the unmagnetised state the domains form themselves into closed loops as shown in Fig. 6.3(a). Placing the material in an externally produced magnetic field causes the domains to re-orientate themselves progressively as the external field strength is increased.

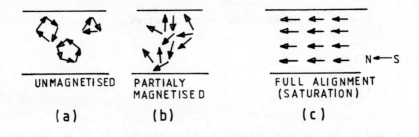

Fig. 6.3

Suppose the current in the coil of N turns in Fig. 6.1 is gradually increased from zero. The torroid is made from a ferrous material. With no current in the coil the domains are arranged in closed loops as in Fig. 6.3(a). As the current is increased, the magnetising force increases and the closed domain loops are broken as they start to align themselves with the external field. There is, as a consequence, a resultant field in one direction (Fig. 6.3(b)). Further increases in current cause the alignment to progress towards that in Fig. 6.3(c). When all the domains are aligned the material is said to be saturated. The flux in the specimen will only rise at a rate equivalent to that in air for additional magnetising force once saturation has occurred.

Now since $B = \mu_o \mu_r H$ (equation 6.2) and B is not proportional to H due to saturation effects, μ_o being a constant, it follows that μ_r must be variable.

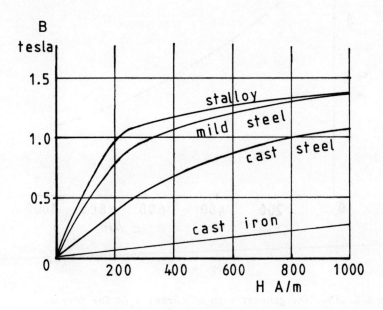

Fig. 6.4

Figure 6.4 shows some typical B/H curves. From the curve for mild steel as an example two points may be obtained which demonstrate how the value of μ_r changes. (a) B = 0.8 tesla, H = 200 A/m, (b) B = 1.35 tesla, H = 1000 A/m.

For case (a), $B = \mu_o \mu_r H$. Therefore

$$\mu_r = \frac{B}{\mu_o H} = \frac{0.8}{4\pi \times 10^{-7} \times 200}$$

$$= 3183$$

For case (b), $\mu_r = \dfrac{1.35}{4\pi \times 10^{-7} \times 1000} = 1074$

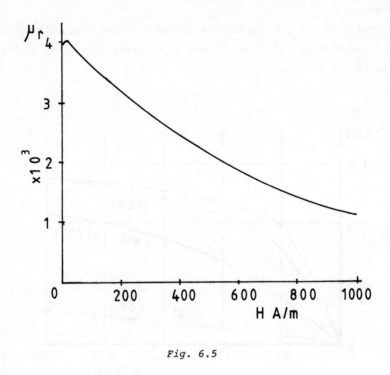

Fig. 6.5

Figure 6.5 shows the general form of curves μ_r/H for ferrous materials.

6.3 HYSTERESIS LOSSES AND THE HYSTERESIS LOOP

Consider a specimen of ferrous material starting in a completely unmagnetised state. It is wound with a coil which can be supplied with a varying current. As the current increases from zero to a value high enough to cause saturation, the flux density increases from 0 to 1. The current is now reduced gradually to zero. The alignment of the domains survives in part after the applied field is removed and the flux density falls to point 2. In Fig. 6.6(a), 0–2, marked R, is known as the remanence. Had the material not been driven into saturation before the reduction of the magnetising force to zero, the general term for the flux remaining would be residual flux or remanent flux.

If the current direction is reversed and increased so as to drive

ELECTROMAGNETISM

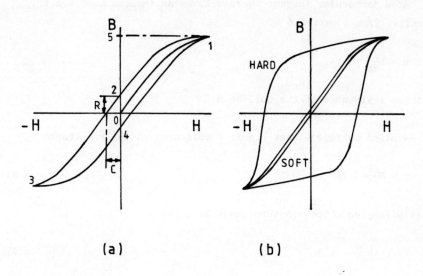

Fig. 6.6

the specimen into saturation with reversed polarity, the curve extends to point 3. The magnitude of reverse magnetising force to cause the flux density to become zero during this procedure is called the coercive force (distance C in Fig. 6.6(a)).

Again reducing the magnetising force to zero brings the curve back to point 4 and then with forward magnetising force to point 1 again.

Supplying the coil with alternating current results in successive excursions through 1, 2, 3, 4. This is known as a hysteresis loop.

Changing the pattern of domains requires the expenditure of energy so that where alternating magnetisation is involved this energy shows up as heat in the material. The energy expended is known as the hysteresis loss. The loss per cycle is a function of the area inside the hysteresis loop so that for minimum loss this area must be small. The softer the material used the narrower is the loop. A permanent magnet material has a very wide loop having high remanence as shown in Fig. 6.6(b).

Let the magnetic circuit shown in Fig. 6.1 be made from a ferrous material. Suppose the current in the coil is increasing. It will be creating a magnetic flux in the circuit.

At a particular instant there will be an induced back e.m.f., e volts. From equation 6.4

$$e = N\frac{d\Phi}{dt} \text{ volts.}$$

Let the resistance of the coil be R Ω.

Applied voltage - back e.m.f. = volt drop in the resistance

$$v - N\frac{d\Phi}{dt} = iR. \tag{6.8}$$

Multiplying equation 6.8 throughout by i gives

$$vi - iN\frac{d\Phi}{dt} = i^2R. \tag{6.9}$$

From the definitions in section 6.1.1

$$\frac{Ni}{\ell} = H \text{ therefore } Ni = H\ell \text{ and } \Phi = BA.$$

Substituting these values in equation 6.9 and transposing gives

$$vi - i^2R = H\ell A\frac{dB}{dt}.$$

Now vi is the power supplied to the circuit and i^2R is the power loss in coil resistance.

Hence $H\ell A\frac{dB}{dt}$ must be the power supplied to the magnetic field.

Energy = power × time. Therefore, in an interval of time dt

$$\text{energy supplied} = H\ell A\frac{dB}{dt} \times dt = H \, dB \times \ell A \text{ joules.} \tag{6.10}$$

Since ℓA is the volume of the magnetic material, the energy supplied per unit volume (per m^3) during time dt seconds is H dB joules.

During time dt, B has increased by dB and the value of H dB is the area of the hatched strip in Fig. 6.7.

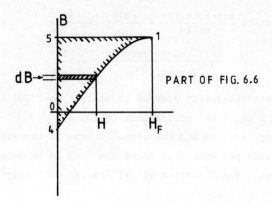

Fig. 6.7

The total energy supplied as H rises from 0 to H_F is therefore the sum of all such strips, Σ H dB. This sum is equal to the total area bounded by the lines 4, 1, 5, 4, the edges of which are cross-hatched.

Reducing the magnetising force to zero so that only the residual flux remains will allow the domains to partially re-orientate themselves so giving energy back to the supply. The corresponding area is that bounded by the lines 2, 1, 5, 2 in Fig. 6.6. The difference between the energy supplied and that returned must be the energy lost and this is represented by the difference between the two areas, i.e. that inside the hysteresis loop. The same loss occurs during the reverse magnetisation of the specimen.

When magnetised by an alternating current at f Hz, the loss occurs f times per second.

Energy loss: (area of loop) × f J/s = W.

The product H × B has units $\dfrac{\text{ampere-turns}}{\text{metre}} \times \dfrac{\text{webers}}{(\text{metres})^2}$ \hfill (i)

From equation 6.4, $e = N\dfrac{d\phi}{dt}$ volts. Transposing: $d\phi = \dfrac{e\ dt}{N}$ \hfill (ii)

The units of (ii) are: webers = $\dfrac{\text{volt seconds}}{\text{turns}}$

Substituting this in (i)

$$HB \text{ has units } \frac{\text{ampere turns}}{\text{metre}} \times \frac{\text{volt seconds}}{\text{turns (metres)}^2}$$

$$= \frac{\text{volt-ampere seconds}}{(\text{metres})^3} = J/m^3.$$

For a hysteresis loop plotted to scales (say) 100 A/m = 1 cm and 0.1 tesla = 1 cm, 1 cm^2 corresponds to $100 \times 0.1 = 10$ J/m^3. Knowing the total area of the loop and the actual volume of the core will enable the energy loss per excursion round the loop to be determined.

In general, for H plotted at (x) A/m per unit length and B plotted at (y) T per unit length,

$$\text{power loss} = x \times y \times \text{area of the hysteresis loop in (units)}^2 \times \text{volume of the core} \times \text{frequency watts.} \quad (6.11)$$

6.4 ENERGY STORED IN A MAGNETIC SYSTEM IN TERMS OF ITS SELF-INDUCTANCE

A circuit is said to possess a self-inductance of one henry if, when a current is changing at a rate of one ampere per second in it, an e.m.f. of one volt is induced in that circuit.

Written as a formula this is

$$e = L\frac{di}{dt} \text{ volts} \quad (6.12)$$

where L is the self-inductance in henry.

Consider a magnetic circuit as shown in Fig. 6.1. The current in the coil is increasing so creating a magnetic flux in the circuit. At a particular instant there will be a back e.m.f., e volts, induced (see also section 6.3). The relevant equations are

(6.4) $e = N\frac{d\phi}{dt}$ volts   (6.12) $e = L\frac{di}{dt}$ volts

(6.9) $vi - iN\frac{d\phi}{dt} = i^2R.$

Equations 6.4 and 6.12 express e in terms of different quantities. Equating these

$$N\frac{d\phi}{dt} = L\frac{di}{dt}.$$

Hence

$$L = N\frac{d\phi}{dt} \times \frac{dt}{di} = N\frac{d\phi}{di}. \qquad (6.13)$$

Transposing equation 6.9 gives

$$vi - i^2R = iN\frac{d\phi}{dt}$$

In words: $\frac{\text{power}}{\text{supplied}} - \frac{\text{power}}{\text{loss}}$ = power to the magnetic field.

This is an alternative expression to that in equation 6.10.

During time interval dt, energy supplied to the magnetic field = $iN\frac{d\phi}{dt} \times dt$ J.

Transposing equation 6.13: L di = N dϕ. Substituting in the equation above gives

energy supplied to the magnetic field in time dt = i × L di J.

Total energy supplied to the field as the current increases from zero to a final steady state value of I amperes is found by integration between the limits 0 and I.

$$\text{Energy to the field} = \int_0^I Li\,di = L\left[\frac{i^2}{2}\right]_0^I = \tfrac{1}{2}LI^2 \text{ J}. \qquad (6.14)$$

<u>Worked example 6.1</u>   A non-magnetic former of a toroid (Fig. 6.1) has a mean circumference of 0.2 m and a uniform cross-sectional area of 3.5 cm^2. The coil consists of 650 turns and carries a steady current of 5 A.

Calculate (a) the m.m.f., (b) the magnetising force, (c) the reluctance, (d) the magnitude of the flux and flux density, (e) the circuit inductance, and (f) the energy stored in the magnetic field.

(a) m.m.f. = N × I = 650 × 5 = 3250 A.

(b) Magnetising force = $\dfrac{\text{m.m.f.}}{\text{length of the flux path }(\ell)} = \dfrac{3250}{0.2} = 16250$ A/m.

(c) Reluctance $S = \dfrac{\ell}{\mu_o A} = \dfrac{0.2}{4\pi \times 10^{-7} \times 3.5 \times 10^{-4}} = 4.547 \times 10^8$ A/Wb.

This means that it would require 4.547×10^8 A to set up a flux of 1 Wb in this torroid.

(d) Flux $\Phi = \dfrac{\text{m.m.f.}}{S} = \dfrac{3250}{4.547 \times 10^8} = 7.148 \times 10^{-6}$ Wb.

Flux density $B = \dfrac{\Phi}{A} = \dfrac{7.148 \times 10^{-6}}{3.5 \times 10^{-4}} = 0.02$ T (tesla = Wb/m^2).

Using equation 6.2 an alternative method is possible. $B = \mu_o H$ and $BA = \Phi$. Therefore

$\Phi = \mu_o HA$
$= 4\pi \times 10^{-7} \times 16250 \times 3.5 \times 10^{-4}$
$= 7.148 \times 10^{-6}$ Wb.

(e) Inductance = $N\dfrac{d\Phi}{di}$ \hfill (equation 6.13).

Since $B = \mu_o H$, the flux is proportional to current. Any size of change in current produces a proportional change in flux. Hence $\dfrac{\Delta\Phi}{\Delta i}$ may be written for $\dfrac{d\Phi}{di}$.

$\Delta i = (0 - 5)$A and $\Delta\Phi = (0 - 7.148 \times 10^{-6})$ Wb; i.e. the changes in each quantity from zero to final conditions.

$L = \dfrac{650 \times 7.148 \times 10^{-6}}{5} = 929.12$ μH.

(f) Energy stored = $\tfrac{1}{2}LI^2$ J \hfill (equation 6.14)
$= \tfrac{1}{2} \times 929 \times 10^{-6} \times 5^2 = 0.0116$ J

<u>Worked example 6.2</u>   A steel toroid has a mean length of 20 cm and cross-sectional area 0.8 cm^2. It has a radial slot 4 mm wide cut

through the steel.

Determine the value of current required in a 400-turn coil wound on the toroid to set up a flux of 72 μWb in the steel and air gap. (The relative permeability of the steel at the working flux density = 1800.)

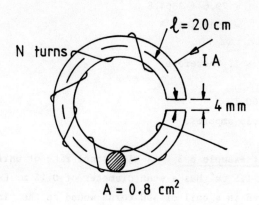

Fig. 6.8

<u>Considering the steel</u>
Flux density $B = \dfrac{\Phi}{A} = \dfrac{72 \times 10^{-6}}{0.8 \times 10^{-4}} = 0.9$ T.

$B = \mu_o \mu_r H$ T. Transposing, $H = \dfrac{B}{\mu_o \mu_r}$

$= \dfrac{0.9}{4\pi \times 10^{-7} \times 1800} = 397.9$ A/m.

$H = \dfrac{NI}{\ell}$ A/m. Transposing: $H\ell = NI \; (= \text{m.m.f.})$

$= 397.9 \times 0.2 = 79.6$ A.

<u>Considering the air gap</u>
Assuming no leakage, the flux and flux density in the air gap will be the same as in the steel. In air $B = \mu_o H$. Therefore

$H = \dfrac{0.9}{4\pi \times 10^{-7}} = 716197$ A/m.

NI = H × length of the gap
   = 716197 × 4 × 10^{-3}
   = 2864.8 A.

Total m.m.f. = (m.m.f. for steel) + (m.m.f. for air)
   = 79.6 + 2864.8
   = 2944.4 A.

Now m.m.f. = N × I. Therefore

2944.4 = 400 I
I = 7.36 amperes.

<u>Self-assessment example 6.3</u>   A mild steel ring of uniform cross-sectional area 1.5 cm^2 has a mean diameter of 0.15 m. Estimate the current required in a coil of 300 turns wound on the ring to establish a flux of 195 μWb in the steel using the data given in the table.

For mild steel

B (tesla)	1.0	1.15	1.3	1.45
H (A/m)	400	500	800	1650

<u>Worked example 6.4</u>   The hysteresis loop for a magnetic material used in the core of a transformer is drawn to the following scales

100 A/m ≡ 1 cm on the H scale
0.05 tesla ≡ 1 cm on the B scale.

The area of the hysteresis loop is 158 cm^2. The volume of the transformer core is 0.015 m^3. The operating frequency is 50 Hz.
  Determine the power loss due to hysteresis in the core.

Using equation 6.11, 100 × 0.05 = 5. This tells us that each square centimetre of the hysteresis loop represents a loss of 5 joules for each cubic metre of the core for each cycle of the input voltage.
  Therefore with 0.015 m^3 and a frequency of 50 Hz, the power loss is given by

ELECTROMAGNETISM

power loss = (5 × 0.015 × 50) × 158
= 592.5 W.

Self-assessment example 6.5
(a) An iron toroid with mean length 25 cm and a cross-sectional area 3 cm^3 is wound with a coil of 120 turns. A current in the coil is increased from zero to 10 A when the flux in the toroid increases from zero to 480 μWb. For a coil current of 10 A, determine (i) the value of H, (ii) the value of B, (iii) the mean value of inductance over the range $(N\frac{\Delta\phi}{\Delta i})$.

(b) During the increase in current in the coil in part (a), the curve shown in Fig. 6.9 was plotted. The scales chosen are such that (x) = (y) = 16 cm. The hatched section has an area of 98 cm^2.

Determine the amount of energy involved in orienting the domains during the magnetisation period.

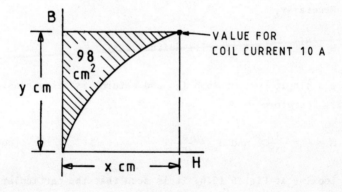

Fig. 6.9

6.5 THE INDUCTANCE OF CONCENTRIC CYLINDERS (CO-AXIAL CABLE)

The core of the cable in Fig. 6.10(a) carries I amperes away from the viewer and to the right while the sheath carries the return current. Since the magnetic fields produced by the go and return currents will be equal and in opposite directions, there will be no magnetic field outside the sheath. A magnetic field exists within the core and between the core and sheath. The magnetic field within the conductor is very small and is ignored at this stage.

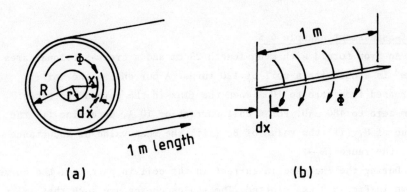

Fig. 6.10

Considering the dielectric, from radius r to radius R. At any radius x, a line of magnetic force has length $2\pi x$ metres.

Generally,

$$H = \frac{NI}{\ell} \quad \text{(section 6.1.1 definitions).}$$

Since a single current loop (go and return) is being considered, N = 1. Therefore

$$H = \frac{I}{\ell} = \frac{I}{2\pi x} \quad \text{and} \quad B_x = \frac{\mu_o I}{2\pi x} \text{ T} \qquad \text{(equation 6.3).}$$

Looking at Fig. 6.10(b) it is seen that the particular lines of force along the one metre length pass through an area $dx \times 1 \text{ m}^2$. Hence the increment of flux, $d\Phi$, through this small area is

$$d\Phi = B_x \times (dx \times 1) = \frac{\mu_o I}{2\pi x} dx \; .$$

The total flux between r and R is found by integration between these limits.

$$\int d\Phi = \int_r^R \frac{\mu_o I}{2\pi} \frac{dx}{x}$$

$$\Phi = \frac{\mu_o I}{2\pi} \left[\log_e x \right]_r^R = \frac{\mu_o I}{2\pi} \log_e \frac{R}{r} \text{ Wb.}$$

ELECTROMAGNETISM

Inductance = flux linkages per ampere $\left(N\frac{d\phi}{di}\right)$ (equation 6.13).

For N = 1, in a non-magnetic medium (see worked example 6.1) the following may be written

$$L = \frac{\Delta\Phi}{\Delta I}.$$

In the cable, as i rises from zero to final value I, the flux increases from zero to

$$\frac{\mu_o I}{2\pi} = \log_e \frac{R}{r} \text{ Wb}.$$

Therefore

$$\frac{\text{change in flux}}{\text{change in current}} = L = \frac{\mu_o}{2\pi} \log_e \frac{R}{r} \text{ H/metre length}.$$

6.6 THE INDUCTANCE OF A PARALLEL PAIR OF TWIN CONDUCTORS REMOTE FROM EARTH

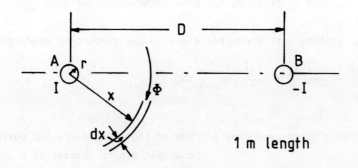

Fig. 6.11

In Fig. 6.11, conductor A carries current away from the viewer, the return current being in conductor B. The distance D between the conductors is much greater than their radii, r. A 1 m length of the twin conductor system will be considered.

Conductor A

At any radius x from A, the magnetic flux density is

$$B_x = \frac{\mu_o I}{2\pi x} \text{ T} \qquad \text{(equation 6.3).}$$

Theoretically, there is no limit to the distance from the conductor at which a magnetic field will be created. In practice, at only a few metres distance the effect is undetectable. Let the radius at which no effect is felt be P metres. The increment of flux passing through the area $dx \times 1 \text{ m}^2$ (see Fig. 6.10(b)) is

$$d\Phi = \frac{\mu_o I}{2\pi x} dx.$$

The flux linking with conductor A due to its own current $= \int_r^P \frac{\mu_o I}{2\pi} \frac{dx}{x}$.

$$\Phi = \frac{\mu_o I}{2\pi} \log_e \frac{P}{r} \text{ Wb}.$$

Similarly, the flux which links with (passes round) conductor A due to the current -I in conductor B will be the same integral, but between the limits P and D. Note that at radii less than D from conductor B, lines of force do not pass round conductor A.

The flux linking with conductor A due to the current in conductor B

$$= \int_D^P \frac{\mu_o(-I)}{2\pi} \frac{dx}{x} = \frac{-\mu_o I}{2\pi} \log_e \frac{P}{D} \text{ Wb}.$$

Total flux linking conductor A = sum of that due to its own current and that due to current in B

$$= \frac{\mu_o I}{2\pi} \left(\log_e \frac{P}{r} - \log_e \frac{P}{D} \right).$$

Subtracting logarithms performs the division function

$$\text{total flux linking conductor A} = \frac{\mu_o I}{2\pi} \log_e \left(\frac{\frac{P}{r}}{\frac{P}{D}} \right) = \frac{\mu_o I}{2\pi} \log_e \frac{D}{r}.$$

Therefore, by similar reasoning to that in section 6.5

ELECTROMAGNETISM

$$L = \frac{\mu_o}{2\pi} \log_e \frac{D}{r} \text{ H/m length.}$$

This is the inductance of conductor A.

By repeating the process for conductor B an identical expression will be arrived at.

The total inductance of the circuit (both conductors) is therefore

$$L = 2\left(\frac{\mu_o}{2\pi} \log_e \frac{D}{r}\right) \text{ H/m}$$

$$= \frac{\mu_o}{\pi} \log_e \frac{D}{r} \text{ H/m length.}$$

6.7 THE STEINMETZ INDEX

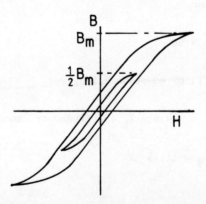

Fig. 6.12

When a magnetic material is not operated up to saturation value but rather to some other lesser value, the area of the hysteresis loop is reduced. From a consideration of Fig. 6.12 it may be seen that reducing the maximum flux density from B_m to $\tfrac{1}{2}B_m$ reduces the area of the loop by a factor considerably greater than 2.

Charles Steinmetz, an American engineer, established an empirical formula connecting the hysteresis losses for the same core to different peak working flux densities. His formula is power loss = $k_h B_m^n f$ W. (6.15)

where k_h is a constant for the material, the hysteresis coefficient, and n is called the Steinmetz index. This has values between 1.6 and 3 generally, depending on the material.

As an example, consider a transformer core with Steinmetz index 1.9 and a hysteresis loss of 150 W at 50 Hz when operating at a maximum flux density of 1.4 tesla.

It is required to estimate what the hysteresis loss would be if the peak flux density were reduced to 1.2 tesla.

The original conditions give

$$150 = k_h \times 1.4^{1.9} \times 50 \qquad (i)$$

For the new condition

$$P = k_h \times 1.2^{1.9} \times 50 \qquad (ii)$$

Dividing equation (i) by equation (ii) gives

$$\frac{150}{P} = \left(\frac{1.4}{1.2}\right)^{1.9} \qquad \text{(since } k_h \text{ and frequency cancel)}$$

$$P = \frac{150}{(1.166)^{1.9}} = 111.9 \text{ W.}$$

6.8 EDDY CURRENT LOSSES

Consider a coil wound on a conducting core as shown in Fig. 6.13(a). The current in the coil is increasing so that it will produce an increasing flux which will induce e.m.f.s in the coil itself and in the core. The e.m.f. induced in the coil inhibits the rate of rise of that current, while that in the core drives what are called eddy currents which heat the core since it has resistance. When the coil current is alternating, the e.m.f.s are alternating, the phase relationship between coil current and induced e.m.f. being $90°$. (See section 6.1.2, in which the flux wave was a sine function, and the induced e.m.f. a cosine function.)

To restrict the flow of eddy currents and hence limit the heat

ELECTROMAGNETISM

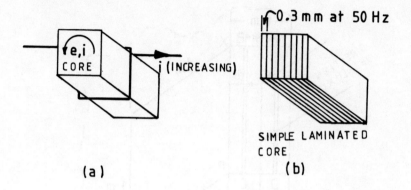

Fig. 6.13

loss, the cores of transformers and motors, etc. are laminated, that is to say they are made up from a large number of very thin strips which are cut from sheet, cleaned and polished, varnished or anodised on one side and then pressed together to form the correct core shape (Fig. 6.13(b)). Large cores are held together using core bolts which are fully insulated from the laminations.

Consider a section of a single lamination situated in a sinusoidally-varying magnetic field. The flux enters the narrow side of the lamination parallel to lines XY and PQ, which may be considered to be drawn on the surface of the lamination (Fig. 6.14).

A loop A, B, C, D within the lamination has a width 2x metres and is 1 metre in height. Its cross-sectional area is $2x$ m^2. Sinusoidal alternating flux linking with this loop will induce an e.m.f. expressed by equation 6.5

$$E = 4.44 B_m AfN \text{ volts}$$
$$= 4.44 B_m (2x) f \text{ volts} \quad \text{(since N = one loop).}$$

This e.m.f. drives current round the loop A, B, C, D which extends 1 metre into the lamination. Looking down on to the top of the lamination it will be seen that the area through which current must pass is dx thick and 1 metre long. (Distance XY × dx) (i)

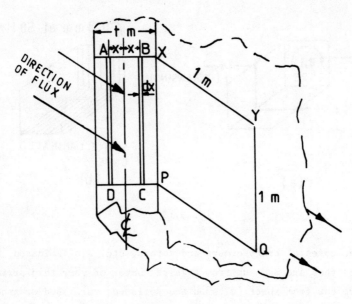

Fig. 6.14

The length of the current path = AB + BC + CD + DA metres.

Now since the thickness of the lamination is likely to be a fraction of 1 mm, AB + CD may be ignored compared with BC (= 1 m) and DA (= 1 m).

Length of current path = 2 m (closely). (ii)

The resistance of the path, $R = \dfrac{\ell}{\sigma A}$ (section 6.1.1)

or since $\rho = \dfrac{1}{\sigma}$

$$R = \dfrac{\rho \ell}{A} \qquad \text{(iii)}$$

where ρ = resistivity of the lamination material.

ELECTROMAGNETISM

Substituting (i) and (ii) in (iii) gives

$$R = \frac{2\rho}{dx}.$$

Power dissipated in this resistance $= \frac{E^2}{R}$ watts

$$= \frac{(4.44 B_m (2x) f)^2}{\frac{2\rho}{dx}} = \frac{(8.88 B_m f x)^2 dx}{2\rho}.$$

The total loss over the whole thickness is obtained by integrating between the limits $x = 0$ (at the centre line) and $x = t/2$ (at the edge).

$$\text{Total loss} = \frac{(8.88 B_m f)^2}{2\rho} \int_0^{t/2} x^2 \, dx$$

$$= \frac{(8.88 B_m f)^2}{2\rho} \left[\frac{x^3}{3}\right]_0^{t/2} = 13.1 \, B_m^2 f^2 \rho^{-1} [x^3]_0^{t/2}$$

$$= 13.1 \, B_m^2 f^2 \rho^{-1} \frac{t^3}{8}. \tag{iv}$$

The volume of this section of lamination $= t \times 1 \times 1 = t \, m^3$.

Thus, dividing equation (iii) above by $t \, m^3$ will yield an expression for the power loss per m^3. Hence

$$\text{power loss per } m^3 = \frac{13.1}{8} B_m^2 f^2 t^2 \rho^{-1} \text{ watts.}$$

The fraction $\frac{13.1}{8}$ cannot be relied upon since the current path has been idealised. The equation is generally expressed as

$$P = k_e (B_m f t)^2 \rho^{-1} \text{ W} \tag{6.16}$$

where k_e is a constant for a particular core (it takes into account the material and its volume).

Where the flux does not vary sinusoidally, as in the case of forced magnetisation for example (chapter 3, section 3.2.2), the induced voltage will not be sinusoidal and its form factor will have to be taken into account.

ELECTRICAL PRINCIPLES FOR HIGHER TEC

With form factor k_f, equation 6.16 becomes

$$P = k_e (k_f B_m f t)^2 \rho^{-1} \text{ watts.}$$

Worked example 6.6   A core of a choke has a hysteresis loss of 25 W and an eddy current loss of 15.6 W when operated at 50 Hz.

Assuming that the flux density remains unchanged, determine the separate losses if (a) the frequency is increased to 60 Hz, (b) a new core is built using the same material but using laminations one half the thickness of those in (a). (The operating frequency is 50 Hz.)

(a) The total core loss = hysteresis loss + eddy current loss
$$= k_h B_m^n f + k_e B_m^2 f^2 t^2 \rho^{-1} \quad \text{(equations 6.15 and 6.16).}$$

Since the flux density is to remain constant, B_m may be combined with the existing constants

$$P = k_1 f + k_2 f^2 t^2 \text{ W} \quad (k_1 = k_h B_m^n \text{ and } k_2 = k_e B_m^2 \rho^{-1}).$$

<u>Hysteresis loss</u>   Original conditions, $k_1 \times 50 = 25$ W

$$k_1 = \frac{25}{50}.$$

At 60 Hz, $k_1 \times 60 = P$ (hysteresis).

Substituting for k_1

$$P_H = \frac{25}{50} \times 60 = 30 \text{ W.}$$

<u>Eddy current loss</u>   Original conditions, $k_2 \times 50^2 t^2 = 15.6$

$$k_2 t^2 = \frac{15.6}{50^2}.$$

At 60 Hz, $k_2 t^2 \times 60^2 = P$ (eddy currents).

Substituting for $k_2 t^2$

$$P_E = \frac{15.6}{50^2} \times 60^2 = 22.46 \text{ W.}$$

(b) The hysteresis loss is independent of lamination thickness so that at one half of the lamination thickness the hysteresis loss is 25 W.

Eddy current loss = $k_2 f^2 t^2$ W.

Original conditions, $15.6 = k_2 \times 50^2 \times t^2$. \hfill (i)

New conditions, $P_E = k_2 \times 50^2 \times (\tfrac{1}{2}t)^2$. \hfill (ii)

Dividing equation (i) by equation (ii) gives

$$\frac{15.6}{P_E} = \frac{t^2}{(\tfrac{1}{2}t)^2} = 2^2$$

$$P_E = \frac{15.6}{4} = 3.9 \text{ W.}$$

<u>Self-assessment example 6.7</u> A transformer core has an eddy current loss of 125 W at 50 Hz. The laminations are 0.34 mm thick. The design is to be modified so that it will operate at a different voltage and at 400 Hz with the same eddy current loss. At the new voltage the flux density will be one half the original value. What is the necessary thickness of the laminations assuming all other parameters remain unchanged?

6.9   DETERMINATION OF THE SEPARATE HYSTERESIS AND EDDY CURRENT LOSSES IN THE FERROUS CORE OF A COIL OR TRANSFORMER

Consider the condition of free magnetisation of an iron core (flux varies sinusoidally). The total core loss,

$$P_C = k_h B_m^n f + k_e B_m^2 f^2 t^2 \rho^{-1} \text{ W} \qquad \text{(equations 6.15 and 6.16).}$$

Dividing both sides of the equation by the frequency f gives

$$\frac{P_C}{f} = k_h B_m^n + k_e B_m^2 f t^2 \rho^{-1} \hfill (6.17)$$

If the coil is fed from a variable-voltage variable-frequency supply, it is possible to measure the total core loss for a number of different frequencies while maintaining the flux density constant. The conditions for constant flux density can be realised from the transformer e.m.f. equation 6.5.

$$E = 4.44 B_m A f N \text{ volts}.$$

Dividing both sides of this equation by f yields

$$\frac{E}{f} = 4.44 B_m A N.$$

Since 4.44, A and N are constant for a particular device,

$$\frac{E}{f} = k_1 B_m \quad (k_1 = 4.44 A N).$$

Hence, if the frequency is varied while keeping the ratio E/f constant, B_m will be constant.

For a constant flux density B_m, and realising that k_h, k_e, t and ρ are all constants for a particular device, equation 6.17 reduces to

$$\frac{P_C}{f} = k_2 + k_3 f \qquad \text{where:} \qquad (6.18)$$

$$k_2 = k_h B_m^n$$

$$k_3 = k_e B_m^2 t^2 \rho^{-1}.$$

This is of the form y = c + mx, which is the equation of a straight line.

To separate the hysteresis and eddy current losses, firstly the d.c. resistance of the coil under test is determined. The coil is then connected as shown in Fig. 6.15. The variable supply may be an oscillator with sufficient range and power or an alternator set, the speed and output voltage of which can be varied as required.

For a particular ratio of E/f, a number of readings of input power are taken. Ignoring any wattmeter errors, the core loss P_C in each case is the wattmeter indication less the coil power loss $I^2 R_{d.c.}$.

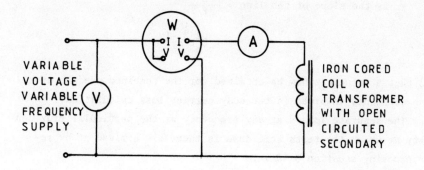

Fig. 6.15

Plotting P_C/f against a base of the frequency f produces a straight line as shown in Fig. 6.16.

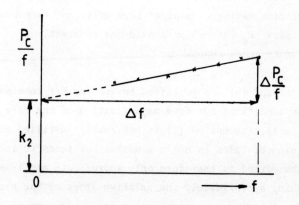

Fig. 6.16

Now, in equation 6.18

$$\frac{P_C}{f} = k_2 + k_3 f$$

k_2 is the intercept on the P_C/f axis at $f = 0$. But $k_2 = k_h B_m^n$, so that this value is now known.

k_3 is the slope of the line = $\dfrac{\Delta P_C/f}{\Delta f}$

$$= k_e B_m^2 t^2 \rho^{-1}$$

so that a value can now be obtained for the combined product of all the constants involved in the eddy current loss calculations.

The separate losses at any frequency at the particular flux density at which the tests were done is therefore arrived at by re-transposing equation 6.18 into

$$P_C = k_2 f \text{ (the hysteresis loss)} + k_3 f^2 \text{ (the eddy current loss)}. \tag{6.19}$$

The assumptions made were

(a) constant flux density
(b) flux variation having sinusoidal form or if not then constant in each of the tests (k_f ignored or considered constant)
(c) the Steinmetz index constant.

Assumptions (a) and (b) are justified by keeping E/f constant and employing the conditions for free magnetisation of the core (chapter 3, section 3.2.1). Assumption (c) is not really justified since, in fact, the Steinmetz index is not a constant but tends to increase with frequency. The method is therefore only approximate and is suitable for establishing approximately the relative sizes of the hysteresis and eddy current losses in a specimen of iron, the core of a transformer or a choke.

<u>Worked example 6.8</u> The following test results refer to a sample of steel used as the core of a coil. The connections were as in Fig. 6.13.

Determine the separate hysteresis and eddy current losses at (a) 50 Hz, (b) 60 Hz.

E (volts)	200	250	300	500
f (Hz)	40	50	60	100
P_C (watts)	40	56.5	73.8	180

Divide P_C by f for each result so giving the following table

f (Hz)	40	50	60	100
P_C/f	1	1.132	1.23	1.8

Plot the graph as shown in Fig. 6.17.

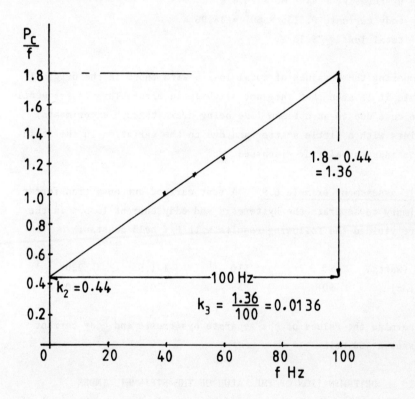

Fig. 6.17

The equation of the line is

$$\frac{P_C}{f} = 0.44 + 0.0136f.$$

Transposing to the form of equation 6.19 gives

$$P_C = 0.44f + 0.0136f^2.$$

The separate losses at 50 Hz = $0.44 \times 50 + 0.0136 \times 50^2$
$$= 22 \quad\quad + 34$$

The hysteresis loss = 22 W. The eddy current loss = 34 W. Total loss = 56 W. At 60 Hz, the losses are

hysteresis, $0.44 \times 60 = 26.4$ W
eddy current, $0.0136 \times 60^2 = 48.96$ W
total loss = 75.36 W.

Comparing these values of total losses with those in the original table it is seen that they are slightly in error. This is frequently the case due to an optimum line being drawn through experimental points with a little scatter and due to the variation in the Steinmetz index as already discussed.

<u>Self-assessment example 6.9</u> A test carried out on a transformer primary to separate the hysteresis and eddy current losses in the core yielded the following results with E/f held constant

P_C (watts)	92.5	110	171.5	220
f (Hz)	50	55	70	80

Determine the values of the separate hysteresis and eddy current losses at 50 Hz.

6.10 DETERMINATION OF THE VALUE OF THE STEINMETZ INDEX

If the value of the Steinmetz index is required it may be obtained by carrying out two sets of tests as outlined previously (section 6.7).

One test is carried out at normal flux density and the other at a fraction of normal. For ease of calculation the second test is often carried out at one half of the normal value. The values of P_C/f are plotted against f for each test.

Let the intercept on the P_C/f axis for normal flux density (B_m) have a value $k_{2(1)}$ and that for the test at one half normal flux density ($\tfrac{1}{2}B_m$) have a value $k_{2(2)}$. From equation 6.18

ELECTROMAGNETISM

$$k_{2(1)} = k_h B_m^n \qquad \text{(i)}$$

and $k_{2(2)} = k_h (½B_m)^n$ \qquad (ii)

Dividing equation (i) by equation (ii) gives

$$\frac{k_{2(1)}}{k_{2(2)}} = \frac{B_m^n}{(½B_m)^n} = 2^n. \qquad (6.20)$$

The Steinmetz index may therefore be calculated.

<u>Worked example 6.10</u> Two tests were carried out on a transformer primary to determine the Steinmetz index for the core material. The results were

<u>Test 1</u>. With $E/f = 5$, which gives normal flux density,

| P_C (watts) | 62.5 | 204 |
| f (Hz) | 50 | 100 |

<u>Test 2</u>. With $E/f = 2.5$, which gives one half normal flux density,

| P_C (watts) | 15.15 | 50 |
| f (Hz) | 50 | 100 |

Determine the value of the Steinmetz index.

Firstly it is worth observing that from equation 6.5, $E = 4.44 B_m A f N$ volts. Dividing both sides by f gives

$$\frac{E}{f} = 4.44 B_m A N = k_1 B_m \qquad \text{(as in section 6.9).}$$

Thus, reducing E/f from 5 to 2.5 reduces B_m in the same proportion.

From the results the following table is produced, as in worked example 6.8.

	$\dfrac{P_C}{f}$	
f (Hz)	Test 1	Test 2
50	1.25	0.303
100	2.04	0.5

A graph drawn to scale may be used to determine the values of the two intercepts. Alternatively it may be argued that since, in test 1, the value of P_C/f falls by $2.04 - 1.25 = 0.79$ as the frequency falls from 100 Hz to 50 Hz, it will fall by the same amount again as the frequency falls from 50 Hz to zero. Hence

$$k_{2(1)} = 1.25 - 0.79 = 0.46.$$

By similar reasoning

$$k_{2(2)} = 0.106.$$

The graph is shown in Fig. 6.18.
From equation 6.20

$$\frac{k_{2(1)}}{k_{2(2)}} = 2^n$$

$$\frac{0.46}{0.106} = 2^n.$$

Taking logarithms to base 10

$$\log_{10} 4.3396 = n \times \log_{10} 2$$

$$0.637 = n \times 0.3010$$

$$n = 2.12$$

ELECTROMAGNETISM

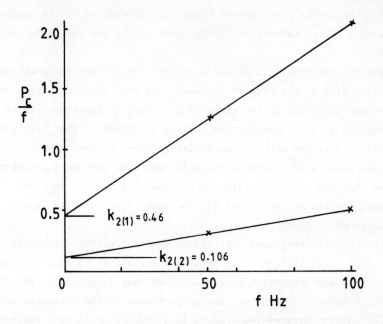

Fig. 6.18

<u>Self-assessment example 6.11</u>   The results of two tests on the core of a choke are as follows

<u>Test 1</u>. With $E/f = 4$: at 30 Hz, $P_C/f = 0.96$, at 60 Hz, $P_C/f = 1.32$.
<u>Test 2</u>. With $E/f = 2$: at 35 Hz, $P_C/f = 0.265$, at 70 Hz, $P_C/f = 0.37$.

Determine the values of (a) the hysteresis loss for the flux density in test 1 at 50 Hz, (b) the eddy current loss for the flux density in test 2 at 40 Hz, (c) the Steinmetz index for the core material.

6.11   LOW RETENTIVITY FERRO-MAGNETIC MATERIALS

In electrical machinery, transformers, motors, etc., a high flux density is required to minimise the cross-sectional area and hence the weight of the core or yoke of the apparatus. This high flux density should be produced with the minimum magnetising force or the current and hence power loss (I^2R) loss will be large. Where the flux

239

reverses periodically, the hysteresis and eddy current losses must be small. The hysteresis losses are determined by the softness of the material while the eddy current losses are determined by the resistance of the core material which is increased by the use of laminations.

Almost pure iron is used and for very high quality equipment it is cold-rolled to the correct thickness and then annealed, which has the effect of orienting the grain of the steel in the direction of rolling. It is easy to magnetise in this direction so that high flux density is achieved with low magnetising force. It does however saturate more rapidly than normal mild steel and care must be taken to see that the permitted value of maximum flux density is not exceeded (in the region of 1.8 T). The flux paths must lie in the direction of rolling.

Cheaper equipment uses laminations of silicon iron which contains 0.5% to 4% silicon. This increases resistance to eddy currents but reduces the saturation flux density so that larger cores result.

Since eddy current losses are proportional to the thickness of the laminations squared (equation 6.16), reductions in eddy current losses are achieved by using nickel-iron alloy strip as thin as 3.5×10^{-6} m wound in a spiral to form a core, when it is used up to about 5 MHz. Above this frequency individual dust particles of nickel-iron or steel covered with an insulating material and then bound together to form the correct shape may be used.

Magnetite or ferrous ferrite was the original lode stone which the early mariners used as a compass. Magnetite, which is a compound of ferric and ferrous oxides, is now combined with oxides of other materials such as manganese, nickel and iron to form the manufactured ferrites. These have extremely high resistances and so do not suffer eddy current losses at all until the frequency is in excess of about 100 MHz. They do not have such a high relative permeability as steel and saturate at a much lower value of flux density (in the region of 0.4 T). The volume of a ferrite core in consequence is three to four times that of a steel core for the same total flux. Ferrite cores are used in chokes and small transformers in audio work.

ELECTROMAGNETISM

Further problems

6.12. A hysteresis loop is plotted with scales: horizontally, 1000 A/m = 1 cm, vertically, 0.25 T = 1 cm. The area of the loop is 3.1 cm^2. The maximum vertical height (from zero) is 6.35 cm.

Calculate (a) the maximum flux density in the material, (b) the energy lost in the material in joules per m^3 once round the loop, (c) the power loss in a core for a choke made from this material which has a mass of 2.87 kg. The density of the material is 7800 kg/m^3. The operating frequency is 100 Hz.

6.13. A transformer core has a mass of 6.9 kg. The hysteresis loss at 50 Hz is 15.75 W. The density of the material is 7800 kg/m^3. The hysteresis loop for the material has an area of 7.12 cm^2 when the horizontal axis is scaled at 250 A/m per cm. The maximum height of the loop (from zero) is 7.25 cm.

Determine (a) the value of the maximum flux density, (b) the induced e.m.f. in the primary coil which has 540 turns if the core cross-sectional area is 25 cm^2.

6.14. A coil comprising 500 turns is wound on a torroidal former with mean length 0.26 m and a cross-sectional area 3 cm^2. The former is non-magnetic and the coil carries a current of 2 A.

Determine (a) the value of the magnetising force, H, (b) the flux density within the torroid, (c) the inductance of the coil, (d) the energy stored in the magnetic field, (e) the value of the induced e.m.f. in the coil if the current is reversed, taking 0.06 second to complete the reversal.

6.15. An iron ring 18 cm in diameter and 5 cm^2 in cross-sectional area is wound with 250 turns of wire. It is to operate at a flux density of 1.1 T when the relative permeability of the material is 2175.

Determine (a) the steady coil current required, (b) the inductance of the coil, (c) the energy stored.

6.16. A ring made of magnetic material is wound with 200 turns. When the current is increased from 1.5 A to 2.5 A the flux in the ring increases from 150 μWb to 300 μWb. Calculate the inductance of the coil.

If an alternating current flows in the coil with peak value 1.5 A, what is the mean value of inductance? (Hint: what is the total flux change as the current changes from +1.5 A to -1.5 A?)

6.17. Derive the relationship which exists between the supply frequency, the maximum flux density in the core, the core cross-section, the number of winding turns and the induced e.m.f., in a single-phase transformer. Hence or otherwise explain the effect upon the core flux density of supplying a transformer at its rated voltage but at twice the normal frequency.

A single-phase transformer is to operate at 240 V : 110 V. The core flux density is to be close to, but not exceeding, 1.5 T when the cross-sectional area is 2500 mm^2 and the frequency is 50 Hz. Determine to the nearest whole number the number of turns required on both primary and secondary windings.

6.18. A concentric cable has a core radius of 0.5 cm. It is to have an inductance of 2×10^{-7} H/metre length. Determine the required inner radius of the sheath.

6.19. A single-phase power line comprises two conductors each with radius 0.75 cm, spaced 1.5 m apart in air. Determine the inductance of the line per metre length.

What change in conductor radius with spacing 1.5 m would have the same effect on the inductance as using the original conductors at 1.25 m spacing?

6.20. (a) Describe a method of separating the eddy current loss and the hysteresis loss from the total core loss in a transformer core.

(b) The following results were noted during a test on a small transformer

Applied e.m.f. (volts)	200	180	160	140	120	100	80
Frequency (Hz)	50	45	40	35	30	25	20
Power input (W)	70	62	51	44	36	29	21

(The I^{2}R loss in the coil is small enough to be ignored.)

Draw a suitable graph from these results and hence determine the separate hysteresis and eddy current losses at (i) 25 Hz, and (ii) 50 Hz.

6.21. The total iron losses in the laminated core of an alternator are 240 W at 50 Hz and 300 W at 60 Hz, the ratio E/f being constant for the two tests. Find the separate hysteresis and eddy current losses at 60 Hz given that the Steinmetz index is 1.6.

Determine the corresponding values with a 30% increase in flux density and a 50% increase in the thickness of the laminations.

6.22. What are the requirements for magnetic materials for the following applications?

(a) the core of a high-quality power transformer operating at 50 Hz,
(b) the core of a radio tuning coil operating at 100 MHz.

Specify in each case a suitable material.

6.23. In a 440 V, 50 Hz, single-phase transformer the total iron loss is 2500 W. When the applied voltage is reduced to 220 V at a frequency of 25 Hz the corresponding loss is 850 W. Calculate the value of the eddy current loss at normal voltage and frequency.

6.24. Two sets of tests were carried out on a transformer in order to determine the Steinmetz index for the core material. The results were

Test 1. With $E/f = 5$.

Core loss (W)	17	37.5	65	120
Frequency (Hz)	30	50	70	100

Test 2. With $E/f = 2.5$.

Core loss (W)	3.9	8.8	15.4	28.8
Frequency (Hz)	30	50	70	100

Determine the separate hysteresis and eddy current losses at 50 Hz for each flux density and the value of the Steinmetz index.

6.25. What losses occur in a transformer core? How is each of these minimised? A single-phase, 50 Hz transformer is to be connected to a 440 V supply. The output from the transformer is to be 110 V on no load. The 110 V winding is to have 800 turns while the maximum core flux density is to be 1.3 T.

Calculate the cross-sectional area of the core and the number of turns on the 440 V winding. Prove any formulae used.

CHAPTER 6   ANSWERS

6.3. B = 1.3 T; H = 800 A/m; I = 1.257 A

6.5. (a) (i) 4800 A/m (ii) 1.6 T (iii) 5.76 mH   (b) Scales: 300 A/m = 1 cm; 0.1 T = 1 cm; volume of core = 75×10^{-6} m^3; energy = 0.2205 J

6.7. 0.085 mm

6.9. 17.5 W; 75 W

6.11. (a) 30 W   (b) 4.8 W   (c) 1.907

6.12. (a) 1.588 T   (b) 1587.5 J/m^3 (c) 58.4 W

6.13. (a) 1.45 T   (b) 435 V

6.14. (a) 3846 A/m   (b) 4.833 mWb/m^2 (c) 362.5 μH   (d) 725 μJ   (e) 0.024 V

6.15. (a) 0.91 A   (b) 0.151 H   (c) 0.063 J

6.16. 0.03 H; 0.02 H on a.c.

6.17. Halves B_m; ratio 289 : 133

6.18. 1.36 cm

6.19. 2.12 μH/m; new conditions: 2.046 μH/m; radius 0.9 cm

6.20. (b)(i) H 22.95 W; E 6.82 W   (ii) H 45.8 W; E 27.2 W

6.21. H 228 W; E 72 W; new: H 346.9 W; E 273.8 W

6.23. 1600 W

6.24. E/f = 5: H 15 W; E 22.5 W; E/f = 2.5: H 3.15 W; E 5.6 W; n = 2.25

6.25. A = 0.000476 m^2; primary 3200 turns

7 Permanent Magnets

In some pieces of equipment it is necessary to have a steady magnetic field in an air gap, an example being the moving-coil d.c. meter. A radial magnetic field is produced by either machining the pole faces of a permanent magnet into a cylindrical form or by the use of soft-iron pole shoes. The name permanent magnet means that the material used exhibits magnetism without the need for excitation by a current-carrying coil. Such magnets need to have high remanent flux and considerable resistance to demagnetisation by external fields, i.e. a large coercive force. Both these factors mean that the hysteresis loop has a large area (see chapter 6, section 6.3). The coercive force of steel is increased by adding alloying elements such as tungsten, nickel, chromium and cobalt. Some such alloys are so hard as to be unrollable when cold and therefore are formed into rectangular blocks or cylinders, the ends of which are ground to receive pole pieces. Some ferrites may be made permanently magnetic. They are not generally so powerful as the alloy magnets but are extremely difficult to demagnetise. With their excellent eddy current performance (chapter 6, section 6.11) they are eminently suitable for use in the region of strong alternating magnetic fields.

Figure 7.1 shows two permanent magnet arrangements for a moving-coil d.c. meter.

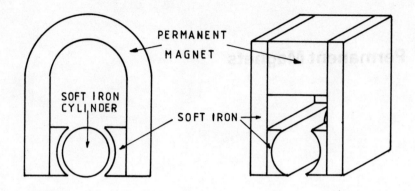

Fig. 7.1

7.1 THE DEMAGNETISATION CURVE

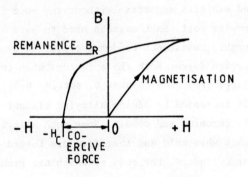

Fig. 7.2

During the manufacture of an alloy permanent magnet, the steel is situated in a loop of low reluctance and is heated to above the Curie temperature (the temperature at which ferro-magnetic materials lose their strongly paramagnetic properties). They are allowed to cool in a sufficiently strong external magnetic field to drive them into saturation. Upon removal of this field the flux density falls to its remanence value, B_R (Fig. 7.2). If the magnetising coil current is now reversed in direction so producing a demagnetising force (-H A/m), the demagnetising curve may be plotted. The current is increased in

PERMANENT MAGNETS

stages and for each current the resultant flux density is measured. At a demagnetising force $-H_C$, the resultant flux density is zero, the flux density of the permanent magnet being matched exactly by that created by the coil. This establishes the effective m.m.f. of the permanent magnet since it creates a magnetising force of $+H_C$. For length of permanent magnet steel ℓ_s, its effective m.m.f. is $H_C \ell_s$ amperes. In calculations it may be considered to act as though it were in fact an electromagnet with this value of m.m.f.

If the permanent magnet is now removed from its low reluctance situation and is caused to set up flux in a high reluctance air gap, the flux density will fall from B_R to some lower value. The simplest magnetic circuit is shown in Fig. 7.3. (See also Fig. 7.1).

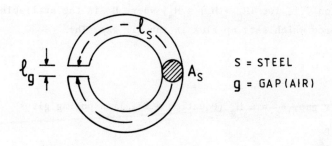

Fig. 7.3

In the case of the electromagnet setting up flux in steel and an air gap

total m.m.f. supplied = m.m.f. for steel + m.m.f. for air.

(See also chapter 6, section 6.1.1 and worked example 6.2.)
Transposing gives

total m.m.f. supplied - m.m.f. for steel - m.m.f. for air = 0.

For the permanent magnet

$$\text{m.m.f. supplied} = H_C \ell_s$$
$$\text{m.m.f. required by magnet steel} = H_s \ell_s$$
$$\text{m.m.f. required by air gap} = H_g \ell_g$$

Hence $H_C \ell_s - H_s \ell_s - H_g \ell_g = 0$. Or regrouping

$$H_C \ell_s - H_s \ell_s = H_g \ell_g$$
$$\ell_s (H_C - H_s) = H_g \ell_g \,. \tag{7.1}$$

$(H_C - H_s)$ is the difference in magnetising force between what is provided by the magnet and what is required to set up flux in itself. The analogy is in the electric circuit where the external potential difference from a generator or battery is equal to the e.m.f. less any internal drops in voltage. In symbols

$$V = E - I_a R_a \text{ volts.}$$

In equation 7.1, let $(H_C - H_s) = H_m$, where H_m is the available magnetising force which sets up flux in the air gap. Then

$$H_m \ell_s = H_g \ell_g \,. \tag{7.2}$$

In the air gap, $B_g = \mu_o H_g$ (equation 6.2). Transposing gives

$$H_g = \frac{B_g}{\mu_o} \,. \tag{7.3}$$

Substituting equation 7.3 in equation 7.2 gives

$$H_m \ell_s = \frac{B_g}{\mu_o} \ell_g \,. \tag{7.4}$$

The corresponding flux density in the steel is B_m.

Figure 7.4 shows H_C and H_m, the difference between them being H_s. They are shown without the negative sign (which was used in Fig. 7.2) which indicated the value of demagnetising force needed to arrive at a particular point on the curve. The permanent magnet provides external magnetising force, a positive effect, so that this negative sign will no longer appear.

PERMANENT MAGNETS

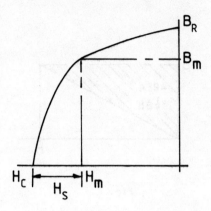

Fig. 7.4

7.2 OPTIMUM OPERATING POINT

The cross-sectional area of the air gap shown in Fig. 7.3 is the same as that of the steel since it is created simply between the two ends of the steel formed into a loop. In reality there would be some leakage flux, i.e. that flux taking paths other than directly from one pole face to the other, but this will be neglected as it will be assumed to be small for short gap lengths.

From equation 6.10

energy stored in a magnetic field during an increment of change in flux density (dB) = $HdB\ell A$ joules.

In Fig. 6.7 the sum of all such increments HdB is equal to the area between the B-axis and the magnetisation curve.

In air, $B = \mu_o H$, so that the B/H graph is a straight line as shown in Fig. 7.5. The area corresponding to 4, 1, 5, 4 in Fig. 6.7 is the area of the hatched triangle ($=\tfrac{1}{2}BH$).

Therefore, the energy stored in an air gap of length ℓ_g and cross-sectional area A m^2 is

$$\tfrac{1}{2} B_g H_g \ell_g A_g \text{ joules} \qquad (g = \text{gap (air), see fig. 7.3}). \qquad (7.5)$$

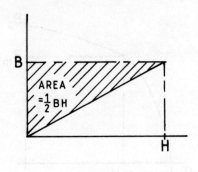

Fig. 7.5

Now $H_m \ell_s = H_g \ell_g$ (equation 7.2), and with no leakage $\Phi_g = \Phi_m$. Hence

$$B_g A_g = B_m A_s . \qquad (7.6)$$

Substituting equations 7.6 and 7.2 in equation 7.5 gives

energy stored in the air gap = $\tfrac{1}{2} B_m H_m A_s \ell_s$ joules.

But $A_s \ell_s$ = volume of steel, therefore

energy stored in the air gap = $\tfrac{1}{2} B_m H_m \times$ volume of steel.

Transposing

$$\text{volume of steel} = \frac{\text{energy stored in the air gap}}{\tfrac{1}{2} B_m H_m} . \qquad (7.7)$$

Hence for a fixed value of energy to be stored in the air gap it follows that a magnet with the smallest possible volume will be obtained when the $B_m H_m$ product is a maximum. The optimum operating point is that point on the demagnetising characteristic which gives the maximum $B_m H_m$ product.

How the $B_m H_m$ product varies with B_m will now be investigated using an example.

PERMANENT MAGNETS

<u>Worked example 7.1</u> The demagnetising curve for a certain permanent magnet steel is given in the following table

B (tesla)	1.25	1.2	1.1	1.0	0.9	0.8	0.6	0.4	0
-H (A/m)	0	18000	32100	35800	38200	40000	41500	42500	43500

Determine the optimum operating point for the material. Working at this flux density determine the volume of a magnet which will provide a stored energy of 0.05 J in an air gap 3 mm long. (Assume $A_s = A_g$.)

The first step is to multiply B_m by its associated H_m and compile a new table of B_m against the $B_m H_m$ product.
  In the first column, 1.25 × 0 = 0. In the second column, 1.2 × 18000 = 21600, etc.

B_m (tesla)	1.25	1.2	1.1	1.0	0.9	0.8	0.6	0.4	0
$B_m H_m$	0	21600	35310	35800	34380	32000	24900	17000	0

These figures are plotted together with the original values of B_m against H_m in Fig. 7.6.
  It is seen that the maximum value of the $B_m H_m$ product lies in the region of 1.0 to 1.1 tesla. To establish exactly where the maximum occurs it is necessary to take one or two more points from the original B/H curve. For example, when B = 1.02 tesla, H scales 35250 A/m. The product is 1.02 × 35250 = 35955. This is in fact the maximum value. The reader is advised to plot these curves to a larger scale than is possible in this book and so to verify the value of $B_m H_m$ (max).

  For a minimum volume of steel, the flux density should be 1.02 tesla when H_m = 35250 A/m. From equation 7.4

$$H_m \ell_s = \frac{B_g}{\mu_0} \ell_g.$$

The length of the air gap equals 3 mm, and since $A_s = A_g$, $B_m = B_g$ (equation 7.6). Substituting these values into the equation gives

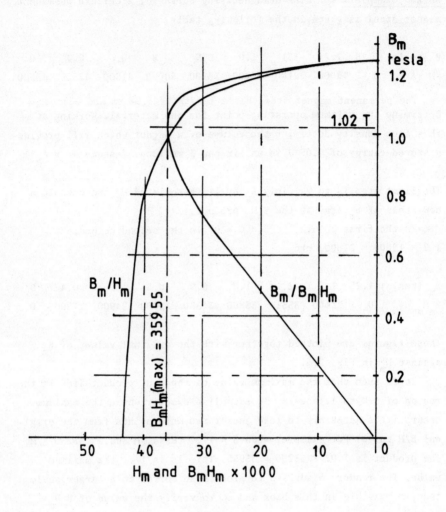

Fig. 7.6

$$35250 \, \ell_s = \frac{1.02}{4\pi \times 10^{-7}} \times 3 \times 10^{-3}$$

$$\ell_s = 0.0691 \text{ m} \quad (6.91 \text{ cm}).$$

The energy stored in the air gap = $\frac{1}{2} B_g H_g A_g \ell_g$ J (equation 7.5).

$H_g = \dfrac{B_g}{\mu_o}$ (equation 7.3), and $A_s = A_g$. Therefore

PERMANENT MAGNETS

$$\text{energy} = 0.05 = \tfrac{1}{2} \times 1.02 \times \frac{1.02}{4\pi \times 10^{-7}} \times A_g \times 3 \times 10^{-3}$$

$$A_g = \frac{0.05 \times 4\pi \times 10^{-7} \times 2}{1.02 \times 1.02 \times 3 \times 10^{-3}} = 40.26 \times 10^{-6} \text{ m}^2 \ (0.4026 \text{ cm}^2).$$

The permanent magnet steel has a length of 6.91 cm and a cross-sectional area of 0.4026 cm^2. Its volume is 6.91 × 0.4026 = 2.782 cm^3.

The length of the air gap is 3 mm and it too has a cross-sectional area of 0.4026 cm^2.

7.3 THE EFFECT OF POLE SHOES

Almost invariably the flux density required in the air gap is not the same as that required in the steel to allow minimum volume to be used. To reconcile the two requirements, pole shoes are used. These are pieces of very low-reluctance soft iron fixed to the poles of the permanent magnet so as to change the effective area over which the magnetic flux is distributed.

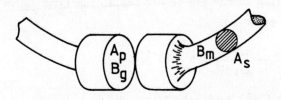

Fig. 7.7

In Fig. 7.7

A_p = face area of the pole shoe
A_s = cross-sectional area of magnet steel
B_g = flux density at the pole face = flux density in the air gap
B_m = flux density in the magnet steel.

Assuming no leakage, the total flux in the magnet is equal to the total flux in the air gap. Hence

$$B_g A_p = B_m A_s. \tag{7.8}$$

<u>Worked example 7.2</u> Calculate the dimensions of a permanent magnet of minimum volume to maintain a flux density of 0.2 tesla in an air gap 5 mm long and having a cross-sectional area of 6 cm^2. Soft iron pole shoes are used, the reluctance of which may be ignored. For the permanent magnet, the optimum operating point occurs when B_m = 0.96 tesla and $B_m H_m$ = 3024.

Since B_m = 0.96 and $B_m H_m$ = 3024, at the optimum operating point

$$H_m = \frac{3024}{0.96}$$
$$= 3150 \text{ A/m}.$$

$$H_m \ell_s = \frac{B_g}{\mu_o} \ell_g \quad \text{(equation 7.4)}.$$

The magnet steel works at its optimum point, hence H_m = 3150 A/m. The flux density in the gap is 0.2 tesla. Substituting values in equation 7.4 gives

$$3150 \, \ell_s = \frac{0.2}{4\pi \times 10^{-7}} \times 5 \times 10^{-3}$$

$$\ell_s = 0.253 \text{ m} \quad (25.3 \text{ cm}).$$

$$B_g A_p = B_m A_s \quad \text{(equation 7.8)}.$$

The area of the gap = area of the pole faces = 6 cm^2.

The steel is to work at 0.96 tesla.

Substituting values in equation 7.8 gives

$$0.2 \times 6 \times 10^{-4} = 0.96 \times A_s$$
$$A_s = 1.25 \times 10^{-4} \text{ m}^2 \quad (1.25 \text{ cm}^2).$$

The magnet steel is 25.3 cm in length and has a cross-sectional

PERMANENT MAGNETS

area of 1.25 cm^2. Its volume is 25.3 × 1.25 = 31.63 cm^3.

<u>Self-assessment example 7.3</u> For the same permanent magnet material as in worked example 7.2, determine the minimum volume of steel necessary to set up a flux density of 0.65 tesla in an air gap 2 mm long and with a cross-sectional area of 4.5 cm^2. The reluctance of the pole shoes used may be ignored.

<u>Self-assessment example 7.4</u> An air gap with a cross-sectional area of 3 cm^2 and length 10 mm is to be supplied with a flux density of 0.42 tesla. The required flux density can be provided in two ways

(i) using a piece of permanent magnet steel with cross-sectional area 3 cm^2 and forming it into a loop when $A_s = A_g$.
(ii) using a piece of permanent magnet steel of minimum volume in conjunction with pole shoes.

The magnet steel has a demagnetising characteristic given in the following table:

B (tesla)	0.25	0.5	0.75	0.9	1.0	1.25	1.4
-H (A/m)	4910	4494	3933	3520	3146	1742	0

Determine (a) the operating point for minimum volume, (b) the dimensions of the magnet described in (i) above, (c) the dimensions of the magnet described in (ii) above, (d) whether the volume for (ii) is less than that for (i). What percentage change has occurred?

<u>Worked example 7.5</u> A permanent magnet is required to maintain a flux density of 0.4 tesla in an air gap of length 4 mm and cross-sectional area 2.75 cm^2. The demagnetisation characteristic for the material is given in worked example 7.1.
(a) For $A_s = A_g$, determine (i) the required length of magnet steel, (ii) the volume of magnet steel, (iii) the value of energy stored in the air gap.
(b) Determine the reduction in magnet steel volume which could be achieved if it is re-designed to work under optimum conditions, the

gap flux density and dimensions remaining unchanged.

(a)(i) $H_m \ell_s = \dfrac{B_g}{\mu_o} \ell_g$ (equation 7.4).

$B_g = B_m = 0.4$ T, and from the characteristics, $H_m = 17000$ for this value of B_m. Hence

$$17000 \, \ell_s = \dfrac{0.4}{4\pi \times 10^{-7}} \times 4 \times 10^{-3}$$

$$\ell_s = 0.0749 \text{ m} \quad (7.49 \text{ cm}).$$

(ii) $A_g = A_s = 2.75 \text{ cm}^2$.

Volume = $A_s \ell_s = 2.75 \times 7.49 = 20.6 \text{ cm}^3$.

(iii) Energy stored in the gap = $\tfrac{1}{2} B_g H_g \ell_g A_g$ J (equation 7.5)

$$= \tfrac{1}{2} \times 0.4 \times \dfrac{0.4}{4\pi \times 10^{-7}} \times 4 \times 10^{-3} \times 2.75 \times 10^{-4}$$

$$= 0.07 \text{ J}.$$

(b) From worked example 7.1, for optimum conditions, $B_m = 1.02$ T, $H_m = 35250$ A/m. The gap is unchanged so that

$$35250 \, \ell_s = \dfrac{0.4}{4\pi \times 10^{-7}} \times 4 \times 10^{-3} \quad \text{(equation 7.4)}$$

$$\ell_s = 0.036 \text{ m} \quad (3.6 \text{ cm}).$$

But $B_m A_s = B_g A_g$ (equation 7.8), therefore

$$1.02 \times A_s = 0.4 \times 2.75 \times 10^{-4}$$
$$A_s = 1.078 \times 10^{-4} \text{ m}^2$$
$$= 1.078 \text{ cm}^2.$$

Volume = $1.078 \times 3.6 = 3.88 \text{ cm}^3$.

PERMANENT MAGNETS

The volume is reduced from 20.6 cm^3 to 3.88 cm^3 by the use of pole shoes.

Further problems

7.6. Explain the meaning of the terms 'hard' and 'soft' as applied to magnetic materials. What are the main requirements of materials used for (a) the core of a mains transformer, (b) the core of a smoothing choke, (c) a permanent magnet.

Calculate the dimensions of a permanent magnet of minimum volume to maintain a flux of 450 µWb in an air gap 5 mm long and having an effective cross-sectional area of 5 cm^2. (The reluctance of the soft iron pole pieces may be ignored.)

Data for the magnet material: BH(max.) = 42500; B_m at BH(max.) = 0.72 tesla.

7.7. The demagnetisation curve of a sample of magnet steel is given by the following data

B (tesla)	1.2	1.1	1.0	0.9	0.8	0.6	0.4	0.2	0
-H (A/m)	17600	31900	38000	41250	43400	45600	47000	47300	47850

Neglecting magnetic leakage, determine the dimensions of a permanent magnet of minimum volume which will store energy of 0.35 J in an air gap 5 mm long and 1 cm^2 in cross-sectional area. (The air gap flux density is the same as that in the steel, i.e. no pole shoes are used.)

7.8. A flux density of 0.2 tesla is required in an air gap 3 mm in length having a cross-sectional area of 500 mm^2. Calculate the dimensions of a permanent magnet of minimum volume which will perform this function given that: B_m at $B_m H_m$(max.) = 1.1 T; H_m at $B_m H_m$(max.) = 42000. (The pole shoes have a reluctance equal to 0.5 mm of extra air gap. In calculations use ℓ_g = 3 mm + 0.5 mm.)

CHAPTER 7 ANSWERS

7.3. ℓ_s = 32.84 cm; A_s = 3.05 cm^2; volume = 100 cm^3

7.4. (a) 0.9 T, BH = 3168; H = 3520 (b) B_s = 0.42 T where H = 4610; $H_s \ell_s = (B_g/\mu_0)\ell_g$, ℓ_s = 72.5 cm, A_s = 3 cm^2; volume = 217.5 cm^3
(c) B_g = 0.42 T; B_s = 0.9 T; ℓ_s = 94.95 cm; A_s = 1.4 cm^2; volume = 132.9 cm^3 (d) reduction 84.6 cm^3 on 217.5 cm^3:

39% reduction
7.6. $\ell_s = 6.06$ cm; $A_s = 6.25$ cm^2
7.7. BH(max.) = 38122 at B = 0.98 T; volume = 18.36 cm^3 (from energy = $\tfrac{1}{2} B_m H_m$ J/m^3); $A_s = 1.836$ cm^2
7.8. $\ell_s = 1.33$ cm; $A_s = 91$ mm^2

8 Two-Port Networks

A two-port network is a network with two input terminals and two output terminals. The network may contain an active device such as a transistor or comprise passive components only. In this chapter the two-port network will be treated as a transmission element containing passive devices only; our concern being with input and output impedance of the network and the proportion of the input which appears at the output.

There are several conventions for use with two-port networks, the two most common being shown in Fig. 8.1. This book uses that in Fig. 8.1(a). If that in Fig. 8.1(b) is preferred, all that is required is a sign change for I_2 i.e. I_2 in Fig. 8.1(b) = $-I_2$ in Fig. 8.1(a).

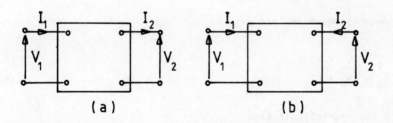

Fig. 8.1

8.1 GENERAL EQUATIONS FOR V_1 AND I_1 IN TERMS OF V_2 AND I_2

For any four-terminal network comprising linear components there will be a linear relationship between input voltage and input current (V_1 and I_1) and output voltage and output current (V_2 and I_2). The relationships are expressed as follows

$$V_1 = AV_2 + BI_2 \quad (8.1)$$
$$I_1 = CV_2 + DI_2 \quad (8.2)$$

where A and D are dimensionless constants, C is a constant with the dimensions of admittance, and B is a constant with the dimensions of impedance. These constants are known as the 'four-terminal' constants or parameters and may be determined using the open- and short-circuit tests.

8.2   FOUR-TERMINAL NETWORK PARAMETERS FROM OPEN- AND SHORT-CIRCUIT TESTS

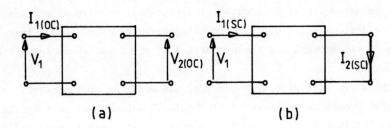

Fig. 8.2

The general equations are

$$V_1 = AV_2 + BI_2 \quad (8.1)$$
$$I_1 = CV_2 + DI_2 \quad (8.2)$$

8.2.1 The open-circuit test

With the output terminals open-circuited, as in Fig. 8.2(a), a voltage V_1 is applied to the input terminals when a current $I_{1(oc)}$ enters the network. The output current is zero ($I_2 = 0$). Hence from equation 8.1

$$V_1 = AV_{2(oc)} + 0$$

TWO-PORT NETWORKS

$$A = \frac{V_1}{V_{2(oc)}} .$$

From equation 8.2

$$I_1 = CV_{2(oc)} + 0$$

$$C = \frac{I_1}{V_{2(oc)}} .$$

8.2.2 The short-circuit test (Fig. 8.2(b))

The output terminals are short-circuited through a low-impedance ammeter. The network is supplied with a voltage V_1 and a current $I_{1(sc)}$ enters the network. The potential difference across the output terminals is zero ($V_2 = 0$). Hence, from equation 8.1

$$V_1 = 0 + BI_{2(sc)}$$

$$B = \frac{V_1}{I_{2(sc)}} .$$

From equation 8.2

$$I_{1(sc)} = 0 + DI_{2(sc)}$$

$$D = \frac{I_{1(sc)}}{I_{2(sc)}} .$$

8.3 FOUR-TERMINAL PARAMETERS FOR FOUR NETWORKS

8.3.1 Series impedance (Z Ω)

With the current convention chosen, V_1 and I_1 are input voltage and input current respectively and V_2 and I_2 the corresponding outputs.
Using Kirchhoff's first law, $I_1 = I_2$.
Using Kirchhoff's second law, $V_1 - I_1Z = V_2$, or since $I_1 = I_2$,
$V_1 - I_2Z = V_2$.
Transposing gives $V_1 = V_2 + ZI_2$
Compare this with equation 8.1 $V_1 = AV_2 + BI_2$.
Hence A = 1 and B = Z. (8.3)

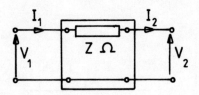
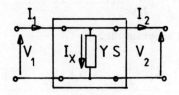

Fig. 8.3   Fig. 8.4

From above: $I_1 = I_2$.

Compare with equation 8.2: $I_1 = CV_2 + DI_2$.

Hence $C = 0$ and $D = 1$.   (8.4)

8.3.2 Shunt admittance (Y S)
In Fig. 8.4, since there is no series impedance, $V_1 = V_2$.
Comparing this with equation 8.1, $V_1 = AV_2 + BI_2$.
Hence $A = 1$ and $B = 0$.   (8.5)
By Kirchhoff's first law, $I_2 = I_1 - I_x$.   (i)
Now $I_x = YV_1 = YV_2$. Substituting for I_x in (i) and transposing gives
$$I_1 = YV_2 + I_2.$$
Comparing this with equation 8.2, $I_1 = CV_2 + DI_2$.
Hence $C = Y$ and $D = 1$.   (8.6)

8.3.3 The T-network
In Fig. 8.5(a)

Current I_2 flows in Z_2, hence: $V_x = V_2 + I_2 Z_2$ (i)
$\phantom{Current I_2 \text{ flows in } Z_2, \text{ hence: }} I_x = V_x Y$ (ii)
$\phantom{Current I_2 \text{ flows in } Z_2, \text{ hence: }} I_1 = I_2 + I_x$. (iii)

Substituting (ii) in (iii): $I_1 = I_2 + V_x Y$.   (iv)

TWO-PORT NETWORKS

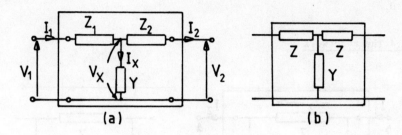

Fig. 8.5

Substituting (i) in (iv): $I_1 = I_2 + (V_2 + I_2 Z_2)Y$
                          $= I_2 + V_2 Y + I_2 Z_2 Y$.

Re-grouping: $I_1 = YV_2 + (Z_2 Y + 1) I_2$.
Compare with equation 8.2, $I_1 = CV_2 + \quad\quad DI_2$.

Hence $C = Y$ and $D = (1 + Z_2 Y)$. (8.7)

Referring back to Fig. 8.5(a): $V_1 = V_x + I_1 Z_1$ (v)

Substituting (iv) in (v): $V_1 = V_x + (I_2 + V_x Y)Z_1$. (vi)

Substituting (i) in (vi): $V_1 = V_2 + I_2 Z_2 + (I_2 + (V_2 + I_2 Z_2)Y)Z_1$
                           $V_1 = V_2 + I_2 Z_2 + I_2 Z_1 + V_2 YZ_1 + I_2 Z_2 YZ_1$.

Re-grouping: $V_1 = (1 + YZ_1)V_2 + (Z_2 + Z_1 + Z_2 Z_1 Y)I_2$.
Compare with equation 8.1, $V_1 = \quad A V_2 + \quad\quad B I_2$.

Hence $A = (1 + YZ_1)$ and $B = (Z_2 + Z_1 + Z_2 Z_1 Y)$. (8.8)

If the network is symmetrical, as shown in Fig. 8.5(b), $Z_1 = Z_2 = Z$ and the four-terminal parameters reduce to

$$A = (1 + YZ) \quad B = (2Z + Z^2 Y) \quad C = Y \quad D = (1 + YZ). \quad (8.9)$$

Making the network symmetrical has made the two constants A and D equal. This is true generally. In any symmetrical, four-terminal network which does not contain sources of e.m.f., i.e. a passive network, $A = D$.

8.3.4 The π-network

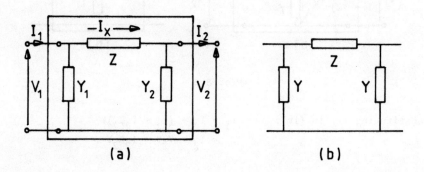

Fig. 8.6

Current in Y_2 is V_2Y_2, and I_x = current in $Y_2 + I_2$.
Hence $\qquad\qquad I_x = V_2Y_2 + I_2.$ \hfill (i)

Also $\qquad\qquad V_1 = V_2 + I_xZ.$ \hfill (ii)

Substituting (i) in (ii): $V_1 = V_2 + (V_2Y_2 + I_2)Z = V_2 + V_2Y_2Z + I_2Z.$

Re-arranging $\qquad\qquad V_1 = (1 + Y_2Z)V_2 + ZI_2$ \hfill (iii)
Compare with equation 8.1, $V_1 = \qquad\quad A V_2 + B I_2$

Hence $A = (1 + Y_2Z)$ and $B = Z$. \hfill (8.10)

Again, in Fig. 8.6(a): current in $Y_1 = V_1Y_1$ \hfill (iv)
$\qquad\qquad$ and I_1 = current in $Y_1 + I_x$. \hfill (v)

Substituting (i) and (iv) in (v): $I_1 = V_1Y_1 + (V_2Y_2 + I_2).$ \hfill (vi)

Substituting (iii) for V_1 in (vi): $I_1 = [(1 + Y_2Z)V_2 + ZI_2]Y_1 + (V_2Y_2 + I_2)$
$\qquad\qquad\qquad\qquad\qquad = V_2Y_1 + Y_2Y_1ZV_2 + Y_1ZI_2 + V_2Y_2 + I_2$
$\qquad\qquad\qquad\qquad\qquad = (Y_1 + Y_2 + Y_1Y_2Z)V_2 + (1 + Y_1Z)I_2$
Compare with equation 8.2, $\qquad I_1 = \qquad\qquad C V_2 + \qquad D I_2.$

TWO-PORT NETWORKS

Hence $C = (Y_1 + Y_2 + Y_1 Y_2 Z)$ and $D = (1 + Y_1 Z)$. (8.11)

For a symmetrical π-network, as shown in Fig. 8.6(b), $Y_1 = Y_2 = Y$ and the four-terminal parameters reduce to

$$A = (1 + YZ) \quad B = Z \quad C = (2Y + Y^2 Z) \quad D = (1 + YZ). \quad (8.12)$$

Symmetry results in the two constants A and D being equal as in the case of the symmetrical-T-network.

8.4 AD - BC = 1

A general property of a four-terminal passive network, whether symmetrical or not, is that $AD - BC = 1$. This will be examined for the asymmetrical π-network shown in Fig. 8.6(a).

Using equations 8.10 and 8.11

$$AD = (1 + Y_2 Z) \times (1 + Y_1 Z) = 1 + Y_1 Z + Y_2 Z + Y_1 Y_2 Z^2$$
$$BC = Z \times (Y_1 + Y_2 + Y_1 Y_2 Z) = Y_1 Z + Y_2 Z + Y_1 Y_2 Z^2$$
Hence, $AD - BC = 1$.

Self-assessment example 8.1 Using the ABCD parameters for the asymmetric T-network in Fig. 8.5(a), deduce that $AD - BC = 1$.

Worked example 8.2 A four-terminal symmetrical π-network has a series impedance of $100\underline{/65^\circ}$ Ω. Each of the shunt admittances is $2 \times 10^{-3}\underline{/85^\circ}$ S, all at a particular frequency.

Calculate (a) the values of the constants A, B, C and D, (b) the values of the input voltage and current (V_1 and I_1) if, when a 50 Ω resistor is connected across the output terminals, a current of 2.5 mA flows in it.

(a) Using equations 8.12

$$\begin{aligned}
A = D = (1 + YZ) &= 1 + 2 \times 10^{-3}\underline{/85^\circ} \times 100\underline{/65^\circ} \\
&= 1 + 0.2\underline{/150^\circ} \\
&= 1 - 0.1732 + j0.1 \\
&= 0.827 + j0.1 = 0.833\underline{/6.9^\circ}.
\end{aligned}$$

$B = Z = 100\underline{/65°}$

$$C = 2Y + Y^2Z = 2 \times 2 \times 10^{-3}\underline{/85°} + (2 \times 10^{-3})^2 \times 100\underline{/65°}$$
$$= 4 \times 10^{-3}\underline{/85°} + 4 \times 10^{-4}\underline{/85° + 85° + 65°}$$
$$= 3.486 \times 10^{-4} + j3.985 \times 10^{-3} + (-2.29 \times 10^{-4} - j3.28 \times 10^{-4})$$
$$= 1.19 \times 10^{-4} + j3.66 \times 10^{-3} = 3.66 \times 10^{-3}\underline{/88.13°}.$$

(b) The potential difference across the output terminals is $I_2 R_L$. Hence

$$V_2 = 2.5 \times 10^{-3} \times 50 = 0.125 \text{ V}.$$

Using V_2 as reference, i.e. writing $V_2 = 0.125\underline{/0°}$ V, then $I_2 = 2.5 \times 10^{-3}\underline{/0°}$ A, since in a resistor, voltage and current are in phase.

Using equation 8.1

$$V_1 = AV_2 + BI_2$$
$$= 0.833\underline{/6.9°} \times 0.125\underline{/0°} + 100\underline{/65°} \times 2.5 \times 10^{-3}\underline{/0°}$$
$$= 0.103 + j0.0125 + 0.106 + j0.227$$
$$= 0.209 + j0.239 = 0.318\underline{/48.9°} \text{ V}.$$

Using equation 8.2

$$I_1 = CV_2 + DI_2$$
$$= 3.66 \times 10^{-3}\underline{/88.13°} \times 0.125\underline{/0°} + 0.833\underline{/6.9°} \times 2.5 \times 10^{-3}\underline{/0°}$$
$$= 14.93 \times 10^{-6} + j457.3 \times 10^{-6} + 2.067 \times 10^{-3} + j250.2 \times 10^{-6}$$
$$= 2.082 \times 10^{-3} + j707.4 \times 10^{-6} = 2.2\underline{/18.76°} \text{ mA}.$$

Self-assessment example 8.3 A single-phase power transmission line may be represented by the symmetrical T-network shown in Fig. 8.7. $V_2 = 158770\underline{/0°}$ V and $I_2 = 126\underline{/-36.87°}$ A.

Determine (a) the values of the four-terminal network parameters, (b) the magnitude and phase of V_1, (c) the magnitude and phase of I_1.

TWO-PORT NETWORKS

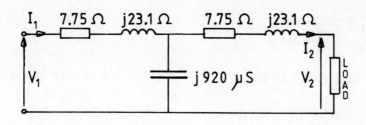

Fig. 8.7

8.5 GENERAL EQUATIONS FOR V_2 AND I_2 IN TERMS OF V_1 AND I_1

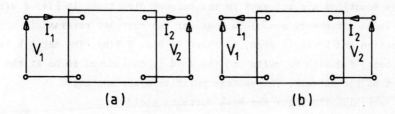

Fig. 8.8

For the network as shown in Fig. 8.8(a) (identical to Fig. 8.1(a)), equation 8.1: $V_1 = AV_2 + BI_2$, equation 8.2: $I_1 = CV_2 + DI_2$. Multiplying equation 8.1 by D and equation 8.2 by B gives

$$DV_1 = ADV_2 + BDI_2 \qquad \text{(i)}$$
$$BI_1 = BCV_2 + BDI_2. \qquad \text{(ii)}$$

Subtracting (ii) from (i)

$$DV_1 - BI_1 = (AD - BC)V_2.$$

From section 8.4, $AD - BC = 1$, therefore

$$V_2 = DV_1 - BI_2. \qquad (8.13)$$

267

Now multiplying equation 8.1 by C and equation 8.2 by A gives

$$CV_1 = CAV_2 + BCI_2 \qquad \text{(iii)}$$
$$AI_1 = CAV_2 + ADI_2. \qquad \text{(iv)}$$

Subtracting (iv) from (iii)

$$CV_1 - AI_1 = (BC - AD)I_2.$$

Since $AD - BC = 1$, $BC - AD = -1$, therefore

$$-I_2 = CV_1 - AI_1. \qquad (8.14)$$

These equations are relevant to the current directions in Fig. 8.8(a).

It is often more useful to consider the currents reversed in direction and hence in sign, as shown in Fig. 8.8(b). The network is now being fed with V_2 volts and the load is considered to be at the front of the network. The voltage polarities are unchanged.

Reversing the signs for both currents yields

$$V_2 = DV_1 + BI_1 \qquad (8.15)$$
$$I_2 = CV_1 + AI_1. \qquad (8.16)$$

For symmetrical networks, $A = D$, and feeding the network from either end results in the same relationships between the sending and receiving conditions. For asymmetric networks, reversing the network will change its performance.

8.6 IMAGE IMPEDANCE

From the maximum power transfer theorem (chapter 2, section 2.5), it is known that for maximum power transfer from a generator with internal impedance Z_I into a load with impedance Z_L, the condition $|Z_I| = |Z_L|$ must be satisfied.

Where Z_L is not equal to Z_I, the load may be matched to the generator using either a transformer or a four-terminal network. A network which is designed to satisfy this condition is said to be image-matched.

TWO-PORT NETWORKS

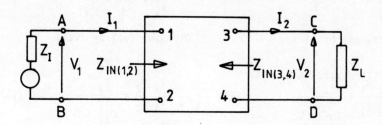

Fig. 8.9(a)

For image matching, the impedance viewed from terminals AB in Fig. 8.9(a) in either direction will be the same. Looking to the left the impedance is Z_I and to the right it is $Z_{IN(1,2)}$.

$$Z_{IN(1,2)} = Z_I.$$

The same is true for the load end; the impedance viewed either way from terminals CD will be the same.

$$Z_{IN(3,4)} = Z_L.$$

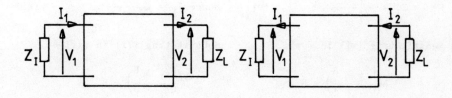

Fig. 8.9(b)   *Fig. 8.9(c)*

Equation 8.1: $V_1 = AV_2 + BI_2$.   Equation 8.15: $V_2 = DV_1 + BI_1$.
Equation 8.2: $I_1 = CV_2 + DI_2$.   Equation 8.16: $I_2 = CV_1 + AI_1$.
Dividing throughout by I_2 Dividing throughout by I_1

$$\frac{V_1}{I_2} = \frac{AV_2}{I_2} + \frac{BI_2}{I_2} \quad \text{(i)} \qquad \frac{V_2}{I_1} = \frac{DV_1}{I_1} + \frac{BI_1}{I_1} \quad \text{(vi)}$$

$$\frac{I_1}{I_2} = \frac{CV_2}{I_2} + \frac{DI_2}{I_2} \, . \qquad \text{(ii)} \qquad\qquad \frac{I_2}{I_1} = \frac{CV_1}{I_1} + \frac{AI_1}{I_1} \, . \qquad \text{(vii)}$$

Dividing (i) by (ii)   Dividing (vi) by (vii)

$$\frac{I_2}{I_2} \frac{V_1}{I_1} = \frac{A\frac{V_2}{I_2} + B}{C\frac{V_2}{I_2} + D} \qquad \text{(iii)} \qquad\qquad \frac{I_1}{I_1} \frac{V_2}{I_2} = \frac{D\frac{V_1}{I_1} + B}{C\frac{V_1}{I_1} + A} \qquad \text{(viii)}$$

$$\frac{V_2}{I_2} = Z_L . \qquad \text{(iv)} \qquad\qquad \frac{V_1}{I_1} = Z_I . \qquad \text{(ix)}$$

Substituting (iv) in (iii)   Substituting (ix) in (viii)

$$\frac{V_1}{I_1} = \frac{AZ_L + B}{CZ_L + D} \, . \qquad \text{(v)} \qquad\qquad \frac{V_2}{I_2} = \frac{DZ_I + B}{CZ_I + A} \, . \qquad \text{(x)}$$

For image matching, $\frac{V_1}{I_1} = Z_{IN(1,2)}$ For image matching, $\frac{V_2}{I_2} = Z_{IN(3,4)}$

$$\qquad\qquad\qquad\qquad = Z_I . \qquad\qquad\qquad\qquad\qquad\qquad = Z_L .$$

Hence $Z_I = \dfrac{AZ_L + B}{CZ_L + D}$.   (vi)   Hence $Z_L = \dfrac{DZ_I + B}{CZ_I + A}$.   (xi)

Substituting (xi) in (vi)   Substituting (vi) in (xi)

$$Z_I = \frac{A\left(\dfrac{DZ_I + B}{CZ_I + A}\right) + B}{C\left(\dfrac{DZ_I + B}{CZ_I + A}\right) + D} \, . \quad (8.17) \qquad Z_L = \frac{D\left(\dfrac{AZ_L + B}{CZ_L + D}\right) + B}{C\left(\dfrac{AZ_L + B}{CZ_L + D}\right) + A} \, . \quad (8.18)$$

Either of the two final equations may be solved. The solution to 8.17 will now be examined.

In the numerator, to bring B over the common denominator $(CZ_I + A)$, the two quantities are multiplied together. Similarly in the denominator, multiplying D by $(CZ_I + A)$ brings it over the common denominator. Multiplying out the brackets yields

$$Z_I = \frac{ADZ_I + AB + BCZ_I + AB}{CDZ_I + CB + CDZ_I + AD} \times \frac{CZ_I + A}{CZ_I + A}.$$

Cross-multiplying

$$CDZ_I^2 + CBZ_I + CDZ_I^2 + ADZ_I = ADZ_I + AB + BCZ_I + AB$$

$$2CDZ_I^2 = 2AB$$

$$Z_I = \sqrt{\left(\frac{AB}{CD}\right)}. \tag{8.19}$$

Solving equation 8.18 in the same manner yields the result

$$Z_L = \sqrt{\left(\frac{DB}{CA}\right)} \tag{8.20}$$

The result may be deduced by a consideration of section 8.5 in which it is seen that viewing the network from the V_2I_2-end results in changing the positions of D and A in the general equations.

As an alternative to using the four-terminal network parameters the values of the image impedances may be determined using the open- and short-circuit tests. (See section 8.2 and Fig. 8.2.)

In Fig. 8.9(a), supplying terminals 1 and 2, and with terminals 3 and 4 open-circuited, then $I_2 = 0$.

Equation 8.1 reduces to: $V_1 = AV_2$. (i)
Equation 8.2 reduces to: $I_1 = CV_2$. (ii)

Dividing (i) by (ii) : $\dfrac{V_1}{I_1} = \dfrac{AV_2}{CV_2} = \dfrac{A}{C}$

$$\frac{V_1}{I_1} = Z_{IN(1,2)(oc)}$$

(where subscript 1,2 = at terminals 1 and 2, and subscript oc = output terminals open-circuited).

Next supplying terminals 1 and 2, with terminals 3 and 4 short-circuited

equation 8.1 reduces to: $V_1 = BI_2$. (iii)
equation 8.2 reduces to: $I_1 = DI_2$. (iv)

Dividing (iii) by (iv): $\dfrac{V_1}{I_1} = \dfrac{B}{D}$

$$\dfrac{V_1}{I_1} = Z_{IN(1,2)(sc)}.$$

(sc = output terminals short-circuited).

Therefore, $Z_{IN(1,2)(oc)} \times Z_{IN(1,2)(sc)} = \dfrac{A}{C} \times \dfrac{B}{D}$.

Now from equation 8.19: $Z_I = \sqrt{\left(\dfrac{AB}{CD}\right)}$

hence, $Z_I = \sqrt{\left(Z_{IN(1,2)(oc)} \times Z_{IN(1,2)(sc)}\right)}.$ (8.21)

Similarly, viewed from terminals 3 and 4, open- and short-circuiting terminals 1 and 2 yields a similar result.

$$Z_L = \sqrt{\left(Z_{IN(3,4)(oc)} \times Z_{IN(3,4)(sc)}\right)}.$$ (8.22)

<u>Worked example 8.4</u>   Determine the values of Z_A and Z_B which, when connected to terminals 1, 2 and 3, 4 respectively in Fig. 8.10, give image matching.

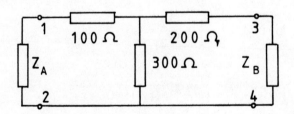

Fig. 8.10

The two methods available will be examined: (a) using the ABCD parameters, (b) using the open- and short-circuit tests.

(a) Using equations 8.7 and 8.8

$$A = 1 + YZ_1 = 1 + \frac{100}{300} = 1.333$$

$$B = Z_1 + Z_2 + Z_1Z_2Y = 100 + 200 + \frac{100 \times 200}{300} = 366.66$$

$$C = Y = \frac{1}{300}$$

$$D = 1 + YZ_2 = 1 + \frac{200}{300} = 1.66.$$

From equation 8.19: $Z_I = Z_A = \sqrt{\left(\frac{AB}{CD}\right)} = \sqrt{\left(\frac{1.33 \times 366.66 \times 300}{1.66}\right)}$

$$= 296.6 \ \Omega.$$

From equation 8.20: $Z_L = Z_B = \sqrt{\left(\frac{DB}{CA}\right)} = \sqrt{\left(\frac{1.66 \times 366.66 \times 300}{1.33}\right)}$

$$= 370.8 \ \Omega.$$

(b)

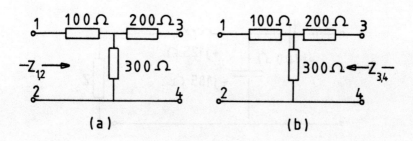

Fig. 8.11

In Fig. 8.11(a), with an open circuit at terminals 3 and 4

$$Z_{1,2} = 100 + 300$$
$$= 400 \ \Omega.$$

A short circuit across terminals 3 and 4 places the 200 Ω resistor in parallel with the 300 Ω resistor

$$Z_{1,2} = 100 + \frac{200 \times 300}{200 + 300} = 220 \ \Omega.$$

Using equation 8.21: $Z_A = \sqrt{(400 \times 220)} = 296.6 \ \Omega$.

In Fig. 8.11(b), with an open circuit at terminals 1 and 2

$$Z_{3,4} = 200 + 300$$
$$= 500 \ \Omega.$$

A short circuit across terminals 1 and 2 places the 100 Ω resistor in parallel with the 300 Ω resistor

$$Z_{3,4} = 200 + \frac{100 \times 300}{100 + 300} = 275 \ \Omega.$$

Using equation 8.22: $Z_B = \sqrt{(500 \times 275)} = 370.8 \ \Omega$.

<u>Self-assessment example 8.5</u>   Determine the value of Z_L in Fig. 8.12 which will give image matching.

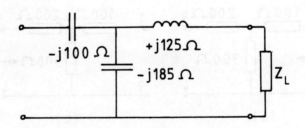

Fig. 8.12

8.7   ITERATIVE IMPEDANCE

If the value of an impedance connected to terminals 3 and 4 of the four-terminal network in Fig. 8.13(a) is equal to the input impedance $Z_{1,2}$, then this value is known as the iterative impedance (from 're-iterate'—to state again).

For iterative matching, $Z_{1,2} = Z_B$.

Similarly in Fig. 8.13(b), looking from terminals 3 and 4, itera-

TWO-PORT NETWORKS

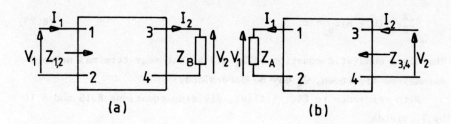

Fig. 8.13

tive matching occurs when $Z_{3,4} = Z_A$.

Using equations 8.1 and 8.2 which apply to Fig. 8.13(a), dividing both equations by I_2 gives

$$\frac{V_1}{I_2} = A\frac{V_2}{I_2} + B\frac{I_2}{I_2} \tag{i}$$

$$\frac{I_1}{I_2} = C\frac{V_2}{I_2} + D\frac{I_2}{I_2} . \tag{ii}$$

Dividing (i) by (ii) gives

$$\left(\frac{I_2}{I_2}\right)\frac{V_1}{I_1} = \frac{A\dfrac{V_2}{I_2} + B}{C\dfrac{V_2}{I_2} + D} .$$

But $\dfrac{V_1}{I_1} = Z_{1,2} = Z_B$

and $\dfrac{V_2}{I_2}$ is also equal to Z_B. Therefore

$$Z_B = \frac{AZ_B + B}{CZ_B + D} .$$

Cross-multiplying gives

$$CZ_B^2 + DZ_B = AZ_B + B.$$

Re-arranging

$$CZ_B^2 + (D - A)Z_B = B. \tag{8.23}$$

This is a quadratic equation in Z_B and as the four-terminal network parameters are known, Z_B may be determined.

With reference to Fig. 8.13(b), dividing equations 8.15 and 8.16 by I_1 yields

$$\frac{V_2}{I_1} = D\frac{V_1}{I_1} + B\frac{I_1}{I_1} \tag{iii}$$

$$\frac{I_2}{I_1} = C\frac{V_1}{I_1} + A\frac{I_1}{I_1}. \tag{iv}$$

Dividing (iii) by (iv) and substituting $\frac{V_1}{I_1} = Z_A$, $\frac{V_2}{I_2} = Z_{3,4} = Z_A$ gives

$$\frac{I_1}{I_1}\frac{V_2}{I_2} = \frac{DZ_A + B}{CZ_A + A}.$$

Cross-multiplying and re-arranging: $CZ_A^2 + (A - D)Z_A = B.$ (8.24)

Again, knowing the four-terminal network parameters, the quadratic may be solved. Where the values of the circuit components are known it is often more direct to work from first principles and this will be demonstrated in worked example 8.6.

<u>Worked example 8.6</u>   Determine the value of Z_A in Fig. 8.14 to give iterative matching.

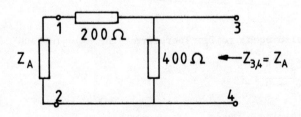

Fig. 8.14

TWO-PORT NETWORKS

The 200 Ω resistor and the 400 Ω resistor form Z_1 and Y in a T-network with Z_2 equal to zero (Fig. 8.5(a)).

Using equations 8.7 and 8.8

$$A = 1 + YZ_1 = 1 + \frac{200}{400} = 1.5$$
$$B = Z_2 + Z_1 + Z_2 Z_1 Y = 0 + 200 + 0 \times \frac{200}{400} = 200$$
$$C = Y = \frac{1}{400}$$
$$D = 1 + Z_2 Y = 1 + 0 \times \frac{1}{400} = 1.$$

Using equation 8.24: $CZ_A^2 + (A - D)Z_A = B$

$$\frac{1}{400} Z_A^2 + (1.5 - 1)Z_A = 200.$$

Multiplying by 400 gives

$$Z_A^2 + 200 Z_A = 80000.$$

The quadratic may be solved by any known method. One way is to complete the square

$$Z_A^2 + 200 Z_A + 100^2 = 80000 + 100^2 = 90000$$
$$Z_A + 100 = \pm \sqrt{90000}$$
$$Z_A = \pm 300 - 100 = 200 \text{ Ω (or } -400 \text{ Ω)}.$$

The value (-400 Ω) is a mathematical solution but not achievable using passive components. Therefore $Z_A = 200$ Ω.

The same result is achievable working from first principles. The impedance, looking in at terminals 3 and 4 in Fig. 8.14, has to be equal to Z_A.

The impedance $Z_{3,4}$ comprises a resistor of value 400 Ω in parallel with $(200 + Z_A)$ Ω, hence

$$Z_{3,4} = Z_A = \frac{400(200 + Z_A)}{400 + 200 + Z_A}$$

cross-multiplying:

$$600Z_A + Z_A^2 = 80000 + 400Z_A$$

transposing gives: $Z_A^2 + 200Z_A = 80000$

which is the same quadratic as that achieved using the four-terminal network parameters.

Self-examination example 8.7   Determine the value of an impedance Z_B which when connected to terminals 3 and 4 in Fig. 8.14 will give iterative matching.

8.8   CHARACTERISTIC IMPEDANCE

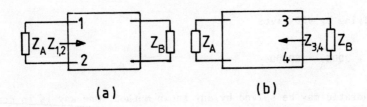

Fig. 8.15

In Fig. 8.15(a), for image impedance

$$Z_{1,2} = Z_A = \sqrt{\left(\frac{AB}{CD}\right)} \qquad \text{(equation 8.19)}.$$

For a symmetrical network, A = D, hence

$$Z_A = \sqrt{\left(\frac{B}{C}\right)}.$$

In Fig. 8.15(b), for iterative impedance

$$Z_{3,4} = Z_A.$$

From equation 8.24

$$CZ_A^2 + (A - D)Z_A = B.$$

TWO-PORT NETWORKS

For a symmetrical network, $A = D$, and equation 8.24 reduces to

$$CZ_A^2 = B.$$

Transposing: $Z_A = \sqrt{(B/C)}$.

Hence for a symmetrical network, image and iterative impedances Z_A are equal. Looking the other way, this is found to be true for Z_B, and because of the symmetry, $Z_A = Z_B$.

This unique impedance, where image and iterative impedances have the same value looking in either direction, is called the characteristic impedance, symbol Z_o.

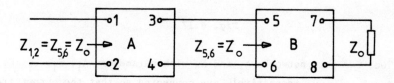

Fig. 8.16

In Fig. 8.15(a), let $Z_B = Z_o$. Now $Z_{1,2} = Z_o$. Instead of an actual impedance Z_o, an identical network terminated in Z_o could be used. In Fig. 8.16, network B is terminated in Z_o so that $Z_{5,6} = Z_o$. Connecting this to network A results in $Z_{1,2}$ being equal to $Z_{5,6}$, which is in turn equal to Z_o.

The process could be continued until an infinite number of identical networks were connected in series, the final one being terminated in Z_o. In fact the more networks there are in series, the smaller is the effect of the termination on the input impedance.

Following this reasoning, characteristic impedance is sometimes defined as the impedance of an infinite number of identical symmetrical two-port networks connected in series.

Terminating a symmetrical network or a number of identical such networks in series with an impedance Z_o ensures maximum power transfer and no reflection of power from the termination towards the source. The system is said to be correctly terminated. (See further work on transmission lines, chapter 11.)

8.9 CASCADE CONNECTION OF NETWORKS

Networks are said to be connected in cascade when the output from the first network becomes the input to the second, and so on, as shown in Fig. 8.17.

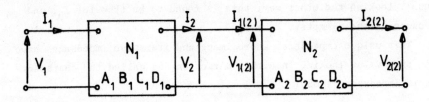

Fig. 8.17

Two, two-port networks N_1 and N_2 with parameters A_1, B_1, C_1, D_1 and A_2, B_2, C_2, D_2 respectively are connected so that the output from N_1 becomes the input to N_2.

$$V_2 = V_{1(2)} \quad \text{and} \quad I_2 = I_{1(2)}. \tag{i}$$

Using equation 8.1, in network N_1: $\quad V_1 = A_1 V_2 + B_1 I_2$.

Substituting for V_2 and I_2 from (i): $V_1 = A_1 V_{1(2)} + B_1 I_{1(2)}$. (ii)

For network N_2, equation 8.1 becomes: $V_{1(2)} = A_2 V_{2(2)} + B_2 I_{2(2)}$ (iii)

and equation 8.2 becomes: $I_{1(2)} = C_2 V_{2(2)} + D_2 I_{2(2)}$. (iv)

Substituting for $V_{1(2)}$ from (iii) and for $I_{1(2)}$ from (iv) both in (ii) gives

$$\begin{aligned} V_1 &= A_1(A_2 V_{2(2)} + B_2 I_{2(2)}) + B_1(C_2 V_{2(2)} + D_2 I_{2(2)}) \\ &= A_1 A_2 V_{2(2)} + A_1 B_2 I_{2(2)} + B_1 C_2 V_{2(2)} + B_1 D_2 I_{2(2)} \\ &= (A_1 A_2 + B_1 C_2) V_{2(2)} + (A_1 B_2 + B_1 D_2) I_{2(2)}. \end{aligned} \tag{8.25}$$

TWO-PORT NETWORKS

This formula relates the input voltage V_1 of the first network to $V_{2(2)}$ and $I_{2(2)}$, the output quantities from the second network.

Now using equation 8.2 in network N_1: $\quad I_1 = C_1 V_2 + D_1 I_2$.

Substituting values for V_2 and I_2 from (i): $I_1 = C_1 V_{1(2)} + D_1 I_{1(2)}$.

(v)

Substituting (iii) for $V_{1(2)}$ and (iv) for $I_{1(2)}$ in (v) gives

$$I_1 = C_1(A_2 V_{2(2)} + B_2 I_{2(2)}) + D_1(C_2 V_{2(2)} + D_2 I_{2(2)})$$
$$= C_1 A_2 V_{2(2)} + C_1 B_2 I_{2(2)} + D_1 C_2 V_{2(2)} + D_1 D_2 I_{2(2)}$$
$$= (C_1 A_2 + D_1 C_2) V_{2(2)} + (C_1 B_2 + D_1 D_2) I_{2(2)}. \tag{8.26}$$

This formula relates the input current I_1 of the first network to $V_{2(2)}$ and $I_{2(2)}$, the output quantities from the second network.

The effect of the two cascaded networks is the same as that of a single network with four-terminal network parameters A, B, C and D

when $V_1 = AV_{2(2)} + BI_{2(2)}$ \hfill (vi)

and $I_1 = CV_{2(2)} + DI_{2(2)}$. \hfill (vii)

Comparing (vi) with (8.25): $A = A_1 A_2 + B_1 C_2$ \hfill (8.27)

and $B = A_1 B_2 + B_1 D_2$. \hfill (8.28)

Comparing (vii) with (8.26): $C = C_1 A_2 + D_1 C_2$ \hfill (8.29)

and $D = C_1 B_2 + D_1 D_2$. \hfill (8.30)

(Note: if the reader is familiar with matrix algebra, it will be realised that these results come from the product of the two individual matrices for the networks N_1 and N_2.)

<u>Worked example 8.8</u> Two networks N_1 and N_2 are to be connected in cascade by joining terminals 3 and 5, and 4 and 6, as shown in Fig. 8.18. Determine the overall values of the four-terminal network parameters for the combined network and hence determine the required input conditions for a potential difference of 2 V to be developed across a 1.6 k Ω resistor connected across the output terminals 7 and 8.

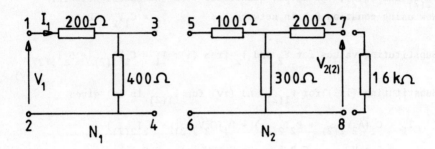

Fig. 8.18

From worked example 8.6, for network N_1: $A_1 = 1.5$, $B_1 = 200$

$$C_1 = \frac{1}{400}, \quad D_1 = 1.$$

From worked example 8.4, for network N_2: $A_2 = 1.33$, $B_2 = 366.66$

$$C_2 = \frac{1}{300}, \quad D_2 = 1.66.$$

Equation 8.27: $A = A_1 A_2 + B_1 C_2 = 1.5 \times 1.33 + 200 \times \frac{1}{300} = 2.666$

Equation 8.28: $B = A_1 B_2 + B_1 D_2 = 1.5 \times 366.66 + 200 \times 1.66 = 883.33$

Equation 8.29: $C = C_1 A_2 + D_1 C_2 = \frac{1}{400} \times 1.33 + 1 \times \frac{1}{300} = 0.006667$

Equation 8.30: $D = C_1 B_2 + D_1 D_2 = \frac{1}{400} \times 366.66 + 1 \times 1.66 = 2.583$

$V_1 = AV_{2(2)} + BI_{2(2)}$ and $I_{2(2)} = \frac{2}{1600} = 1.25$ mA

$= 2.666 \times 2 + 883.33 \times 1.25 \times 10^{-3}$

$= 6.437$ V.

$I_1 = CV_{2(2)} + DI_{2(2)}$

$= 0.006667 \times 2 + 2.583 \times 1.25 \times 10^{-3}$

$= 16.56$ mA.

Self-assessment example 8.9 Determine the values of the combined ABCD parameters for the networks N_1 and N_2 connected in cascade as shown in Fig. 8.19.

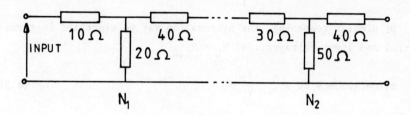

Fig. 8.19

8.10 UNITS OF ATTENUATION AND GAIN

The bel or decibel (after Alexander Graham Bell)

Gain or attenuation in this system is given in terms of power. If a network is considered with passive components only, the power in R_L (Fig. 8.20) will be less than the input power.

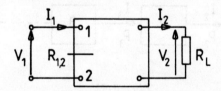

Fig. 8.20

The quantity $\log_{10} \frac{\text{power out}}{\text{power in}}$ is defined as the attenuation in bels.

Multiplying by 10 gives the sub-unit, the decibel (dB).

Now, power output = $I_2^2 R_L$ W and power input = $I_1^2 R_{1,2}$ W.

Therefore, attenuation = $10 \log_{10} \left(\dfrac{I_2^2 R_L}{I_1^2 R_{1,2}} \right)$ dB

$= 20 \log_{10} \dfrac{I_2}{I_1} + 10 \log_{10} \dfrac{R_L}{R_{1,2}}$ dB.

Now, if R_L is equal in value to $R_{1,2}$, as in the case of iterative matching, then the attenuation expression reduces to

$$\text{attenuation} = 20 \log_{10} \frac{I_2}{I_1} \text{ dB.} \qquad (8.31)$$

By similar reasoning, the attenuation may be expressed in terms of output and input voltages. For $R_L = R_{1,2}$

$$\text{attenuation} = 20 \log_{10} \frac{V_2}{V_1} \text{ dB.} \qquad (8.32)$$

The neper

In the neper system the attenuation is expressed in terms of voltage or current using natural logarithms.

$$\text{Attenuation} = \log_e \frac{I_2}{I_1} \text{ neper (N)}$$

or

$$\text{attenuation} = \log_e \frac{V_2}{V_1} \text{ neper.}$$

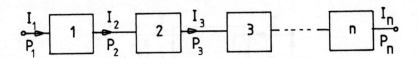

Fig. 8.21

For several networks connected in series as shown in Fig. 8.21, the overall attenuation may be expressed using either system. For power, the attenuation over the first network is expressed as

$$\text{attentuation} = 10 \log_{10} \frac{P_2}{P_1} \text{ dB}$$

and for the second network as

$$\text{attenuation} = 10 \log_{10} \frac{P_3}{P_2} \text{ dB, etc.}$$

Hence the overall attenuation of n networks in cascade is

TWO-PORT NETWORKS

$$10\left(\log_{10}\frac{P_2}{P_1} + \log_{10}\frac{P_3}{P_2} + \log_{10}\frac{P_4}{P_3} + \ldots + \log_{10}\frac{P_n}{P_{(n-1)}}\right) dB$$

$$= 10\left(\log_{10}\frac{P_2}{P_1} \times \frac{P_3}{P_2} \times \frac{P_4}{P_3} \times \ldots \times \frac{P_n}{P_{(n-1)}}\right)$$

$$= 10 \log_{10}\frac{P_n}{P_1} dB. \tag{8.33}$$

If all the networks are iteratively matched, this may be expressed in terms of output and input current (or voltage)

$$\text{attenuation} = 20 \log_{10}\frac{I_n}{I_1} dB. \tag{8.34}$$

In the neper system, the overall current attenuation will be

$$\text{attenuation} = \log_e \frac{I_n}{I_1} N.$$

The overall attenuation in both systems is the sum of the individual attenuations in logarithmic form.

8.11 INSERTION LOSS RATIO

The insertion loss ratio of a four-terminal network is defined as the ratio of voltage or current at the load without the network in circuit to the corresponding voltage or current with the network in circuit.

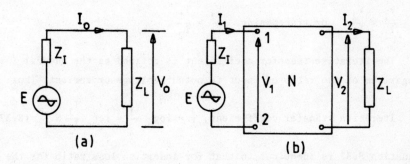

Fig. 8.22

In Fig. 8.22(a), the load voltage without the network is V_0 volts. In Fig. 8.22(b), the load voltage after the insertion of the network is V_2 volts.

$$\text{Insertion loss ratio} = \log_e \frac{V_0}{V_2} = \log_e \frac{I_0}{I_2}.$$

Generally V_0/V_2 or I_0/I_2 will be a complex quantity and there will be an insertion phase shift of $(\arg V_0 - \arg V_2)$.

If the network is terminated in its iterative impedance, the input impedance to the network is equal to Z_L, $V_0 = V_1$ and $I_0 = I_1$ (Fig. 8.22(b)). Hence

$$\text{insertion loss ratio} = \log_e \frac{V_1}{V_2} = \log_e \frac{I_1}{I_2}. \qquad (8.35)$$

Using decibels, for the same condition

$$\text{insertion loss ratio} = 20 \log_{10} \frac{V_1}{V_2} = 20 \log_{10} \frac{I_1}{I_2} \text{ dB}. \qquad (8.36)$$

For iterative matching the loss is therefore a function of the network alone being concerned with $V_1(I_1)$ and $V_2(I_2)$, which leads to the definitions of the transfer coefficient, γ (gamma).

8.12 TRANSFER COEFFICIENT, γ

8.12.1 Iterative transfer coefficient

When a network is terminated in its iterative impedance, $Z_{1,2} = Z_L$ (Fig. 8.22(b)). Hence

$$\frac{V_1}{I_1} = \frac{V_2}{I_2} \quad \text{or transposing:} \quad \frac{V_1}{V_2} = \frac{I_1}{I_2}.$$

The iterative transfer coefficient is defined as the natural logarithm of the ratio of input to output voltage or current. Thus

$$\text{iterative transfer coefficient, } \gamma = \log_e \frac{V_1}{V_2} = \log_e \frac{I_1}{I_2} \text{ N.} \qquad (8.37)$$

Equation 8.37 is identical to that for insertion loss ratio for the iteratively terminated network.

TWO-PORT NETWORKS

8.12.2 Propagation coefficient

Where a network is symmetrical, both iterative impedances are the same and are known as the characteristic impedance of the network (section 8.8). In this case the constant is known as the propagation constant or coefficient. It has particular significance in work on transmission lines. The propagation coefficient is defined by equation 8.37. Using the current relationship, since

$$\gamma = \log_e \frac{I_1}{I_2}, \text{ then } e^\gamma = \frac{I_1}{I_2}.$$

Generally I_2 differs from I_1 in both magnitude and phase so the transfer coefficient is complex.

$$\gamma = \alpha + j\beta, \quad \text{therefore } e^\gamma = e^{\alpha + j\beta} = e^\alpha \times e^{j\beta}.$$

But $e^{j\beta} = (\cos \beta + j \sin \beta)$ (Euler's formula).

Therefore, $e^\gamma = e^\alpha (\cos \beta + j \sin \beta)$.
In polar form, $e^\gamma = e^\alpha \underline{/\beta}$. \hfill (8.38)

α is the attenuation coefficient and it determines the relationship between the moduli of input and output currents.

$$e^\alpha = \frac{|I_1|}{|I_2|}.$$

β is the phase-change coefficient and is the phase angle between I_1 and I_2. For example,

$$I_1 = 5\underline{/0^\circ} \text{ and } I_2 = 2.5\underline{/-45^\circ} \text{ A}$$

$$\frac{I_1}{I_2} = \frac{5}{2.5\underline{/-45^\circ}} = 2\underline{/45^\circ}.$$

$$e^\gamma = 2\underline{/45^\circ}; \quad e^\alpha = 2, \ \beta = 45^\circ.$$

Worked example 8.10

(a) An oscillator with internal resistance 600 Ω and e.m.f. 5 V feeds

a 600 Ω load directly. Determine the power developed in the load.
(b) It is required to attenuate the load power by 3 dB using an iterative network of the form shown in Fig. 8.23. Determine (i) the values of R_1 and R_2, (ii) the value of the load power with the attenuator in circuit.

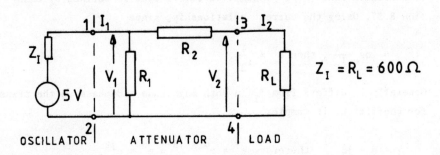

Fig. 8.23

(a) Without attenuator: load current $= \dfrac{E}{R_L + R_I} = \dfrac{5}{600 + 600} = 4.166$ mA.

Load power $= I^2 R_L = (4.166 \times 10^{-3})^2 \times 600 = 0.0104$ W.
Load voltage $= IR_L = 2.5$ V.

(b)(i) Since the network is to be iterative, the resistance looking to the right from terminals 1 and 2 is 600 Ω. Similarly looking from terminals 3 and 4 to the left the resistance is also 600 Ω.

From terminals 1 and 2 the resistance comprises R_1 in parallel with $(R_2 + 600)$ and this is itself to be 600 Ω. Hence

$$600 = \dfrac{R_1(R_2 + 600)}{R_1 + R_2 + 600} \quad \text{.} \tag{i}$$

For 3 dB attenuation: $20 \log_{10} \dfrac{V_1}{V_2} = 3$ (equation 8.32)

Transposing: $\log_{10} \dfrac{V_1}{V_2} = \dfrac{3}{20}$

$$10^{3/20} = \dfrac{V_1}{V_2}$$

$$\frac{V_1}{V_2} = 1.4126. \qquad (ii)$$

To determine V_2, consider R_L and R_2 as a potential divider, the voltage across which is the voltage V_1. Therefore

$$V_2 = \left(\frac{R_L}{R_L + R_2}\right)V_1$$

Transposing: $\dfrac{V_1}{V_2} = \dfrac{R_L + R_2}{R_L}$. $\qquad (iii)$

But $V_1/V_2 = 1.4126$ (equation (ii)). Substituting this value and R_L in (iii) gives

$$1.4126 = \frac{600 + R_2}{600}$$

$$847.56 = 600 + R_2$$

$$R_2 = 247.56 \ \Omega.$$

Substituting for R_2 in (i) above gives

$$600 = \frac{R_1(247.56 + 600)}{R_1 + 247.56 + 600}$$

$$847.56 R_1 = 600 R_1 + 508536$$

$$R_1 = 2054.2 \ \Omega.$$

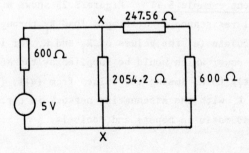

Fig. 8.24

(ii) The circuit is now as shown in Fig. 8.24. The resistance of the network and load resistor to the right of points XX is 600 Ω. (The reader might care to check whether 2054.2 Ω in parallel with (247.56 + 600) Ω is indeed 600 Ω.)

The current leaving the oscillator = $\frac{5}{600 + 600}$ = 4.166 mA as in (a).

The voltage $V_{XX} = V_1$ (in fig. 8.23) = IR_{XX} = 4.166 × 10^{-3} × 600 = 2.5 V (again, as in (a)).

Current in R_L = $\frac{V_{XX}}{247.56 + 600}$

= $\frac{2.5}{847.56}$ = 2.95 mA.

Load power = $I^2 R_L$ = $(2.95 \times 10^{-3})^2$ × 600 = 5.22 mW.

This is exactly one half of the value in (a) which is the requirement for 3 dB attenuation since

$10 \log_{10} \frac{P_{(a)}}{\frac{1}{2}P_{(a)}}$ = $10 \log_{10} 2$ = 3 dB ($P_{(a)}$ = power in part (a)).

It should be observed that although the load power is reduced, the oscillator is still delivering the original power into the same resistance as it did when the 600 Ω resistor was connected directly to its output terminals. It is the proportion of that power which reaches the load which is changed by the attenuator.

<u>Self-assessment example 8.11</u>   Figure 8.25 shows a source of e.m.f. with internal resistance R_I driving a load R_L through an attenuating network. Calculate (a) the values of R_I and R_L for iterative matching, (b) the power which would be supplied by the source when connected directly to R_L (using the values from (a)), (c) the power delivered to R_L with the attenuating network in circuit, (d) the insertion loss ratio in nepers and decibels.

TWO-PORT NETWORKS

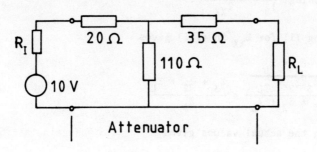

Fig. 8.25

Worked example 8.12   Determine the values of the attenuation coefficient α and phase-change coefficient β for the network shown in Fig. 8.26.

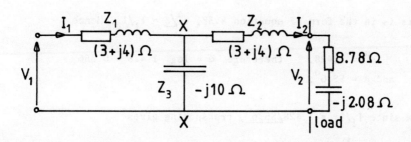

Fig. 8.26

Impedance to the right of terminals XX = $Z_{XX} = \dfrac{Z_3(Z_2 + Z_L)}{Z_3 + Z_2 + Z_L}$.   (i)

$V_{XX} = I_1 Z_{XX}$ (ii)

and $I_2 = \dfrac{V_{XX}}{Z_2 + Z_L}$.

Substituting (ii) for V_{XX} gives

$I_2 = \dfrac{I_1 Z_{XX}}{Z_2 + Z_L}$.

Transposing: $\dfrac{I_1}{I_2} = \dfrac{Z_2 + Z_L}{Z_{XX}}$. (iii)

Substituting (i) for Z_{XX} in (iii) gives

$$\dfrac{I_1}{I_2} = \dfrac{\dfrac{Z_2 + Z_L}{Z_3(Z_2 + Z_L)}}{Z_3 + Z_2 + Z_L} = \dfrac{Z_3 + Z_2 + Z_L}{Z_3} .$$

Introducing the actual values gives

$$\dfrac{I_1}{I_2} = \dfrac{-j10 + (3 + j4) + (8.78 - j2.08)}{-j10}$$

$$= \dfrac{11.78 - j8.08}{-j10} = \dfrac{14.28\underline{/-34.4^\circ}}{10\underline{/-90^\circ}}$$

$$= 1.428\underline{/55.6^\circ} .$$

This is in the form of equation 8.38, $e^\alpha\underline{/\beta} = I_1/I_2$. Hence

$e^\alpha = 1.428$, therefore, $\alpha = \log_e 1.428 = 0.356$
and $\beta = 55.6^\circ$.

Now since $I_1/I_2 = 1.428\underline{/55.6^\circ}$, transposing gives

$$I_2 = \dfrac{I_1}{1.428\underline{/55.6^\circ}}$$

$$= 0.71_1\underline{/-55.6^\circ} \text{ A}.$$

I_2 has a modulus 0.7 times that of I_1 and lags I_1 by 55.6°.
Further work would show that the same relationship exists between V_1 and V_2.

<u>Self-assessment example 8.13</u>   For the network and load shown in Fig. 8.27, (a) determine the value of Z_L required for iterative matching, (b) calculate the values of α and β for the network.

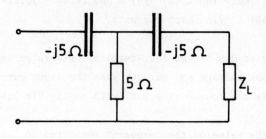

Fig. 8.27

Further problems

8.14. (a) What relationships exist between the ABC and D parameters of a linear two-port network (i) if it is passive, and (ii) if it is symmetrical.

(b) For Fig. 8.28, calculate the two values of impedance necessary for image matching at input and output terminals.

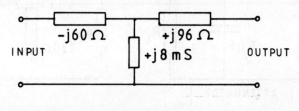

Fig. 8.28

8.15. Tests on a linear two-port network gave the following results.

Open-circuit test: input 10 V and 0.125 A; output 6.25 V.
Short-circuit test: input 10 V and 0.2 A; current in short-circuit 0.12 A.

Determine the values of the four-terminal network parameters A, B, C and D. Confirm that the network is passive.

8.16. Tests on a linear two-port network gave the following results.

Open-circuit test: input (6 + j0) V and (4.8 + j2.4) mA; output

(1.2 - j2.4) V.

Short-circuit test: input (6 + j0) V and (4.62 + j0.924) mA; output (2.77 - j1.848) mA (in short circuit).

Determine the values of the four-terminal network parameters and hence find the output voltage and current when the input current is 7.5$\underline{/-30^o}$ mA from a source of e.m.f. with negligible internal impedance giving 10$\underline{/0^o}$ V.

What is the value of the component connected to the output terminals?

8.17. (a) Define the terms image impedance and iterative impedance as applied to a four-terminal network.

(b) Calculate the insertion loss in decibels or nepers when the LC attenuator shown in Fig. 8.29 is placed between the two resistors R_I and R_L. ($R_I = R_L$, = 141.42 Ω. ω = 1414.2 rad/s.)

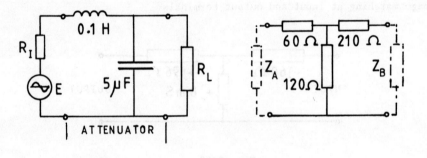

Fig. 8.29   Fig. 8.30

8.18. Calculate, for the network shown in Fig. 8.30, (a) the two iterative impedances, and (b) the insertion loss when the network is inserted between these iterative impedances.

8.19. A voltage source with internal resistance 1000 Ω and e.m.f. 10 V feeds a 1000 Ω load through a network, as shown in Fig. 8.31. It gives iterative matching and an attenuation of 40 dB.

(a) Determine the values of R_1 and R_2.

(b) Determine the load power (i) with the network in circuit, (ii) with the load connected directly to the source.

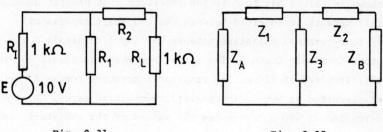

Fig. 8.31   Fig. 8.32

8.20. (a) For the asymmetric network shown in Fig. 8.32, derive the ABCD parameters in terms of Z_1, Z_2 and Z_3.
(b) Given that the values of the components are $Z_1 = -j40$ Ω, $Z_2 = j60$ Ω and $Z_3 = -j200$ Ω, (i) calculate the values of the four-terminal network parameters, (ii) using these parameters, determine the values of the two impedances Z_A and Z_B which will give image matching.

8.21. Two identical passive two-port networks N_1 and N_2 have the following parameters

$$A = (1 + j1), \ B = 50, \ C = j0.02, \text{ and } D = 1.$$

They are connected in cascade. A voltage of $5\underline{/0^o}$ V is applied to the input terminals and a resistive load of 20 Ω is connected to the output terminals.

Calculate (a) the output voltage, (b) the input current.

8.22. The filter shown in Fig. 8.33 is terminated in its characteristic impedance Z_o. Determine (a) the value of Z_o, (b) the value of the attenuation coefficient for the network when terminated in Z_o.

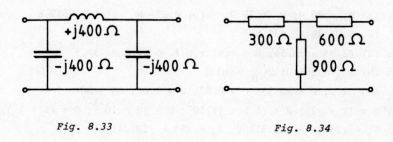

Fig. 8.33   Fig. 8.34

8.23. Calculate the values of the two image impedances for the network shown in Fig. 8.34. What is the insertion loss brought about when this network is inserted between the two image impedances?

8.24. A four-terminal resistive network has input terminals A and B and output terminals C and D. The resistance measured across AB with CD short-circuited is 720 Ω. The resistance measured across AB with CD open-circuited is 900 Ω. The resistance measured across CD with AB open-circuited is 500 Ω. Determine the values of the resistors, assuming a T-configuration.

8.25. (a) Determine the value of the characteristic impedance of the network shown in Fig. 8.35.
(b) Calculate the values of α and β for the network correctly terminated.

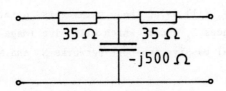

Fig. 8.35

CHAPTER 8   ANSWERS

8.3. (a) $A = D = 0.979\underline{/0.417^\circ}$; $B = 48.2\underline{/71.7^\circ}$; $C = j920 \times 10^{-6}$
(b) $V_1 = 160480\underline{/1.64^\circ}$ (c) $I_1 = 123\underline{/36^\circ}$

8.5. 60 Ω

8.7. 400 Ω

8.9. $A_1 = 1.5$; $B_1 = 70$; $C_1 = 0.05$; $D_1 = 3$; $A_2 = 1.6$; $B_2 = 94$; $C_2 = 0.02$; $D_2 = 1.8$; $A = 3.8$; $B = 267$; $C = 0.14$; $D = 10.1$

8.11. (a) $R_I = 90$ Ω; $R_L = 75$ Ω   (b) 0.275 W   (c) 0.0689 W
(d) 6.02 dB; 0.693 N

8.13. (a) $(3.93 - j6.36)$ Ω   (b) $\alpha = 1.06$ N; $\beta = -51.8^\circ$

8.14. (b) $Z_{in} = 256$ Ω; $Z_{out} = 40$ Ω

8.15. $A = 1.6$; $B = 83.33$; $C = 0.02$; $D = 1.66$; $AD - BC = 1$

8.16. $A = (1 + j2)$; $B = (1.5 + j1)10^3$; $C = j2 \times 10^{-3}$; $D = (1 + j1)$;
$V_2 = (23.49 + j10.87)$ V; $I_2 = (14 + j29.24)$ mA; $Z_L = 798\underline{/-39.57^\circ}$ Ω

TWO-PORT NETWORKS

8.17. (b) 0.97 dB; 0.11 N

8.18. (a) $Z_A = 300\ \Omega$; $Z_B = 150\ \Omega$ (b) 1.39 N; 12 dB

8.19. (a) $R_2 = 99\ k\Omega$; $R_1 = 1010\ \Omega$ (b)(i) 2.5 μW (ii) 0.025 W

8.20. (b)(i) A = 1.2; B = j32; C = j5 × 10^{-3}; D = 0.7 (ii) $Z_A = 104.7\ \Omega$; $Z_B = 61.1\ \Omega$

8.21. A = j3; B = 100 + j50; C = -0.02 + j0.04; D = (1 + j1)
  (a) $V_2 = 0.673\underline{/-47.72^o}$ V (b) $I_1 = 63\underline{/23.8^o}$ mA

8.22. (a) $Z_o = 400\ \Omega$ (b) α = 0 (low pass filter designed for zero attenuation)

8.23. 890 Ω; 1120 Ω; 8.2 dB

8.24. $Z_1 = 600\ \Omega$; $Z_2 = 200\ \Omega$; $Z_3 = 300\ \Omega$

8.25. (a) $Z_o = 187.3\underline{/-44^o}\ \Omega$ (b) α = 0.264 N (2.29 dB); β = 15°

9 Resonance, Q-Factor and Coupled A C Circuits

9.1 SERIES RESONANCE

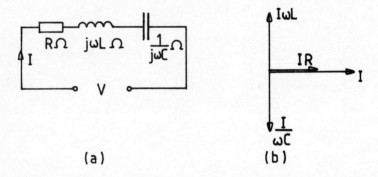

Fig. 9.1

Consider a circuit comprising a resistive coil in series with a capacitor as shown in Fig. 9.1(a). With angular velocity ω rad/s, the impedance Z of the circuit is expressed as

$$Z = \left(R + j\omega L + \frac{1}{j\omega C}\right)$$

$$= \left(R + j\left(\omega L - \frac{1}{\omega C}\right)\right) \Omega.$$

When $\omega L = 1/\omega C$, the reactive terms disappear and the impedance of the circuit is purely resistive. This condition is termed 'series resonance'. $Z = R \ \Omega$, and the angular velocity is written as ω_o rad/s and the current as I_o amperes. Hence

RESONANCE, Q-FACTOR AND COUPLED A C CIRCUITS

$$\omega_o L = \frac{1}{\omega_o C}.$$

$$\omega_o^2 = \frac{1}{LC} \quad \text{or} \quad \omega_o = \sqrt{\left(\frac{1}{LC}\right)} \text{ rad/s.} \tag{9.1}$$

At resonance, the circuit current I_o is V/R A. This is the greatest value the current can achieve since at angular velocities above and below ω_o the impedance increases, so reducing the current. For this reason the circuit is sometimes called an 'acceptor' circuit.

At resonance, the voltage developed across the capacitor is $I_o X_C$ volts.

$$V_C = I_o \times \frac{1}{\omega_o C} = \frac{V}{R \omega_o C} \text{ V.}$$

The ratio V_C/V is the circuit voltage magnification factor, symbol Q. Hence

$$Q = \frac{\frac{V}{R \omega_o C}}{V} = \frac{1}{R \omega_o C}. \tag{9.2}$$

The voltage developed across the inductive part of the coil is $I_o X_L$ volts.

$$V_L = \frac{V}{R} \omega_o L \text{ V.}$$

Again, $$Q = \frac{V_L}{V} = \frac{\frac{V}{R} \omega_o L}{V} = \frac{\omega_o L}{R}. \tag{9.3}$$

If the resistance of the coil is small, the voltage developed across the coil terminals will not be greatly different from the value of $I_o X_L$. Figure 9.2 shows the curves of V_C and V_L plotted against angular velocity for the circuit in Fig. 9.1(a). The voltage V_C is at a maximum at ω_o rad/s, while the maximum value of coil voltage occurs at a slightly higher frequency because of the coil resistance. The two maxima would coincide at $\tilde{\omega}_o$ if the resistance could be eliminated.

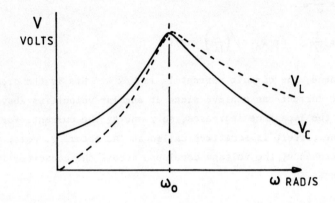

Fig. 9.2

9.2 PARALLEL RESONANCE

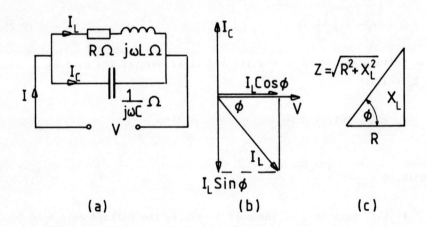

Fig. 9.3

The resistive coil is now connected in parallel with a perfect capacitor as shown in Fig. 9.3(a). The current I_C leads the applied voltage by exactly $90°$. The current I_L in the coil lags the voltage by an angle ϕ which is determined by the ratio of R to X_L in the impedance triangle for the coil shown in Fig. 9.3(c). The complete phasor diagram (Fig. 9.3(b)) shows I_L being resolved into two components, one

RESONANCE, Q-FACTOR AND COUPLED A C CIRCUITS

in phase with the voltage and the other in quadrature with the voltage.

Quadrature current = $I_L \sin \phi$
In-phase current = $I_L \cos \phi$.

When $I_L \sin \phi = I_C$, the two quadrature currents cancel leaving only the in-phase current $I_L \cos \phi$. The circuit appears to be purely resistive and 'parallel resonance' is said to have occurred. The angular velocity is written as ω_o rad/s.

Referring to the impedance triangle in Fig. 9.3(c)

$$\sin \phi = \frac{X_L}{Z} = \frac{\omega_o L}{\sqrt{\left[R^2 + (\omega_o L)^2\right]}} \quad \text{(i)}$$

$$\cos \phi = \frac{R}{Z} = \frac{R}{\sqrt{\left[R^2 + (\omega_o L)^2\right]}} \quad \text{(ii)}$$

$$I_C = \frac{V}{X_C} = \frac{V}{\frac{1}{\omega_o C}} = V\omega_o C \text{ A} \quad \text{(iii)}$$

$$I_L = \frac{V}{Z_{coil}} = \frac{V}{\sqrt{\left[R^2 + (\omega_o L)^2\right]}} \quad \text{(iv)}$$

At resonance, $I_C = I_L \sin \phi$. Substituting for I_C, I_L and $\sin \phi$ from equations (iii), (iv) and (i) above gives

$$V\omega_o C = \frac{V}{\sqrt{\left[R^2 + (\omega_o L)^2\right]}} \times \frac{\omega_o L}{\sqrt{\left[R^2 + (\omega_o L)^2\right]}} = \frac{V\omega_o L}{R^2 + (\omega_o L)^2}.$$

Cancelling $V\omega_o$ on both sides of the equation and then transposing gives

$$C = \frac{L}{R^2 + (\omega_o L)^2}$$

$$R^2 + (\omega_o L)^2 = \frac{L}{C}. \quad \text{(v)}$$

Further transpositions give the result

$$\omega_o = \sqrt{\left(\frac{1}{LC} - \frac{R^2}{L^2}\right)}. \tag{9.4}$$

Notice that parallel resonance does NOT occur when $\omega L = 1/\omega C$ as in the series case. However, when R is small enough for R^2/L^2 to become insignificant compared with $1/LC$, equation 9.4 reduces to that for series resonance (equation 9.1).

The impedance at resonance is called the 'dynamic impedance', Z_D.

$$Z_D = \frac{V}{\text{circuit current}} = \frac{V}{I_L \cos \phi}.$$

Substituting for I_L and $\cos \phi$ from equations (iv) and (ii) gives

$$Z_D = \frac{V}{\frac{V}{\sqrt{[R^2 + (\omega_o L)^2]}} \times \frac{R}{\sqrt{[R^2 + (\omega_o L)^2]}}}$$

$$= \frac{[R^2 + (\omega_o L)^2] V}{VR}.$$

From equation (v), $R^2 + (\omega_o L)^2 = L/C$. Hence

$$Z_D = \frac{\frac{L}{C} \times V}{VR} = \frac{L}{CR} \ \Omega. \tag{9.5}$$

At resonance the circuit has its highest possible impedance so that the current I (Fig. 9.3(a)) is a minimum. This is called the 'make-up' current. If the resistance could be made zero then the dynamic impedance would be infinitely great and no make-up current would be required to maintain the branch currents I_C and I_L.

The parallel circuit exhibits current magnification,

$Q = I_C/I_{\text{make-up}}$.

From equation (iii), $I_C = V\omega_o C$ and $I_{\text{make-up}} = \frac{V}{Z_D}$.

Therefore, $Q = \dfrac{V\omega_o C}{\frac{V}{L/CR}} = \dfrac{V\omega_o C}{V} \dfrac{L}{CR} = \dfrac{\omega_o L}{R}$ as for the series circuit.

RESONANCE, Q-FACTOR AND COUPLED A C CIRCUITS

Also, for the inductive branch, $Q = I_L/I_{\text{make-up}}$.

For small values of resistance, $I_L = \dfrac{V}{\omega_o L}$ (closely).

Hence Q is approximately equal to $\dfrac{\frac{V}{\omega_o L}}{\frac{V}{L/CR}} = \dfrac{VL}{V\omega_o L \cdot CR} = \dfrac{1}{\omega_o CR}$.

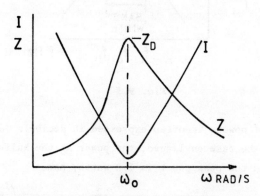

Fig. 9.4

Figure 9.4 shows the variations in impedance and make-up current with angular velocity for the circuit in Fig. 9.3(a).

9.3   Q-FACTOR AND BANDWIDTH

The bandwidth of a circuit is defined arbitrarily as the range of frequencies within which the circuit power remains in excess of 50% of its peak value. Considering the series circuit (Fig. 9.1(a)), the power developed in the resistance at resonance is given by

$$\text{power} = I_o^2 R \text{ watts, where } I_o = \dfrac{V}{R} \text{ amperes.}$$

In order to fall to one half of this value, i.e. to $I_o^2 R/2$ watts, the current must fall to I amperes, when $I^2 R = I_o^2 R/2$. Transposing gives

$$I = \sqrt{\left(\dfrac{I_o^2 R}{2R}\right)} = \dfrac{I_o}{\sqrt{2}}.$$

303

Figure 9.5 shows the two half-power points and the bandwidth.

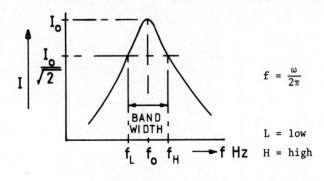

Fig. 9.5

The ratio of powers is often expressed in decibels (dB) (see section 8.10). In the case envisaged, the power at the half-power points is $I^2 R$ watts and the peak power is $I_o^2 R$ watts, hence

$$\text{ratio of powers (in dB)} = 10 \log_{10} \frac{I^2 R}{I_o^2 R} = 10 \log_{10} \frac{\left(\frac{I_o}{\sqrt{2}}\right)^2}{I_o^2}$$

$$= 10 \log_{10} \left(\frac{1}{\sqrt{2}}\right)^2 = 20 \log_{10} \frac{1}{\sqrt{2}}$$

$$= -3 \text{ dB.}$$

There is an attenuation of 3 dB between the peak power and the half-power points, which are often referred to as the 3 dB points.

For the current to fall to $I_o/\sqrt{2}$, the impedance must rise to $\sqrt{2}R$. This occurs at two frequencies

(a) f_H, when $\left(\omega L - \frac{1}{\omega C}\right) = R$. The circuit is predominantly inductive since with the frequency greater than f_o, ωL is greater than $\frac{1}{\omega C}$.

(b) f_L, when $\left(\frac{1}{\omega C} - \omega L\right) = R$. The circuit is predominantly capacitive.

The impedance of the series circuit $= R + j\left(\omega L - \frac{1}{\omega C}\right) \Omega$ \hfill (i)

From equation (i), considering Fig. 9.6, it should be clear how the

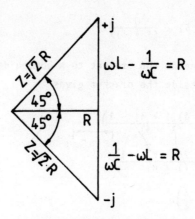

Fig. 9.6

two impedances for half power are realised.

Let the circuit be de-tuned by a fractional amount δ.

$$\delta = \frac{\omega - \omega_o}{\omega_o}.$$

Transposing gives

$$\omega_o \delta = \omega - \omega_o$$

$$\omega = \omega_o (1 + \delta). \tag{ii}$$

For δ positive, ω is higher than ω_o, the resonant angular velocity, and vice-versa.

Two values of ω corresponding to f_L and f_H in Fig. 9.5 are now going to be determined in terms of the circuit Q-factor.

Substituting for equation (ii) in equation (i) gives

$$Z = R + j\left(\omega_o(1 + \delta)L - \frac{1}{\omega_o(1 + \delta)C}\right). \tag{iii}$$

From equation 9.3, $\omega_o L/R = Q$. Transposing gives, $\omega_o L = RQ$. (iv)
From equation 9.2, $1/\omega_o CR = Q$. Transposing gives, $1/\omega_o C = RQ$. (v)

305

Substituting equations (iv) and (v) in equation (iii) gives

$$Z = R + j\left(RQ(1 + \delta) - \frac{RQ}{(1 + \delta)}\right).$$

Bringing the terms inside the bracket to a common denominator $(1 + \delta)$ and the common RQ outside the bracket gives

$$Z = R + jRQ\left(\frac{(1 + \delta)(1 + \delta) - 1}{(1 + \delta)}\right)$$

$$= R + jRQ\left(\frac{(1 + 2\delta + \delta^2 - 1)}{(1 + \delta)}\right)$$

$$= R + jRQ\delta\left(\frac{2 + \delta}{1 + \delta}\right).$$

For circuits with high Q the deviation δ is small to the half-power points and $(2+\delta)/(1+\delta)$ is very nearly equal to 2; then $Z = R + 2jRQ\delta$. For half power, the j term is equal to the resistance R (see Fig. 9.6). Therefore $2RQ\delta = R$.

For f_H, δ is positive. Hence $R = +2RQ\delta$. \hfill (vi)
for f_L, δ is negative. Hence $R = -2RQ\delta$. \hfill (vii)

Divide equation (vi) throughout by R, $1 = +2Q\delta$
divide equation (vii) throughout by R, $1 = -2Q\delta$, so that

$$\delta = \pm \frac{1}{2Q}.$$

But $\delta = \dfrac{\omega - \omega_o}{\omega_o}$. Therefore $\dfrac{\omega_H - \omega_o}{\omega_o} = +\dfrac{1}{2Q}$ \hfill (viii)

$$\text{and } \frac{\omega_L - \omega_o}{\omega_o} = -\frac{1}{2Q}. \hfill \text{(ix)}$$

Subtracting (ix) from (viii)

$$\frac{\omega_H - \omega_o}{\omega_o} - \frac{\omega_L - \omega_o}{\omega_o} = \frac{1}{2Q} - \left(\frac{-1}{2Q}\right)$$

RESONANCE, Q-FACTOR AND COUPLED A C CIRCUITS

$$\frac{\omega_H - \omega_L}{\omega_o} = \frac{2}{2Q}$$

$$\omega_H - \omega_L = \frac{\omega_o}{Q}.$$

But $(\omega_H - \omega_L)$ is the bandwidth expressed in rad/s, therefore

$$\text{bandwidth} = \frac{\omega_o}{Q}, \text{ or in terms of frequency, } \frac{f_o}{Q} \text{ Hz.} \qquad (9.6)$$

Equation 9.6 is applicable also to the parallel resonant circuit. It may be proved in exactly the same manner commencing with the impedance of the circuit for any frequency.

<u>Worked example 9.1</u>   A series resonant circuit has the following parameters: frequency at resonance, $5000/2\pi$ Hz; impedance at resonance, 56 Ω; Q-factor, 25.
(a) Assuming that the capacitor is pure, calculate (i) the capacitance of the capacitor, (ii) the inductance of the inductor.
(b) Assuming that these values are independent of frequency, determine the two frequencies at which the circuit impedance has a phase angle of $\pi/4$ radian.

Since $f_o = \frac{5000}{2\pi}$ Hz, $\omega_o = 2\pi f_o = 5000$ rad/s.

(a) At resonance, the circuit impedance is that of the resistance alone, R = 56 Ω.

Now $Q = \frac{\omega_o L}{R}$ (equation 9.3), therefore

$$25 = \frac{5000 L}{56}$$

$$L = 0.28 \text{ H}.$$

At resonance, $\omega_o L = \frac{1}{\omega_o C}$, therefore

$$5000 \times 0.28 = \frac{1}{5000 C}$$

$$C = \frac{1}{(5000)^2 \times 0.28} = 0.143 \text{ μF}.$$

(b) For a phase angle of $\pi/4$ rad ($45°$), $\left(\omega L - \frac{1}{\omega C}\right) = R$ (Fig. 9.6)

$$\omega \times 0.28 - \frac{1}{\omega \times 0.143 \times 10^{-6}} = 56.$$

Multiplying throughout by $0.143 \times 10^{-6} \omega$ gives

$$4.004 \times 10^{-8} \omega^2 - 1 = 8.008 \times 10^{-6} \omega.$$

Dividing throughout by 4.004×10^{-8} and transposing

$$\omega^2 - 200\omega = 24975025.$$

This quadratic equation may be solved by any known method. Completing the square

$$\omega^2 - 200\omega + (-100)^2 = 24975025 + 100^2$$
$$(\omega - 100)^2 = 24985025$$
$$(\omega - 100) = \pm 4998.5$$
$$\omega = 4998.5 + 100 = 5098.5 \text{ rad/s}$$
$$\text{or} \quad \omega = -4998.5 + 100 = -4898.5 \text{ rad/s}.$$

The second result is impossible practically since it is negative. However starting from $(1/\omega C - \omega L) = R$ to determine the lower frequency, it will be found that the results are the same with the signs changed. The value 4898.5 rad/s is the required lower angular velocity. The reader might care to verify this result for himself.

The two frequencies for $45°$ phase shift and half power (3 dB) points are therefore

$$\frac{5098.5}{2\pi} = 811.4 \text{ Hz} \quad \text{and} \quad \frac{4898.5}{2\pi} = 779.6 \text{ Hz}.$$

An alternative approach is to use equation 9.6.

$$\text{Bandwidth} = \frac{f_o}{Q} = \frac{\frac{5000}{2\pi}}{25} = 31.83 \text{ Hz}.$$

RESONANCE, Q-FACTOR AND COUPLED A C CIRCUITS

Assuming symmetry about f_o, the bandwidth is from $(f_o - 31.83/2)$ up to $(f_o + 31.83/2)$

$$f_L = \frac{5000}{2\pi} - 15.9 = 779.8 \text{ Hz}; \quad f_H = \frac{5000}{2\pi} + 15.9 = 811.7 \text{ Hz}.$$

The slight differences between the answers using the two methods are due to the assumptions made during the proof of equation 9.6.

<u>Worked example 9.2</u> An inductor has a resistance of 20 Ω and when connected across a 120 V, 50 Hz supply draws a current of 1 A. For a supply voltage of 85 V, calculate (a) the frequency of a supply to cause parallel resonance with a capacitor of capacitance 30 μF, (b) the dynamic impedance of the parallel circuit, (c) the value of the make-up current, (d) the capacitor current, and (e) the circuit Q-factor.

The impedance of the inductor at 50 Hz $= \frac{V}{I} = \frac{120}{1} = 120$ Ω.

Now, $Z = \sqrt{(R^2 + X_L^2)}$

$120 = \sqrt{(20^2 + X_L^2)}$, hence, $X_L = 118.32$ Ω.

Therefore, $L = \frac{118.32}{2\pi} = 0.3766$ H.

(a) For parallel resonance,

$$f_o = \frac{1}{2\pi}\sqrt{\left(\frac{1}{LC} - \frac{R^2}{L^2}\right)} \quad \text{(equation 9.4)}$$

$$f_o = \frac{1}{2\pi}\sqrt{\left(\frac{1}{0.3766 \times 30 \times 10^{-6}} - \frac{20^2}{0.3766^2}\right)}$$

$= 46.58$ Hz.

(b) $Z_D = \frac{L}{CR}$ (equation 9.5)

$= \frac{0.3766}{30 \times 10^{-6} \times 20} = 627.7$ Ω.

309

(c) Supply (make-up) current $= \dfrac{V}{Z_D} = \dfrac{85}{627.7} = 0.135$ A.

(d) $I_C = \dfrac{V}{X_C} = V\omega C$ A
$= 85 \times 2\pi \times 46.58 \times 30 \times 10^{-6} = 0.746$ A.

(e) Q-factor $= \dfrac{I_C}{I_{make-up}} = \dfrac{0.746}{0.135} = 5.5$

Self-assessment example 9.3 A coil with inductance 2.5 mH and resistance 15 Ω is connected in parallel with a capacitor of capacitance 0.001 μF and fed from a 1.5 V source. Determine (a) the frequency at which the current drawn from the source is a minimum, (b) the r.m.s. value of the make-up current, (c) the r.m.s. value of the capacitor current, (d) the Q-factor of the circuit, (e) the values of f_L, f_H and the bandwidth.

9.4 COUPLED COILS
9.4.1 Mutual inductance and coupling coefficient

Two circuits are said to possess mutual inductance if, when a current changes in one of them, the resultant flux change links at least in part with a second circuit so inducing a voltage in it. The two circuits are said to have a mutual inductance of one henry if, when a current changes in one circuit at a rate of one ampere per second, one volt is induced in the second circuit. In symbols

$$e_2 = M \dfrac{di_1}{dt}$$

the subscripts 1 and 2 denote the first and second circuit respectively, M is the mutual inductance in henry.

For alternating currents of sine form, $i_1 = I_{m(1)} \sin \omega t$, so that

$$e_2 = M \dfrac{dI_{m(1)} \sin \omega t}{dt} = M\omega I_{m(1)} \cos \omega t \text{ volts.} \tag{i}$$

The maximum value of e_2 is $E_{m(2)}$ and this occurs when $\cos \omega t = 1$.

The voltage wave (equation (i)) is a cosine function whereas the current was a sine function. The voltage wave is therefore 90° ahead of the current wave and this phase shift is indicated by using +j.

Hence

$$E_{m(2)} = jM\omega I_{m(1)}.$$

Dividing both sides of the equation by $\sqrt{2}$ reduces both $E_{m(2)}$ and $I_{m(1)}$ to r.m.s. values when

$$E_2 = j\omega M I_1. \qquad (9.7)$$

This is similar to the familiar result associated with self-inductance, namely

$$E_1 = j\omega L_1 I_1 \quad \text{or} \quad \frac{E_1}{I_1} = j\omega L_1 \ \Omega. \qquad (9.8)$$

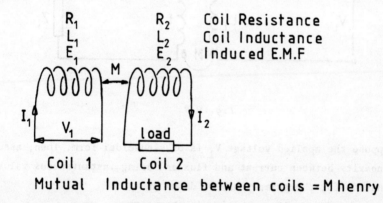

Fig. 9.7

Equations 9.7 and 9.8 are illustrated in Fig. 9.7. Assuming that all the flux created by the current I_1 in coil 1 links with coil 2, i.e. assuming perfect or tight coupling, the mutual inductance has the highest value possible, M_{max}

$$M_{max} = \sqrt{(L_1 L_2)}. \qquad (9.9)$$

If there is leakage flux so that only part of the flux from coil 1 links with coil 2, then

$$M = k\sqrt{M_{max}} = k\sqrt{(L_1 L_2)} \tag{9.10}$$

where k is the coupling coefficient.

For a derivation of equations 9.9 and 9.10, see appendix 3.

9.4.2 The effect of secondary loading on the input impedance of the primary

Figure 9.8 shows the circuit diagram of two mutually coupled coils, including the resistances of both the primary and secondary coils, R_1 and R_2 respectively. The secondary is connected to a load with impedance Z_L.

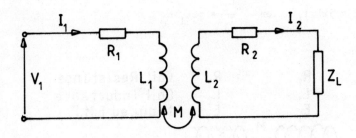

Fig. 9.8

Suppose the applied voltage V_1 is of sinusoidal form. Then, assuming linearity between current and flux and using instantaneous values, the equation for the primary may be drawn up.

There are four voltages acting in the primary

1. $v_1 = V_{m(1)} \sin \omega t$ V. This will drive a current,
$i_1 = I_{m(1)} \sin(\omega t - \phi)$ A.
2. The resistive volt drop, $i_1 R_1$ V.
3. The back e.m.f., e, induced by the self-inductance, L_1

$$e_1 = L_1 \frac{di_1}{dt} \text{ V.}$$

By Lenz's law this will oppose V_1.

4. The mutually induced e.m.f., $e_{1(M)}$, in the primary set up by current i_2 in the secondary

$$e_{1(M)} = M \frac{di_2}{dt}.$$

It is necessary to decide whether the mutually induced voltage helps or opposes V_1. With no current in the secondary, i.e. the load disconnected, there will be no primary mutually induced voltage and I_1 will have a certain value which depends on the primary parameters. Connecting the load and considering the effect of increasing the magnitude of the secondary current, there will be power developed in the secondary so that the power input to the primary must increase. This will occur only if the voltage $e_{1(M)}$ is helping v_1. If $e_{1(M)}$ opposed v_1, for increasing secondary current, the input current would fall, which is clearly a nonsense.

The equation for the primary in instantaneous values becomes

$$v_1 = M \frac{di_2}{dt} - i_1 R_1 - L_1 \frac{di_1}{dt} = 0.$$

Transposing and substituting the steady-state values for sinusoidal input from equations 9.7 and 9.8

$$V_1 = I_1 R_1 + j\omega L_1 I_1 - j\omega M I_2$$
$$V_1 = I_1 (R_1 + j\omega L_1) - j\omega M I_2. \tag{9.11}$$

In the secondary it is necessary to consider a further four voltages

1. The mutually induced voltage created by the current in the primary

$$e_2 = M \frac{di_1}{dt} \text{ V}.$$

This drives current in the secondary and through the load.
2. The voltage drop, $i_2 R_2$ V.
3. The self-induced back e.m.f. $= L_2 \frac{di_2}{dt}$ V.
4. The voltage drop across the load, $i_2 Z_L$ V.

In instantaneous values the secondary equation becomes

$$M\frac{di_1}{dt} - i_2 R_2 - L_2 \frac{di_2}{dt} - i_2 Z_L = 0.$$

Transposing and substituting the steady-state values for sinusoidal input from equations 9.7 and 9.8

$$j\omega M I_1 = I_2 R_2 + j\omega L_2 I_2 + I_2 Z_L$$
$$j\omega M I_1 = I_2 (R_2 + j\omega L_2 + Z_L). \tag{9.12}$$

Note that in equation 9.12, Z_L will have to be expressed in rectangular form and added correctly to R_2 and $j\omega L_2$.

Consider the case where the load is inductive, $Z_L = R_L + j\omega L_L$.

Total secondary impedance = $(R_2 + R_L) + j\omega(L_2 + L_L)$.

Let $(R_2 + R_L) = R_s$ and $(L_2 + L_L) = L_s$. Equation 9.12 becomes

$$j\omega M I_1 = I_2 (R_s + j\omega L_s). \tag{9.13}$$

This yields a pair of simultaneous equations (9.11 and 9.13). A solution by eliminating the I_2 term involves multiplying equation 9.11 by $(R_s + j\omega L_s)$ to form equation 9.14, and equation 9.13 by $j\omega M$ to form equation 9.15.

$$V_1(R_s + j\omega L_s) = I_1(R_1 + j\omega L_1)(R_s + j\omega L_s) - j\omega M I_2 (R_s + j\omega L_s) \tag{9.14}$$

$$j\omega M I_1 \times j\omega M = j\omega M I_2 (R_s + j\omega L_s)$$
$$-\omega^2 M^2 I_1 = j\omega M I_2 (R_s + j\omega L_s). \tag{9.15}$$

Adding equations 9.14 and 9.15 gives

$$V_1(R_s + j\omega L_s) - \omega^2 M^2 I_1 = I_1 (R_1 + j\omega L_1)(R_s + j\omega L_s)$$

dividing throughout by $(R_s + j\omega L_s)$

$$V_1 - \frac{\omega^2 M^2 I_1}{R_s + j\omega L_s} = I_1(R_1 + j\omega L_1)$$

$$V_1 = I_1\left[(R_1 + j\omega L_1) + \frac{\omega^2 M^2}{R_s + j\omega L_s}\right]$$

$$\frac{V_1}{I_1} = (R_1 + j\omega L_1) + \frac{\omega^2 M^2}{R_s + j\omega L_s} \qquad (9.16)$$

$\frac{V_1}{I_1} = Z_{in}$, the input impedance of the primary coil.

Without the effect of mutual inductive coupling to the secondary the input impedance would be simply $(R_1 + j\omega L_1)$.

Hence, bringing a second coil close to the first coil, such as to establish mutual inductance M, has the effect of modifying the input impedance by a term

$$\frac{\omega^2 M^2}{R_s + j\omega L_s} \; .$$

Rationalising this term gives

$$\frac{\omega^2 M^2 (R_s - j\omega L_s)}{R_s^2 + (\omega L_s)^2} \; .$$

Since the secondary inductive reactance is now multiplied by $-j$ it appears to be capacitive at the input terminals. Thus the reactance at the input becomes the difference between

$$j\omega L_1 \quad \text{and} \quad \frac{j\omega L_s \omega^2 M^2}{R_s^2 + (\omega L_s)^2}$$

which results in a lowering of the input impedance from that in the uncoupled state (see worked example 9.4).

<u>Worked example 9.4</u>   Determine the value of input current to the primary coil in the circuit shown in Fig. 9.9.

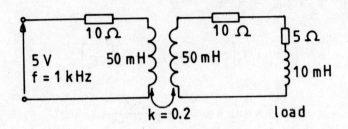

Fig. 9.9

From equation 9.10, $M = k\sqrt{(L_1 L_2)}$

$$\dot{M} = 0.2\sqrt{(50 \times 10^{-3} \times 50 \times 10^{-3})} = 10 \text{ mH}.$$

For each coil, at 1 kHz, $X_L = 2\pi \times 1000 \times 50 \times 10^{-3} = 314.16 \text{ }\Omega$.

For the load, $X_{L(load)} = 2\pi \times 1000 \times 10 \times 10^{-3} = 62.83 \text{ }\Omega$.

Total secondary impedance $= (10 + 5) + j(314.16 + 62.83)$

$$= (15 + j377) \text{ }\Omega.$$

Primary input impedance $= R_1 + j\omega L_1 + \dfrac{\omega^2 M^2}{R_s + j\omega L_s}$ (equation 9.16)

$$= 10 + j314.16 + \dfrac{(2\pi \times 1000)^2 (10 \times 10^{-3})^2}{(15 + j377)}$$

$$= 10 + j314.16 + \dfrac{3947.8(15 - j377)}{15^2 + 377^2}$$

$$= 10 + j314.16 + \dfrac{59217.6 - j1488336}{142354}$$

$$= 10.416 + j303.7 \text{ }\Omega.$$

Observe that this value is less than $(R_1 + j\omega L_1)$.

$$|I_1| = \dfrac{5}{10.416 + j303.7} = \dfrac{5}{303.88} = 16.45 \text{ mA}.$$

Without the presence of the secondary, I_1 would have been

RESONANCE, Q-FACTOR AND COUPLED A C CIRCUITS

$$\frac{5}{10 + j314.16} = 15.9 \text{ mA}.$$

The effect of the secondary is small since the degree of coupling is small ($k = 0.2$).

Generally speaking when a capacitor is connected in the secondary it will be employed to cause the secondary to be purely resistive, i.e. to tune it to resonance. Rarely, if ever, will sufficient capacitance be used so as to cause the secondary impedance to become capacitive overall. The solution of problems will therefore be as outlined in worked example 9.4 or with Z_s purely resistive.

<u>Worked example 9.5</u>   Two similar coils have resistance 20 Ω and inductance 25 mH. They are mutually coupled and $k = 0.6$.

The input voltage to the first coil is 20 V at a frequency of 1 kHz. A load of $50\underline{/-36.87^\circ}$ Ω is connected to the second coil. Determine (a) the power input to the first coil, (b) the power developed by the load.

From equation 9.10, $M = 0.6\sqrt{[(25 \times 10^{-3})(25 \times 10^{-3})]} = 15$ mH.

For both coils, $X_L = 2\pi \times 1000 \times 25 \times 10^{-3} = 157$ Ω.

The load = $50\underline{/-36.87^\circ} = (40 - j30)$ Ω.

Using equation 9.16

$$Z_{in} = 20 + j157 + \frac{(2\pi \times 1000)^2 \times (15 \times 10^{-3})^2}{(20 + j157) + (40 - j30)}$$

$$= 20 + j157 + \frac{8882.6(60 - j127)}{60^2 + 127^2} \quad (\text{*, part (b)(ii)})$$

$$= 47 + j100.2 \text{ Ω}.$$

$$I_1 = \frac{V}{Z_{in}} = \frac{20}{47 + j100.2} = \frac{20}{110.67\underline{/64.87^\circ}} = 0.181\underline{/-64.87^\circ} \text{ A}.$$

317

(a) Power input = VI cos ϕ = 20 × 0.181 × cos 64.87°
$$= 1.54 \text{ W.}$$

OR $|I|^2$ × total effective resistance in the primary = 0.181 × 47
$$= 1.54 \text{ W.}$$

(b) To calculate the power developed by the load there are two ways to progress.
(i) Determine the secondary induced voltage and hence the secondary current and power.

Using equation 9.7, $E_s = j\omega M I_1$
$$= j2\pi \times 1000 \times 15 \times 10^{-3} \times 0.181 \underline{/-64.87°}$$
$$= 17.06 \underline{/25.13°} \text{ V} \quad (j = +90°).$$

The angle is of no significance in this question and will be disregarded.

$$I_2 = \frac{E_s}{Z_s} = \frac{17.06}{(20 + j157) + (40 - j30)}$$

$$|I_2| = \frac{17.06}{140.46} = 0.12 \text{ A.}$$

Power in the load = $|I_2|^2 \times R_{load} = 0.12^2 \times 40 = 0.59$ W.

(ii) The resistance referred into the primary because of the total secondary resistance of 60 Ω is expressed as

$$\frac{8882.6 \times 60}{60^2 + 127^2} = 27 \text{ Ω .} \qquad (\text{*, previously})$$

The power in the primary due to this referred resistance is

$$|I_1|^2 \times 27 = 0.18^2 \times 27 = 0.8845 \text{ W.}$$

The secondary resistance comprises 20 Ω in the coil and 40 Ω in the load.

RESONANCE, Q-FACTOR AND COUPLED A C CIRCUITS

The load resistance = $\frac{40}{60} = \frac{2}{3}$ of the total.

The power developed in the primary due to the load resistance is therefore two-thirds of that due to the total secondary resistance, i.e. two-thirds of 0.8845 = 0.59 W.

<u>Self-assessment example 9.6</u> Two coils each with resistance 20 Ω and inductance 25 mH are mutually coupled so that the coupling coefficient is 0.8. The secondary is connected to a load of (40 - j30) Ω. A capacitor with capacitance 2.87 μF is connected in series with the primary coil. For a supply voltage of 20 V at 1 kHz, determine the power developed in the load resistance.

9.4.3 Critical coupling

From the work on maximum power transfer (chapter 2) it is known that maximum power is transferred from a resistive source of e.m.f. into a resistive load when the load resistance is equal to the source resistance.

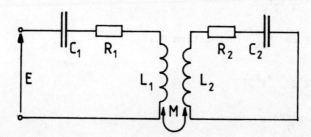

Fig. 9.10

By individually tuning both primary and secondary using capacitors C_1 and C_2 as shown in Fig. 9.10, equation 9.16 for the input impedance reduces to

$$Z_{in} = R_1 + \frac{\omega_o^2 M^2}{R_2}.$$

The primary coil and capacitor may now be considered as a generator with an e.m.f. of E volts and an internal resistance of R_1 ohms.

Maximum power will be transferred into the secondary when the secondary resistance referred into the primary has the same value as R_1. Hence for maximum power transfer

$$\frac{\omega_o^2 M^2}{R_2} = R_1. \qquad \text{(i)}$$

The particular value of mutual inductance which will cause this to occur is known as the critical value M_{CR}. Transposing (i) gives

$$\omega_o^2 M_{CR}^2 = R_1 R_2 \qquad (9.17)$$

but $M = k\sqrt{(L_1 L_2)}$, (equation 9.10)

or $M_{CR} = k_{CR}\sqrt{(L_1 L_2)}$

where k_{CR} is the corresponding value of coupling coefficient to give maximum power transfer.

$$M_{CR}^2 = k_{CR}^2 L_1 L_2.$$

Substituting for M_{CR}^2 in equation 9.17

$$\omega_o^2 k_{CR}^2 L_1 L_2 = R_1 R_2.$$

Transposing gives

$$k_{CR}^2 = \frac{R_1 R_2}{\omega_o L_1 \cdot \omega_o L_2}.$$

Now from equation 9.3

$$Q = \frac{\omega_o L}{R}, \text{ so that } \frac{R}{\omega_o L} = \frac{1}{Q}.$$

Hence $\dfrac{R_1}{\omega_o L_1} = \dfrac{1}{Q_1}$ and $\dfrac{R_2}{\omega_o L_2} = \dfrac{1}{Q_2}$. Therefore

$$k_{CR}^2 = \frac{1}{Q_1} \times \frac{1}{Q_2}$$

$$k_{CR} = \sqrt{\left(\frac{1}{Q_1 Q_2}\right)}. \tag{9.18}$$

Worked example 9.7 Two identical coils each with resistance 10 Ω and self-inductance 75 mH are individually tuned to resonance using capacitors connected as shown in Fig. 9.11. The first coil is supplied with 5 V at a frequency of 2.5 kHz.

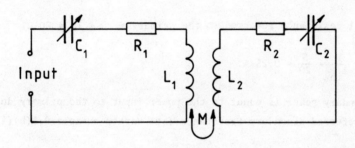

Fig. 9.11

(a) Determine (i) the required mutual inductance between the coils to achieve maximum power transfer into the secondary, (ii) the value of k_{CR} for (i) above, and (iii) the value of the power transferred into the secondary.
(b) Determine the power transferred into the secondary for (i) $k = \tfrac{1}{2}k_{CR}$, (ii) $k = 2k_{CR}$.

(a) (i) From equation 9.17, for critical coupling, $R_1 = \dfrac{\omega_o^2 M_{CR}^2}{R_2}$.

Transposing gives

$$M_{CR} = \sqrt{\left(\frac{R_1 R_2}{\omega_o^2}\right)},$$

or since the coils are identical

$$M_{CR} = \frac{R_1}{\omega_o} = \frac{10}{2\pi \times 2500} = 0.637 \text{ mH}.$$

(ii) From equation 9.10, $M_{CR} = k_{CR}\sqrt{(L_1 L_2)}$, hence

$$k_{CR} = \frac{M_{CR}}{\sqrt{(L_1 L_2)}}$$

$$k_{CR} = \frac{0.637 \times 10^{-3}}{75 \times 10^{-3}} = 8.49 \times 10^{-3}.$$

(iii) Since there is critical coupling, then by definition R_2 appears to be equal to R_1. Therefore

total resistance referred to the primary = 10 + 10 = 20 Ω.

$$I_1 = \frac{V}{Z_{in}} = \frac{5}{20} = 0.25 \text{ A.}$$

The secondary power is equal to the power input to the primary due to the referred secondary resistance (see worked example 9.5(b)(ii)).

Secondary power = $0.25^2 \times 10 = 0.625$ W.

(b)(i) $k = \tfrac{1}{2}k_{CR}$, therefore $M = \tfrac{1}{2}M_{CR} = 0.318 \times 10^{-3}$ H.

Secondary resistance referred into the primary = $\dfrac{\omega^2 M^2}{R_2}$

$$= \frac{(2\pi \times 2500)^2 \times (0.318 \times 10^{-3})^2}{10} = 2.5 \text{ Ω.}$$

$Z_{in} = 10 + 2.5 = 12.5$ Ω.

$$I_1 = \frac{5}{12.5} = 0.4 \text{ A.}$$

Power to secondary = $0.4^2 \times 2.5 = 0.4$ W.

(ii) $k = 2k_{CR}$, therefore $M = 2M_{CR} = 1.273 \times 10^{-3}$ H.

Secondary resistance referred = $\dfrac{(2\pi \times 2500)^2 \times (1.273 \times 10^{-3})^2}{10}$

= 40 Ω.

$$I_1 = \frac{5}{10 + 40} = 0.1 \text{ A.}$$

Secondary power = $0.1^2 \times 40 = 0.4$ W.

This example demonstrates the fact that the power transferred to the secondary falls from a relatively high value at k_{CR} to lower values as k is both increased and decreased. It does not conclusively prove the presence of a maximum value of k_{CR}. This is verified by differentiating the expression for secondary current I_2, with respect to ωM.

$$I_2 = \frac{j\omega M I_1}{R_2} \quad \text{and} \quad I_1 = \frac{E}{R_1 + \frac{\omega^2 M^2}{R_2}}.$$

9.4.4 Coupled tuned circuits and bandwidth

In the series or parallel tuned circuit the bandwidth is quite narrow. The circuits are highly selective, that is to say they have their maximum voltage or current magnification at a particular frequency, the response falling by 3 dB quite rapidly as the frequency deviates either way from the resonant value.

Where, for example, speech, music or television signals are to be transmitted, much wider bandwidths are required.

The ear responds to frequencies between about 40 Hz and 18 kHz. All these frequencies are contained in music either as fundamental or harmonic frequencies. To be able to listen to music in anything like the original clarity from a radio receiver or gramophone record player the equipment must be able to deliver this range of frequencies. From 40 Hz to 18000 Hz is a bandwidth of 18000 - 40 = 17960 Hz. The medium wave radio transmissions have an 8.5 kHz bandwidth whereas VHF transmissions have a bandwidth in excess of 15 kHz. When listening to both of these carrying the same programme, switching from one to the other will show the difference quite clearly.

For television receivers the bandwidth is in the region of 8 MHz since the signal not only provides the capability for high-quality sound but must be capable of adjusting the electron beam intensity with great rapidity as it scans the vision tube forming light and dark squares upon its surface.

At the other end of the bandwidth spectrum there is the telephone system operating with a bandwidth of 3 kHz (400 to 3400 Hz) which is adequate for speech but sometimes leads to recognition problems.

The wide bandwidth required in radio, etc. is achieved by using mutual inductance coupling between two tuned circuits such as has already been considered in this chapter, but with the degree of coupling considered from a bandwidth point of view and not from that for maximum power transfer.

In Fig. 9.11 both primary and secondary are tuned to a common resonant angular velocity, ω_o. This case has already been considered under 'critical coupling' when a single frequency was supplied. Now will be considered the connection of the input to a source which provides a range of frequencies, as would be obtained from a radio aerial for example.

The aerial carries signals of all frequencies from transmitters within its range.

For angular velocities above ω_o, ωL_1 is greater than $1/\omega C_1$, and ωL_2 is greater than $1/\omega C_2$. Both primary and secondary are therefore predominantly inductive. Using equation 9.16

$$Z_{in} = R_1 + j\left(\omega L_1 - \frac{1}{\omega C_1}\right) + \frac{\omega^2 M^2}{R_2 + j\left(\omega L_2 - \frac{1}{\omega C_2}\right)}$$

$$= R_1 + j\left(\omega L_1 - \frac{1}{\omega C_1}\right) + \frac{\omega^2 M^2 \left(R_2 - j\left(\omega L_2 - \frac{1}{\omega C_2}\right)\right)}{R_2^2 + \left(\omega L_2 - \frac{1}{\omega C_2}\right)^2} \quad \text{(i)}$$

Thus the excess inductive reactance in the secondary appears to be capacitive (-j) to the primary. There will be one particular value of ω for which this apparent capacitance cancels the inductance of the primary, i.e. when

$$+j\left(\omega L_1 - \frac{1}{\omega C_1}\right) + \left[\frac{-j\omega^2 M^2\left(\omega L_2 - \frac{1}{\omega C_2}\right)}{R_2^2 + \left(\omega L_2 - \frac{1}{\omega C_2}\right)^2}\right] = 0 \quad \text{(ii)}$$

At this angular velocity the circuit as a whole appears to be resistive and there is resonance at an angular velocity other than ω_o.

At angular velocities less then ω_o, $1/\omega C_1$ is greater than ωL_1 and

$1/\omega C_2$ is greater than ωL_2. Both primary and secondary circuits are predominantly capacitive. However, capacitance in the secondary appears to be inductive to the primary. There will be one particular value of ω for which this apparent inductance cancels the excess capacitance of the primary, resulting in a further resonant frequency. The coupled circuit as a whole therefore has three resonant angular velocities, ω_o, ω_H and ω_L.

As the coil resistances are small they can be neglected and equation (ii) reduces to

$$j\left(\omega L_1 - \frac{1}{\omega C_1}\right) + \frac{-j\omega^2 M^2\left(\omega L_2 - \frac{1}{\omega C_2}\right)}{\left(\omega L_2 - \frac{1}{\omega C_2}\right)^2} = 0$$

Dropping the j in each term and transposing gives

$$\omega L_1 - \frac{1}{\omega C_1} = \frac{\omega^2 M^2}{\left(\omega L_2 - \frac{1}{\omega C_2}\right)} \ .$$

Cross multiplying, $\left(\omega L_1 - \frac{1}{\omega C_1}\right)\left(\omega L_2 - \frac{1}{\omega C_2}\right) = \omega^2 M^2$

$$\omega^2 L_1 L_2 - \frac{L_1}{C_2} - \frac{L_2}{C_1} + \frac{1}{\omega^2 C_1 C_2} = \omega^2 M^2 \ .$$

Multiplying by $\dfrac{C_1 C_2}{\omega^2}$ throughout

$$\frac{\omega^2}{\omega^2} L_1 L_2 C_1 C_2 - \frac{C_1 C_2 L_1}{\omega^2 C_2} - \frac{C_1 C_2 L_2}{\omega^2 C_1} + \frac{C_1 C_2}{\omega^2 \omega^2 C_1 C_2} = \frac{\omega^2 M^2 C_1 C_2}{\omega^2}$$

$$L_1 L_2 C_1 C_2 - \frac{C_1 L_1}{\omega^2} - \frac{C_2 L_2}{\omega^2} + \frac{1}{\omega^4} = M^2 C_1 C_2 . \qquad \text{(iii)}$$

Now $M = k\sqrt{(L_1 L_2)}$ (equation 9.11), therefore $M^2 = k^2 L_1 L_2$ and $\omega_o^2 = 1/LC$ for each circuit (equation 9.1). Therefore

$$\frac{1}{\omega_o^2} = L_1 C_1 = L_2 C_2. \tag{iv}$$

Substitute (iv) in (iii) and also for M^2 gives

$$\frac{1}{\omega_o^2}\frac{1}{\omega_o^2} - \frac{1}{\omega_o^2 \omega^2} - \frac{1}{\omega_o^2 \omega^2} + \frac{1}{\omega^4} = k^2 L_1 L_2 C_1 C_2 = \frac{k^2}{\omega_o^4}$$

$$\frac{1}{\omega_o^4} - \frac{2}{\omega_o^2 \omega^2} + \frac{1}{\omega^4} = \frac{k^2}{\omega_o^4}$$

$$\frac{1}{\omega^4} - \frac{2}{\omega_o^2 \omega^2} = \frac{k^2 - 1}{\omega_o^4}. \tag{v}$$

Let $\frac{1}{\omega^2} = x$, then (v) becomes

$$x^2 - \frac{2x}{\omega_o^2} = \frac{k^2 - 1}{\omega_o^4}.$$

This is a quadratic equation which may be solved using any known method. Completing the square

$$x^2 - \frac{2x}{\omega_o^2} + \left(-\frac{1}{\omega_o^2}\right)^2 = \frac{k^2 - 1}{\omega_o^4} + \left(\frac{1}{\omega_o^2}\right)^2$$

$$\left(x - \frac{1}{\omega_o^2}\right)^2 = \frac{k^2}{\omega_o^4}$$

$$\left(x - \frac{1}{\omega_o^2}\right) = \pm \frac{k}{\omega_o^2}$$

$$x = \frac{k}{\omega_o^2} + \frac{1}{\omega_o^2} = \frac{(k + 1)}{\omega_o^2}$$

$$\text{or } x = -\frac{k}{\omega_o^2} + \frac{1}{\omega_o^2} = \frac{(1 - k)}{\omega_o^2}.$$

But $x = \frac{1}{\omega^2}$, therefore

RESONANCE, Q-FACTOR AND COUPLED A C CIRCUITS

$$\frac{1}{\omega^2} = \frac{(k+1)}{\omega_o^2} \quad \text{or} \quad \omega = \frac{\omega_o}{\sqrt{(1+k)}} \qquad (9.19(a))$$

$$\text{or} \quad \frac{1}{\omega^2} = \frac{(1-k)}{\omega_o^2} \quad \omega = \frac{\omega_o}{\sqrt{(1-k)}} \qquad (9.19(b))$$

Although these two solutions have been obtained from equation (ii) which supposed a higher angular velocity than ω_o, they are both valid. Equation 9.19(b) gives ω_H, the denominator being less than 1, while equation 9.19(a) gives ω_L, the denominator being greater than 1. Reversing the positions of ($1/\omega C_1$ and ωL_1) and of ($1/\omega C_2$ and ωL_2) to establish the lower angular velocity for resonance yields the same roots.

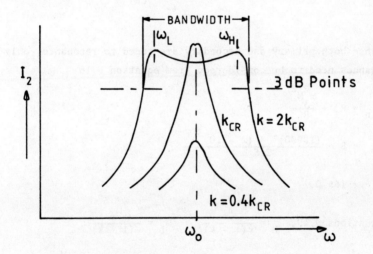

Fig. 9.12

For values of k less than k_{CR}, the secondary current is small and the peak response is small. For values of k from k_{CR} and upwards, the secondary current develops the 'double hump' form shown in Fig. 9.12 which is drawn for two identical coils each with Q in the region of 50. For $k = 2k_{CR}$, the extra bandwidth over the simple response curve (Fig. 9.5) is clear. ω_L and ω_H and the bandwidth in Fig. 9.13 are for this value of coupling coefficient only. Ideally the response curve should be rectangular giving a level response between the two peaks. This is never realised in practice but by careful adjustment of the two resistances and by fine adjustment of the coupling coefficient it may be approached. From worked example 9.7 it can be seen that using $k = 2k_{CR}$ the power transferred to the secondary is less than that for k_{CR}, but for this application power is sacrificed for bandwidth.

<u>Worked example 9.8</u> Two air-cored coils are coupled magnetically, one being connected to an alternating supply and the other short circuited. Calculate (a) the primary input impedance in the coupled condition given that $R_1 = 5\ \Omega$, $R_2 = 2.5\ \Omega$, $L_1 = L_2 = 10$ mH and $M = 2$ mH. Both primary and secondary are tuned using series connected capacitors to a frequency of $\frac{10000}{2\pi}$ Hz. (b) The additional resonant frequencies created by the coupling. (c) The ratio k/k_{CR} for this arrangement of coils.

(a) Since both primary and secondary are tuned to resonance, only the resistances need to be considered. From equation 9.16

$$Z_{in} = R_1 + \frac{\omega_o^2 M^2}{R_2}$$

$$= 5 + \frac{(10000)^2 \times (2 \times 10^{-3})^2}{2.5}$$

$$= 165\ \Omega.$$

(b) Equations 9.19, $\omega_H = \frac{\omega_o}{\sqrt{(1-k)}}$; $\omega_L = \frac{\omega_o}{\sqrt{(1+k)}}$.

Equation 9.10, $M = k\sqrt{(L_1 L_2)}$, transposing gives, $k = \frac{M}{\sqrt{(L_1 L_2)}}$

or for identical inductances, $k = \frac{M}{L} = \frac{2\ \text{mH}}{10\ \text{mH}} = 0.2$.

Therefore, $\omega_H = \frac{10000}{\sqrt{(1-0.2)}} = 11180$ rad/s; $f_H = \frac{11180}{2\pi}$ Hz.

$\omega_L = \frac{10000}{\sqrt{(1+0.2)}} = 9129$ rad/s; $f_L = \frac{9129}{2\pi}$ Hz.

(c) From equation 9.17, for k_{CR}, $R_1 = \frac{\omega_o^2 M^2}{R_2}$. Transposing

RESONANCE, Q-FACTOR AND COUPLED A C CIRCUITS

$$M_{CR} = \sqrt{\frac{R_1 R_2}{\omega_0^2}} = \frac{5 \times 2.5}{10000^2} = 0.35 \times 10^{-3} \text{ H.}$$

$$k_{CR} = \frac{M_{CR}}{L} = \frac{0.35 \times 10^{-3}}{10 \times 10^{-3}} = 0.035$$

Therefore, $\dfrac{k}{k_{CR}} = \dfrac{0.2}{0.035} = 5.66$.

<u>Self-assessment example 9.9</u>   Two coils are coupled by mutual inductance. Both primary and secondary are tuned to resonance. Given that $L_1 = L_2 = 238$ µH, $C_1 = C_2 = 106$ pF and $k = 0.02$, (a) determine the three frequencies, f_o, f_H and f_L at which resonance occurs. (b) What value of k would double the range of frequencies between f_H and f_L?

9.5   THE POWER TRANSFORMER
9.5.1 The ideal transformer

The power transformer comprises two coils wound on a laminated iron core in the simplest single-phase type. The core is designed to have the smallest possible leakage flux and therefore to approach the perfect case of mutual coupling when $k = 1$ and $M = \sqrt{(L_1 L_2)}$ (equation 9.9).

Consider a perfectly coupled pair of coils as shown in Fig. 9.13. The primary coil has N_p turns and has a self-inductance of L_p henry. The secondary has N_s turns and a self-inductance L_s henry. The resistances of both windings are ignored for the moment as they are small compared with ωL_p and ωL_s. The secondary is supplying current to a load with impedance Z_2 ohms.

$$I_1 = \frac{V}{Z_{in}} = \frac{V}{j\omega L_p + \dfrac{\omega^2 M^2}{j\omega L_s + Z_2}} \quad \text{(using equation 9.16 for } Z_{in}\text{)}.$$

Multiply throughout by $(j\omega L_s + Z_2)$

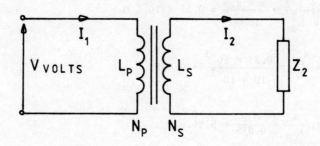

Fig. 9.13

$$I_1 = \frac{V(j\omega L_s + Z_2)}{j\omega L_p(j\omega L_s + Z_2) + \omega^2 M^2}$$

$$= \frac{V(j\omega L_s + Z_2)}{-\omega^2 L_p L_s + j\omega L_p Z_2 + \omega^2 M^2} \quad . \tag{i}$$

This is a general equation, no assumptions at all having been made with respect to coupling, etc. As already stated, with a power transformer tight coupling is obtained, so that $M = \sqrt{(L_p L_s)}$ (very closely), hence

$$M^2 = L_p L_s \quad \text{and} \quad \omega^2 M^2 = \omega^2 L_p L_s \; .$$

Hence in the denominator of equation (i), $-\omega^2 L_p L_s + \omega^2 M^2 = 0$, when

$$I_1 = \frac{V(j\omega L_s + Z_2)}{j\omega L_p Z_2}$$

$$= V\left(\frac{j\omega L_s}{j\omega L_p Z_2}\right) + V\left(\frac{Z_2}{j\omega L_p Z_2}\right)$$

$$= V\left(\frac{L_s}{L_p} \times \frac{1}{Z_2}\right) + V\left(\frac{1}{j\omega L_p}\right)$$

$$= \begin{Bmatrix} \text{a current present} \\ \text{due to the load} \\ \text{impedance } Z_2 \end{Bmatrix} + \begin{Bmatrix} \text{a current lagging V by } 90^0 \\ \text{limited only by } L_p, \text{ i.e. the} \\ \text{magnetising current } I_m \end{Bmatrix} \; .$$

RESONANCE, Q-FACTOR AND COUPLED A C CIRCUITS

$1/Z_2 = Y_2$, the admittance of the load, hence

$$\frac{L_s}{L_p} \times \frac{1}{Z_2}$$

is the admittance of the load referred into the primary. Notice that the effect of Z_2 in the primary is determined by the ratio L_p/L_s. It is no longer necessary to consider $\omega^2 M^2$ with tight-coupled coils. It is general to consider impedances when working with power transformers so that inverting gives

$$\text{effective primary impedance} = Z_2 \frac{L_p}{L_s} \; \Omega.$$

From equation 6.13, $L = N\frac{d\Phi}{di}$ H. Now $d\Phi \propto N \, di$, so that

$$L \propto \frac{N \times N \, di}{di} \quad \text{or} \quad L \propto N^2.$$

Therefore, $L_p \propto N_p^2$ and $L_s \propto N_s^2$, and $\frac{L_p}{L_s} = \left(\frac{N_p}{N_s}\right)^2$.

Hence, the effective impedance in the primary = $Z_2\left(\frac{N_p}{N_s}\right)^2 \; \Omega.$ (9.20)

The equivalent circuit for the ideal transformer is, therefore, as shown in Fig. 9.14.

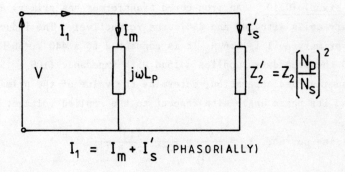

$I_1 = I_m + I'_s$ (PHASORIALLY)

Fig. 9.14

The magnetising current I_m is independent of transformer loading and flows at all the times the transformer is connected to the supply.

I_s' is the secondary current referred to the primary, or balancing current, i.e. the current which flows in the primary because I_s flows in the secondary and in the load.

In Fig. 9.14

$$\text{VA developed in } Z_2' = (I_s')^2 \, Z_2 \left(\frac{N_p}{N_s}\right)^2.$$

This must be the same as that actually developed in the load Z_2 by current I_s if this circuit is to represent the transformer accurately. Therefore

$$I_s^2 Z_2 = (I_s')^2 \, Z_2 \left(\frac{N_p}{N_s}\right)^2.$$

Transposing:

$$\left(\frac{I_s}{I_s'}\right)^2 = \frac{Z_2}{Z_2} \left(\frac{N_p}{N_s}\right)^2.$$

Cancelling Z_2 and taking the square root of both sides gives

$$\frac{I_s}{I_s'} = \frac{N_p}{N_s} \quad \text{and} \quad I_s' = I_s \frac{N_s}{N_p} \text{ A}. \tag{9.21}$$

The current input to the primary is the phasor sum of I_m and I_s'.

<u>Worked example 9.10</u>   An iron-cored transformer has primary and secondary coils with 500 and 250 turns respectively. The inductance of the primary coil is 1.6 H. It is connected to a 440 V, 50 Hz supply and the secondary supplies a load with impedance $(100 + j50) \, \Omega$.

Assuming ideal operation, determine the value of the primary current and its phase angle with respect to the applied voltage.

$$\text{Magnetising current, } I_m = \frac{V}{j\omega L_p} = \frac{440}{j2\pi \times 50 \times 1.6}$$

$$= -j0.875 \text{ A}.$$

Current in primary due to secondary load, I_s'

effective impedance, $Z_2' = \left(\dfrac{N_p}{N_s}\right)^2 \times Z_2 = \left(\dfrac{500}{250}\right)^2 \times (100 + j5)$

$\hspace{4cm} = (400 + j200) \ \Omega.$

$I_s' = \dfrac{440}{(400 + j200)} = \dfrac{440(400 - j200)}{400^2 + 200^2}$

$\hspace{2cm} = (0.88 - j0.44) \ \text{A}.$

$I_1 = I_m + I_s'$
$\hspace{0.6cm} = -j0.875 + (0.88 - j0.44) = (0.88 - j1.315) \ \text{A}$
$\hspace{4cm} = 1.582 \underline{/-56.2^o} \ \text{A}.$

$I_1 = 1.582$ A, the current lags the supply voltage by 56.2^o.

9.5.2 The practical transformer

In the previous section ideal operation was considered. There are a number of imperfections which must be considered.

1. The core being made from iron suffers hysteresis and eddy current losses and so there must be a power input from the supply to provide these. In the equivalent circuit a resistance R_o is drawn in parallel with $j\omega L_p$. This resistor has a value calculated from open-circuit test results on the transformer which determine the core loss. For example, the core loss is equal to 60 W at a full rated voltage of 440 V. It would appear that this loss is taking place in an effective resistance calculated from

$\hspace{1cm} \text{power} = \dfrac{V^2}{R_o}, \hspace{0.5cm} \text{hence } R_o = \dfrac{440^2}{60} = 3227 \ \Omega.$

2. However carefully designed the transformer is, there will in fact be some leakage flux, so that the primary coil is shown in two sections. The main section of N_p turns is considered to be perfectly coupled to the secondary while a very small section is considered to be uncoupled and it is this small section which sets up the leakage flux. The reactance of the uncoupled section is called leakage reactance and is marked X_p in Fig. 9.15. A similar argument may be

applied to the secondary and its leakage reactance is marked X_s in Fig. 9.15. (See also appendix 3 and section 9.4.1.)

3. Each winding has a small resistance and where efficiency and volt-drop calculations are to be performed these resistances have to be considered.

The overall equivalent circuit diagram becomes as shown in Fig. 9.15.

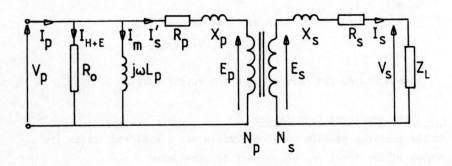

Fig. 9.15

In Fig. 9.15

R_p and X_p = primary winding resistance and leakage reactance respectively.
R_s and X_s = secondary winding resistance and leakage reactance respectively.
E_p = primary induced voltage which opposes V_p.
E_s = secondary induced voltage which drives the secondary current I_s through the load impedance such that the load p.d. equals V_s volts.
I_{H+E} = current to provide hystersis and eddy current losses of the core.
I_p = total input current to the primary = phasor sum of I_s, I_m and I_{H+E}.

It might be argued that R_p should be on the supply side of $j\omega L_p$ and R_o since I_m and I_{H+E} have to flow in the primary winding and hence

RESONANCE, Q-FACTOR AND COUPLED A C CIRCUITS

its resistance. This is indeed true, but in this position, as I_s' flows, the voltage drop across R_p would seem to cause the magnetising current and core losses to fall. These are in fact essentially constant so that the best equivalent circuit is as shown. It gives good results when predicting efficiency and voltage changes with load.

Using equation 9.20, all the secondary values of impedance may be referred into the primary as in Fig. 9.14. However, it is often more convenient to refer the series impedance from the primary into the secondary. Equation 9.20 gives:

$$\text{primary impedance} = \left(\frac{N_p}{N_s}\right)^2 \times \text{secondary impedance}$$

$$\text{primary impedance} \times \left(\frac{N_s}{N_p}\right)^2 = \text{secondary impedance}.$$

Hence: R_p referred to the secondary $= R_p' = R_p\left(\frac{N_s}{N_p}\right)^2$

and X_p referred to the secondary $= X_p' = X_p\left(\frac{N_s}{N_p}\right)^2$

when the equivalent circuit diagram becomes as shown in Fig. 9.16.

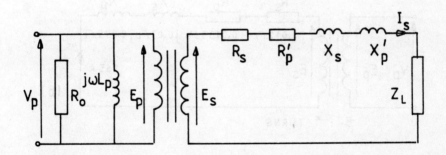

Fig. 9.16

From equation 6.5, the e.m.f. equation for a transformer

$$E = 4.44 B_m A f N \text{ volts}.$$

Therefore, $E_p = 4.44B_m AfN_p$ and $E_s = 4.44B_m AfN_s$.

Dividing E_p by E_s gives: $\dfrac{E_p}{E_s} = \dfrac{N_p}{N_s}$. (9.22)

Worked example 9.11 The ratio of turns of a single-phase transformer is 8 : 1. The resistances of primary and secondary windings are 0.85 Ω and 0.012 Ω respectively. The corresponding leakage reactances are 4.8 Ω and 0.07 Ω.

Determine the value of voltage to be applied to the primary to cause 165 A to flow in a short circuit across the secondary terminals. (Ignore magnetising current and core losses.)

The transformer as described in the question may be drawn as shown in Fig. 9.17(a).

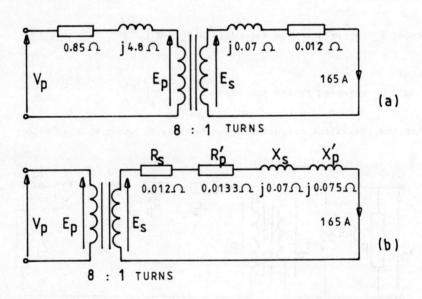

Fig. 9.17

Referring primary resistance and reactance to the secondary yields:

$$R_p' = R_p \times \left(\dfrac{N_s}{N_p}\right)^2 = 0.85 \times \left(\dfrac{1}{8}\right)^2 = 0.0133 \text{ Ω}$$

$$X'_p = X_p \times \left(\frac{N_s}{N_p}\right)^2 = 4.8 \times \left(\frac{1}{8}\right)^2 = 0.075 \; \Omega.$$

Observe that when referring values from a winding with a large number of turns into one with less turns, the impedance values decrease, and vice-versa.

In Fig. 9.17(b), the total impedance of the transformer referred to the secondary $Z_T^{(S)}$ is given by

$$Z_T^{(S)} = \sqrt{[(0.012 + 0.0133)^2 + (0.07 + 0.075)^2]} = 0.147 \; \Omega.$$

$$E_s = IZ_T^{(S)} = 165 \times 0.147 = 24.26 \; V.$$

Note that if the output terminals were not short circuited, the load impedance would have to be included in $Z_T^{(S)}$.

Now, $\dfrac{E_p}{E_s} = \dfrac{N_p}{N_s}$. (equation 9.22)

Therefore, $\dfrac{E_p}{24.26} = \dfrac{8}{1}$

$$E_p = 8 \times 24.26 = 194.1 \; V.$$

Since the primary impedance is zero, $V_p = E_p = 194.1 \; V.$

<u>Worked example 9.12</u> A single-phase transformer with ratio 660 V : 110 V has a core loss of 200 W at full rated voltage. The magnetising current is 2 A in the 660 V winding. The transformer delivers a current of 65 A from its 110 V winding at a power factor of 0.75 lagging.

Determine (a) the iron loss current I_{H+E}, (b) the no-load current and phase angle of the transformer, (c) the balancing current in the primary winding, (d) the total primary current, and (e) the power factor of the primary at this load. (Ignore winding resistance and reactance.)

(a) The core loss is considered to occur in a resistance R_o connected directly across the supply (Fig. 9.15). The value of a current

(I_{H+E}) must be determined which, flowing in a resistor, would give a power of 200 W at 660 V.

In a resistor, V × I = power. Hence

$$I_{H+E} = \frac{200}{660} = 0.303 \text{ A.}$$

(b) On no-load, I_m and I_{H+E} flow simultaneously. I_m lags V_p by $90°$. From Fig. 9.18(a)

$$\text{no-load current, } I_o = \sqrt{\left(I_{H+E}^2 + I_m^2\right)} = \sqrt{(0.303^2 + 2^2)}$$
$$= 2.023 \text{ A.}$$

Also, $\tan \phi_o = \dfrac{I_m}{I_{H+E}}$, hence $\phi_o = 81.38°$.

(c) From equation 9.21

$$I_s' = I_s \frac{N_s}{N_p} = I_s \frac{E_s}{E_p} \quad \text{(using equation 9.22)}$$

$$I_s' = 65 \times \frac{110}{660} = 10.83 \text{ A.}$$

Since at all instants I_s' 'balances' I_s, I_s' has the same phase angle with respect to E_p that I_s has with respect to E_s. The load power factor is 0.75. Therefore I_s lags E_s by $\cos^{-1} 0.75 = 41.4°$ and $E_s = V_s$, since winding impedance is ignored.

It follows that I_s lags E_p (and V_p) by $41.4°$.

(d) The total primary current is equal to the phasor sum of I_m, I_{H+E} and I_s'. In Fig. 9.18(b), resolving I_s' into vertical horizontal components

horizontal component = 10.83 sin 41.4 = 7.16 A
vertical component = 10.83 cos 41.4 = 8.12 A.

Adding horizontal components: 7.16 + 2 = 9.16 A.
Adding vertical components: 8.12 + 0.303 = 8.423 A.

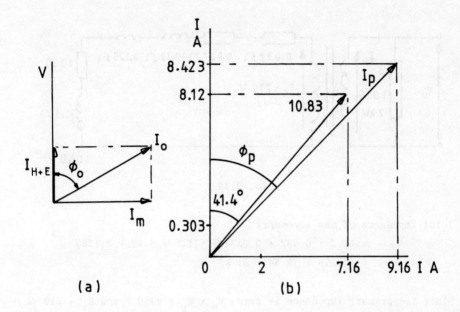

Fig. 9.18

By Pythagoras' theorem: $I_p = \sqrt{(9.16^2 + 8.423^2)} = 12.44$ A.

(e) Power factor $= \cos \phi_p = \dfrac{8.423}{12.44} = 0.677 \quad (\phi_p = 47.4°)$.

<u>Worked example 9.13</u> A single-phase, 50 Hz transformer with voltage ratio 6350 V : 230 V has the following details: $R_p = 20.7\ \Omega$, $R_s = 0.032\ \Omega$, $X_p = 95\ \Omega$, $X_s = 0.12\ \Omega$. The core loss is 2 kW. The secondary is connected to a load of $(2.3 + j1.6)\ \Omega$.

Calculate, for this condition, assuming the primary to be connected to a 6350 V supply, (a) the output terminal voltage, and (b) the operating efficiency.

Refer the primary resistance and reactance into the secondary.

$$R_p' = 20.7\left(\dfrac{230}{6350}\right)^2 = 0.027\ \Omega; \quad X_p' = 95\left(\dfrac{230}{6350}\right)^2 = 0.125\ \Omega.$$

The equivalent circuit is shown in Fig. 9.19.

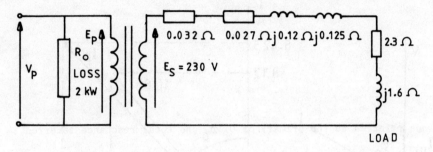

Fig. 9.19

Total impedance of the secondary
$$= (2.3 + 0.032 + 0.027) + j(1.6 + 0.12 + 0.125)$$
$$= (2.359 + j1.845) \ \Omega.$$

Since the primary impedance is zero, $V_p = E_p = 6350$ V and $E_s = 230$ V.

$$I_s = \frac{E_s}{Z_T^{(S)}} = \frac{230}{(2.359 + j1.845)} = \frac{230}{3\underline{/38.03^o}} = 76.8\underline{/-38.03^o} \ A$$

I_s has a modulus of 76.8 A and lags E_s by 38.03^o.

(a) The output terminal voltage $= IZ_{load}$
$$= 76.8\underline{/-38.03^o} \times (2.3 + j1.6)$$
$$= 76.8\underline{/-38.03^o} \times 2.8\underline{/34.82^o}$$
$$= 215.18\underline{/-3.2^o} \ V.$$

V_s has a modulus of 215.18 V and lags E_s by 3.2^o.

(b) Load power $= I_s^2 R_L = 76.8^2 \times 2.3 = 13566$ W.

Power loss in windings $= I_s^2 \times$ winding resistance
$$= 76.8^2 \ (0.032 + 0.027)$$
$$= 348 \ W.$$

Core loss $= 2$ kW.

RESONANCE, Q-FACTOR AND COUPLED A C CIRCUITS

$$\text{Efficiency} = \frac{\text{output}}{\text{input}} = \frac{\text{output}}{\text{output + losses}}$$

$$= \frac{13566}{13566 + 348 + 2000}$$

$$= 0.852 \text{ p.u.}$$

Self-assessment example 9.14   A single-phase transformer has its primary connected to a 220 V supply. It has total equivalent resistance referred to the primary of 0.1 Ω and total reactance referred into the primary of 0.5 Ω. The secondary is connected to a load of resistance 200 Ω and reactance +j100 Ω. The turns ratio is 1 : 5 from primary to secondary. The core loss is 60 W.

Determine (a) the total resistance referred into the secondary, (b) the total reactance referred into the secondary, (c) the secondary current I_s, (d) the secondary terminal voltage, and (e) the efficiency.

There is a considerable amount of further work which can be undertaken on power transformers including some 'short-cuts' in working allowed in the relevant British Standards. These give close approximate answers with substantially reduced working. These are to be considered as applications and not principles.

This chapter has attempted to show that mutual inductance is employed to obtain several different results, maximum power transfer, wide bandwidth, or in the case of power transformers, the greatest possible efficiency while changing voltages from one level to another. Each function overlaps the others to some extent. There is a great tendency to treat mutual inductance for light-current applications as something quite different to that required for heavy-current applications which is, of course, quite unjustified.

Further problems

9.15. A capacitor with capacitance 53 μF is connected in series with a coil of inductance 0.0955 H. The maximum current drawn from a 200 V, variable-frequency supply is 5 A. Determine (a) the circuit current, (b) the power factor, when supplied at 50 Hz.

9.16. What is meant by the magnification factor of a series resonant circuit?

Sketch a graph of current against frequency for a series circuit containing inductance and capacitance to which a constant voltage, variable-frequency supply is connected. Explain how the shape of the curve would be affected by a change in the circuit resistance.

A circuit comprises a capacitor with capacitance 0.02 µF in series with a coil possessing resistance and inductance. The resonant frequency is 5 kHz. The magnification factor Q = 110. Determine (a) the resistance of the coil, (b) the inductance of the coil, (c) the lower 3 dB frequency.

9.17. (a) Explain what is meant by (i) the dynamic impedance of a parallel resonant circuit, (ii) the voltage magnification of a series resonant circuit.

(b) A circuit comprises a capacitor with capacitance 20 µF and a coil with resistance 120 Ω and inductance 0.5 H. Determine the resonant frequency when (i) the capacitor is connected in series with the coil, (ii) the capacitor is connected in parallel with the coil.

9.18. A coil and a capacitor are connected in series to a source of e.m.f. of 10 mV and a frequency of 1 MHz. If the coil has an effective resistance of 20 Ω and an inductance of 159 µH, determine (a) the value of capacitance to tune the circuit to resonance, (b) the value of circuit current at resonance, (c) the p.d. across the capacitor at resonance, and (d) the bandwidth and frequencies for half power.

9.19. An a.c. circuit comprises a coil with inductance 0.1 H and resistance 50 Ω shunted by a capacitor having capacitance 10 µF. The supply frequency is adjusted until the supply current and voltage are in phase. Calculate (a) the supply frequency for this condition, (b) the value of the dynamic impedance of the circuit.

9.20. The secondary of the circuit shown in Fig. 9.20 is at resonance when f = 50 kHz. (M = 0.1 mH and R_s = 10 Ω.) Calculate the effective impedance coupled into the primary from the secondary at this frequency.

9.21. A coil having inductance 5 mH and resistance 20 Ω is connected to a source of 2 V at a frequency of 10 kHz. It is mutually coupled to a second coil having inductance 20 mH and resistance 100 Ω. The coupling coefficient is 0.4. The second coil is connected to a variable capacitor which is adjusted to make the secondary circuit resonant.

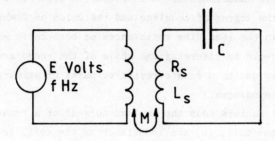

Fig. 9.20

Calculate (a) the value of the primary current, (b) the capacitor voltage under these conditions.

9.22. Two air-cored coils possessing resistance and self-inductance are coupled by mutual inductance. The primary is connected to an alternating power supply while the secondary is short circuited.
(a) Derive an expression for the effective impedance of the primary under these conditions. Given that $R_p = 10\ \Omega$, $L_p = 20$ mH, $R_s = 5\ \Omega$. $L_s = 20$ mH, $k = 0.15$ and $f = 2$ kHz, calculate this value of impedance.
(b) If capacitors are now connected in series with both primary and secondary windings such that both circuits are resonant at $10000/2\pi$ Hz when in isolation, calculate the two additional frequencies created by the mutual coupling.

9.23. (a) Two coils having self-inductance L_1 and L_2 henry respectively have mutual inductance M henry between them. Define the coupling coefficient k, between these two coils in terms of L_1, L_2 and M.
(b) Given that the values of L_1 and L_2 are 0.1 mH and 0.3 mH respectively and that the corresponding resistances are $5\ \Omega$ and $21\ \Omega$, and that each coil is connected in series with a capacitor which causes resonance in each circuit when isolated at 50 kHz, calculate (i) the value of the critical coupling coefficient and the corresponding value of mutual inductance, (ii) the power transferred into the secondary circuit for the condition in (i), the primary being fed at 5 V.

9.24. Two coils coupled by mutual inductance are used to couple a load of impedance $100\underline{/0^{\circ}}\ \Omega$ to a source of e.m.f. having internal impedance of $150\underline{/0^{\circ}}\ \Omega$ operating at 0.8 MHz. The primary winding has inductance

ELECTRICAL PRINCIPLES FOR HIGHER TEC

100 µH and is connected to the source via a capacitor C. The secondary winding with inductance 25 µH is connected directly to the load.

Calculate the degree of coupling and the value of C which gives maximum power in the load. The resistances of both coils may be ignored. (Hint: begin by referring the whole of the secondary impedance into the primary and then equate resistive parts of primary and secondary referred impedances.)

9.25. Upon what factors does the no-load current of a transformer depend?

A transformer draws a magnetising current of 9.9 A and an in-phase current of 1.1 A to supply hysteresis and eddy current losses when fed at its rated voltage. The ratio is 4 : 1 step-down. Determine the value of the total primary current and the power factor when the secondary supplies a load with 200 A at 0.8 power factor lagging.

9.26. A single-phase transformer with ratio 3300 V : 240 V has the following parameters: R_p = 4.5 Ω, X_p = 22.6 Ω, R_s = 0.026 Ω, X_s = 0.12 Ω. The core loss is 325 W. A load of (10 + j5) Ω is connected to the secondary (240 V) terminals.

For the primary connected to a 3300 V supply, determine (a) the secondary current, (b) the balancing current in the primary, (c) the output terminal voltage, and (d) the operating efficiency.

CHAPTER 9   ANSWERS
9.3. (a) 100654 Hz   (b) 9 µA   (c) 0.949 mA   (d) Q = 105.4
      (e) 100176 Hz; 101131 Hz; bandwidth, 955 Hz
9.6. 2.8 W
9.9. (a) f_0 = 1.002 MHz; f_H = 1.012 MHz; f_L = 0.992 MHz
      (b) k = 0.039
9.14. (a) $R_T^{(S)}$ = 2.5 Ω   (b) $X_T^{(S)}$ = 12.5 Ω   (c) I_s = 1100/231.65 $\underline{/29°}$
      = 4.748 $\underline{/-29°}$ A   (d) V_s = 1061 V   (e) Power out = 4508.7 W;
      losses = 56.4 + 60 W; efficiency: 0.975
9.15. (a) 4 A   (b) 0.8
9.16. (a) 14.5 Ω   (b) 0.0507 H   (c) 4.977 kHz
9.17. (b)(i) 50.3 Hz   (ii) 32.7 Hz
9.18. (a) 159 pF   (b) 500 µA   (c) 0.5 V   (d) 20 kHz; 0.99 MHz; 1.01 MHz
9.19. (a) 137.8 Hz   (b) 200 Ω

RESONANCE, Q-FACTOR AND COUPLED A C CIRCUITS

9.20. $98.7 \, \Omega$

9.21. (a) 2.76 mA (b) 8.72 V (I_s = 6.94 mA)

9.22. (a) $(10.1 + j245.7) \, \Omega$ (b) 1484 Hz; 1726 Hz

9.23. (b)(i) k = 0.188; M = 32.6 μH (ii) 1.25 W

9.24. From secondary: $(244.9 - j307.8)k^2 \, \Omega$. To match resistances, k = 0.904. Capacitor: $X_C = -j251.1 \, \Omega$; C = 792 pF

9.25. 57.6 A; 0.71 lag

9.26. (a) 21.18 A (b) 1.54 A (c) 236.7 V (d) 0.928 p.u.

10 Laplace Transforms and the Solution of Circuit Transients

This chapter is not intended to be a comprehensive treatment of Laplace transforms but rather an introduction to the topic for those whose mathematical studies have not reached this stage and as an encouragement to study further this very useful tool for solving differential equations, involved amongst other things in circuit transients and transmission lines with which this book deals. The solution of differential equations by other means is difficult or even impossible.

When a voltage is applied to a circuit comprising elements which store energy, i.e. inductance, capacitance, or a combination of both, there is a period of fluctuation of current and energy preceding the settling down to the final steady-state conditions. Likewise when the voltage is removed or changed there is a period of transient disturbance.

To analyse the transient performance, the Laplace transform is used. This is a development of work carried out by a British electrical engineer, Oliver Heavyside.

10.1 COMPARISON WITH LOGARITHMS

When an operation such as multiplication, division or raising a number to a power is to be undertaken the direct approach may be difficult. For example, consider raising 1.96 to the power 0.54 by direct mathematical methods. The case of division employing logarithms with the aid of a flow diagram will be examined.

The quotient N_1/N_2 is to be determined using logarithms to a base 'b'.

LAPLACE TRANSFORMS AND THE SOLUTION OF CIRCUIT TRANSIENTS

The transforms are:

$\log_b N_1 = x$, hence $b^x = N_1$,

$\log_b N_2 = y$, hence $b^y = N_2$.

Therefore, $\dfrac{N_1}{N_2} = \dfrac{b^x}{b^y} = b^{(x-y)}$.

Hence the logarithm of $\dfrac{N_1}{N_2} = (x - y)$.

To find the value of N_1/N_2 the inverse transform is performed by looking up $(x - y)$ in the anti-logarithm tables.

The procedure is illustrated in the flow diagram (Fig. 10.1).

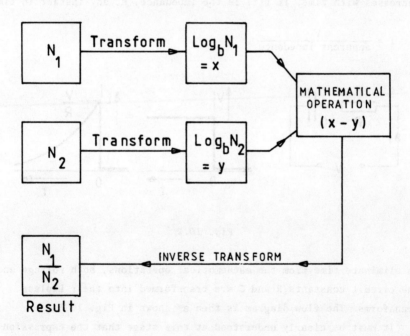

Fig. 10.1

10.2 THE LAPLACE TRANSFORM
10.2.1 Definition
The Laplace transform changes the expression for a quantity from that involving time, $f(t)$, to a complex function, $F(s)$, where s is a

complex quantity, $(\alpha + j\omega)$. α is a real number and ω is the angular velocity in rad/s.

As an example, consider a direct voltage being suddenly applied to a capacitive circuit which is initially uncharged. The current falls exponentially from a maximum value given by $I = V/R$ amperes to zero. The stimulus is a suddenly applied steady voltage, or step voltage as it is termed. The voltage is a function of time in that

at $t < 0$, $V = 0$
$t = 0$, $V = 0$
$t > 0$, $V = V$ volts.

The opposition to current flow or apparent impedance of the circuit increases with time. If $f(t)$ is the impedance, at any instant in time

$$i = \frac{V}{\text{apparent impedance}}.$$

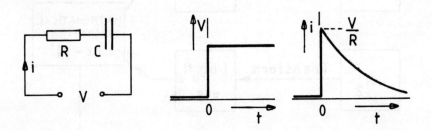

Fig. 10.2

To eliminate time from the mathematical operations, both voltage and the circuit constants R and C are transformed into their Laplace transforms. The flow diagram is then as shown in Fig. 10.3.

It must be clearly understood at this stage that the expression $X_C = (1/\omega C)$ Ω is only valid for a capacitor energised from a constant-voltage constant-frequency alternating source and only then after a period during which conditions settle down to what are known as 'steady-state' conditions. During any period of discontinuity of the supply, conditions are transient and if a transient solution is required this will be quite different from the steady-state solution

LAPLACE TRANSFORMS AND THE SOLUTION OF CIRCUIT TRANSIENTS

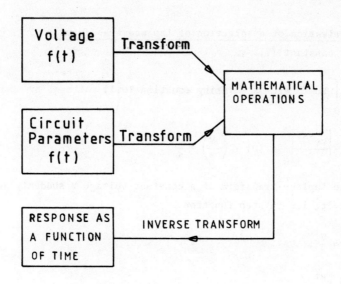

Fig. 10.3

and will involve differential equations which may be solved using Laplace transforms.

If f(t) is a function of time then its Laplace transform $\mathcal{L}[f(t)]$ is defined as

$$\int_0^\infty f(t)e^{-st} \, dt, \text{ and is written as } F(s). \tag{10.1}$$

In words, the Laplace transform of a function is the integral between zero and infinity with respect to time of that function multiplied by natural e raised to the power (-st). The integration eliminates the time variable so that the resulting transform is a function of s. The only requirement for the value of s is that the integral (equation 10.1) must be convergent as $t \to \infty$. If the integral is divergent then the particular time function has no Laplace transform.

Logarithms do not need to be derived from first principles each time they are used as they are readily available in books of tables. It is the same with Laplace transforms; they may be referred to in published works. Some of them are listed at the end of this chapter. However, the derivation of some of them will be examined before progressing to their use.

10.2.2 Derivation of a selection of Laplace transforms
1. $f(t) = \text{constant } (V)$.

$$\mathcal{L}V = \int_0^\infty Ve^{-st} \, dt \qquad \text{(using equation 10.1)}$$

$$= -\left[\frac{Ve^{-st}}{s}\right]_0^\infty = (0) - \left(-\frac{V}{s}\right) = \frac{V}{s} \, .$$

V/s is the Laplace transform of a constant voltage V suddenly applied to a circuit, i.e. a step function.

$$F(s) = \frac{V}{s} \, .$$

2. $f(t) = e^{at}$.

$$\mathcal{L}e^{at} = \int_0^\infty e^{at} e^{-st} \, dt = \int_0^\infty e^{(a-s)t} \, dt.$$

Now since at $t = \infty$, $e^{(a-s)t}$ must converge, s must be larger than a. The integral is therefore expressed as

$$\mathcal{L}e^{at} = \int_0^\infty e^{-(s-a)t} \, dt$$

$$= \left[\frac{e^{-(s-a)t}}{-(s-a)}\right]_0^\infty = (0) - \left(-\frac{1}{s-a}\right).$$

$$F(s) = \frac{1}{s-a} \, .$$

3. $f(t) = \sin \omega t$.

$e^{j\theta} = \cos \theta + j \sin \theta$ \quad (Euler's formula) \hfill (i)
$e^{-j\theta} = \cos \theta - j \sin \theta$. \hfill (ii)

Subtracting (ii) from (i)

LAPLACE TRANSFORMS AND THE SOLUTION OF CIRCUIT TRANSIENTS

$$e^{j\theta} - e^{-j\theta} = 2j \sin \theta$$

$$\sin \theta = \frac{e^{j\theta} - e^{-j\theta}}{2j}.$$

$$\mathcal{L}[\sin \omega t] = \int_0^\infty \left(\frac{e^{j\omega t} - e^{-j\omega t}}{2j}\right) e^{-st} \, dt \qquad \text{(writing } \omega t \text{ for } \theta\text{)}$$

$$= \int_0^\infty \frac{1}{2j}\left[\left(e^{-(s-j\omega)t} - e^{-(s+j\omega)t}\right)\right] dt$$

$$= \frac{1}{2j}\left[\frac{-e^{-(s-j\omega)t}}{(s-j)} + \frac{e^{-(s+j\omega)t}}{(s+j)}\right]_0^\infty$$

$$= \frac{1}{2j}\left[0 + 0 - \left(-\frac{1}{s-j\omega} + \frac{1}{s+j\omega}\right)\right]$$

$$= \frac{1}{2j}\left[\frac{2j\omega}{s^2 + \omega^2}\right] = \frac{\omega}{s^2 + \omega^2}.$$

$$\mathcal{L}[\sin \omega t] = F(s) = \frac{\omega}{s^2 + \omega^2}.$$

4. $f(t) = \sinh \omega t$.

Since $\sinh \omega t = \frac{1}{2}(e^{\omega t} - e^{-\omega t})$

$$\mathcal{L}[\sinh \omega t] = \int_0^\infty \frac{1}{2}(e^{\omega t} - e^{-\omega t})e^{-st} \, dt$$

$$= \frac{1}{2} \int_0^\infty e^{-(s-\omega)t} - e^{-(s+\omega)t} \, dt$$

$$= \frac{1}{2}\left[-\frac{e^{-(s-\omega)t}}{(s-\omega)} + \frac{e^{-(s+\omega)t}}{(s+\omega)}\right]_0^\infty$$

$$= \frac{1}{2}\left(\frac{1}{s-\omega} - \frac{1}{s+\omega}\right) = \frac{1}{2}\left[\frac{s+\omega - (s-\omega)}{s^2 - \omega^2}\right]$$

$$= \frac{\omega}{s^2 - \omega^2}.$$

$$\mathcal{L}[\sinh \omega t] = F(s) = \frac{\omega}{s^2 - \omega^2} .$$

5. The Laplace transform of derivatives

A knowledge of the method of integration by parts is required to understand this section.

Given two different functions of t, u and v

$$\frac{d}{dt}(uv) = v\frac{du}{dt} + u\frac{dv}{dt} \quad \text{(differentiation of a product)}.$$

Transposing gives

$$v\frac{du}{dt} = \frac{d}{dt}(uv) - u\frac{dv}{dt} .$$

Integrating both sides with respect to t and putting in the limits 0 and ∞

$$\int_0^\infty v\frac{du}{dt} dt = \left[uv - \int_0^\infty u\frac{dv}{dt} dt \right]_0^\infty . \tag{i}$$

Now if $y = f(t)$, the Laplace transform of its derivative, using the defining equation 10.1 is

$$\mathcal{L}\left[\frac{df(t)}{dt}\right] = \int_0^\infty e^{-st} \frac{df(t)}{dt} dt \tag{10.1(a)}$$

Let $e^{-st} = v$. Then $\frac{dv}{dt} = -se^{-st}$. (ii),(iii)

Let $\frac{df(t)}{dt} = \frac{du}{dt}$. Then $u = f(t)$. (iv)

Substituting (ii) and (iv) in equation 10.1(a): $\mathcal{L}\left[\frac{df(t)}{dt}\right] = \int_0^\infty v\frac{du}{dt} dt.$ (v)

Multiplying (ii) by (iv): $uv = e^{-st}f(t)$. (vi)

Multiplying (iii) by (iv): $\int_0^\infty u\frac{dv}{dt} dt = \int_0^\infty f(t)(-se^{-st}) dt$

$$= -s \int_0^\infty e^{-st} f(t) dt. \tag{vii}$$

Substituting (vi) for uv and (vii) for $\int_0^\infty u\frac{dv}{dt} dt$ in (i)

$$\int_0^\infty v\frac{du}{dt} dt = \left[e^{-st}f(t) - \left(-s\int_0^\infty e^{-st}f(t) dt\right)\right]_0^\infty.$$

But $\int_0^\infty v\frac{du}{dt} dt = \mathcal{L}\left[\frac{df(t)}{dt}\right]$ (equation (v)).

Therefore, the Laplace transform of the first derivative of $f(t)$ is

$$\left[e^{-st}f(t) + s\int_0^\infty e^{-st}f(t) dt\right]_0^\infty. \qquad \text{(viii)}$$

Now, in (viii), $\int_0^\infty e^{-st}f(t) dt$ should be recognised as the Laplace transform of $f(t)$. Therefore, $s\int_0^\infty e^{-st}f(t) dt = s\mathcal{L}f(t)$.

Also, $\left[e^{-st}f(t)\right]_0^\infty = 0 - f(t)_0$ (i.e. minus the value of the function at $t = 0$).

Therefore, $\mathcal{L}\left[\frac{df(t)}{dt}\right] = -f(t)_0 + s\mathcal{L}f(t)$.

In words, the Laplace transform of the first derivative of a function is equal to s times the Laplace transform of the function itself minus the value of the function itself at time zero.

In symbols: $\mathcal{L}\left[\frac{df(t)}{dt}\right] = \mathcal{L}f'(t) = sF(s) - f(o)$.

Using a similar method the Laplace transform for the second differential can be arrived at

$$\mathcal{L}\left[\frac{d^2f(t)}{dt^2}\right] = \mathcal{L}f''(t) = s\mathcal{L}f'(t) - f'(o)$$
$$= s^2\mathcal{L}f(t) - sf(o) - f'(o)$$
$$= s^2F(s) - sf(o) - f'(o).$$

In words, the Laplace transform of the second derivative of a function is equal to s^2 times the Laplace transform of the function itself, minus s times the value of the function at time zero, minus the value of the first differential of the function at time zero.

10.3 THE EXPONENTIAL FACTOR (SHIFTING THEOREM)

By definition (equation 10.1), $\mathcal{L}f(t) = F(s) = \int_0^\infty f(t)e^{-st}\,dt$.

Multiplying $f(t)$ by e^{-kt} gives

$$\mathcal{L}\left[f(t) \times e^{-kt}\right] = \int_0^\infty f(t) \times e^{-kt} \times e^{-st}\,dt$$

$$= \int_0^\infty f(t)e^{-(s+k)t}\,dt.$$

To examine what this means, suppose that $f(t)$ is a step function V. Then

$$\mathcal{L}\left[Ve^{-kt}\right] = \int_0^\infty Ve^{-(s+k)t}\,dt = \left[\frac{Ve^{-(s+k)t}}{-(s+k)}\right]_0^\infty$$

$$= \frac{V}{s+k}.$$

It has already been deduced that $\mathcal{L}V = V/s$ (section 10.2.2). The effect of multiplying V by e^{-kt} is to add a k to the s in the Laplace transform. This is a general theorem, a further example being the case of $\mathcal{L}\sin\omega t$.

$$\mathcal{L}[\sin\omega t] = \frac{\omega}{s^2 + \omega^2} \quad \text{and} \quad \mathcal{L}[\sin\omega t \times e^{-kt}] = \frac{\omega}{(s+k)^2 + \omega^2}.$$

10.4 EXAMPLES OF FINDING THE ORIGINAL FUNCTION FROM A GIVEN LAPLACE TRANSFORM (using the table of transforms at the end of this chapter)

1. $F(s) = \dfrac{3}{s+2} = 3\left(\dfrac{1}{s+2}\right)$. Find $f(t)$.

Laplace pair no. 3: $\mathcal{L}e^{-at} = \dfrac{1}{s+a}$.

LAPLACE TRANSFORMS AND THE SOLUTION OF CIRCUIT TRANSIENTS

Therefore, the inverse (working backwards) of $1/s + 2 = e^{-2t}$ (a = 2)

and of $3\left(\dfrac{1}{s+2}\right) = 3e^{-2t}$.

Therefore $f(t) = 3e^{-2t}$.

2. $F(s) = \dfrac{3s + 8}{s^2 + 4} = \dfrac{3s}{s^2 + 4} + \dfrac{8}{s^2 + 4}$. (i)

Laplace pair no. 6: $\mathcal{L}[\cos \omega t] = \dfrac{s}{s^2 + \omega^2}$.

Laplace pair no. 5: $\mathcal{L}[\sin \omega t] = \dfrac{\omega}{s^2 + \omega^2}$.

It is necessary to re-arrange (i) so that the denominators are the sum of s^2 and (a number)^2, thus

$$F(s) = 3\left(\dfrac{s}{s^2 + 2^2}\right) + 4\left(\dfrac{2}{s^2 + 2^2}\right).$$

Observe that in the second term it has been necessary to leave a 2 in the numerator to match that in the denominator in order that it is similar to Laplace pair no. 5.

With $\omega = 2$, $f(t) = 3 \cos 2t + 4 \sin 2t$.

3. $F(s) = \dfrac{2}{s^2 - 5^2} = 2\left(\dfrac{1}{s^2 - 5^2}\right)$.

Laplace pair no. 12: $\mathcal{L}[\sinh at] = \dfrac{a}{s^2 - a^2}$.

To give an inverse sinh at, $F(s)$ needs to have a 5 as its numerator to match that in the denominator.

Multiply inside the bracket by 5 and divide outside by 5 so that the overall value remains unchanged

$$F(s) = \dfrac{2}{5}\left(\dfrac{1 \times 5}{s^2 - 5^2}\right).$$

$f(t) = \dfrac{2}{5} \sinh 5t$ \qquad (a = 5).

4. $F(s) = \dfrac{s + 1}{s(s^2 + 4s + 8)}$.

The expression is divided into partial fractions

$$\dfrac{s + 1}{s(s^2 + 4s + 8)} = \dfrac{A}{s} + \dfrac{Bs + C}{s^2 + 4s + 8} \tag{i}$$

$$= \dfrac{A(s^2 + 4s + 8) + (Bs + C)s}{s(s^2 + 4s + 8)}$$

$$= \dfrac{As^2 + 4As + 8A + Bs^2 + Cs}{s(s^2 + 4s + 8)}.$$

Equating numerators

$$s + 1 = As^2 + 4As + 8A + Bs^2 + Cs.$$

For equality, each type of term must balance, numbers, s terms and s^2 terms, hence

s^2 terms: $0 = A + B$ (ii)
s terms: $1 = 4A + C$ (iii)
numbers: $1 = 8A$. (iv)

From (iv), $A = \dfrac{1}{8}$.

Substituting the value for A in (iii): $1 = 4\left(\dfrac{1}{8}\right) + C$

$$C = \dfrac{1}{2}$$

Substituting the value of A in (ii): $B = -\dfrac{1}{8}$

Substituting the values of A, B and C in (i):

$$F(s) = \dfrac{\tfrac{1}{8}}{s} + \dfrac{-\tfrac{1}{8}s + \tfrac{1}{2}}{s^2 + 4s + 8}.$$

From Laplace pair no. 1 the inverse for the first term is obtained

$$\mathcal{L}\frac{1}{8} = \frac{\frac{1}{8}}{s}, \text{ therefore } f(t) = \frac{1}{8} \qquad (v)$$

Now consider the second term. There are no transforms with fractional values of s in them. Multiply numerator and denominator by -8 to produce a single s

$$F(s) = -\frac{1}{8}\left[\frac{-8\left(-\frac{1}{8} + \frac{1}{2}\right)}{s^2 + 4s + 8}\right] = -\frac{1}{8}\left[\frac{(s-4)}{s^2 + 4s + 8}\right]. \qquad (vi)$$

To give a perfect inverse it is necessary to have a term similar to $(s^2 + \omega^2)$ or $((s+k)^2 + \omega^2)$ as the denominator.

Now $(s+2)^2 = s^2 + 4s + 4$
so that $(s^2 + 4s + 8) = s^2 + 4s + 4 + 4 = (s+2)^2 + 4$
$$= (s+2)^2 + 2^2.$$

Equation (vi) becomes: $F(s) = -\frac{1}{8}\left[\frac{(s-4)}{(s+2)^2 + 2^2}\right]. \qquad (vii)$

Laplace pair no. 8: $\mathcal{L}[e^{-kt} \cos \omega t] = \frac{s+k}{(s+k)^2 + \omega^2}.$

Laplace pair no. 7: $\mathcal{L}[e^{-kt} \sin \omega t] = \frac{\omega}{(s+k)^2 + \omega^2}.$

From Laplace pair no. 8, it is observed that to give a perfect inverse

$$F(s) = \frac{s+2}{(s+2)^2 + 2^2}.$$

In equation (vii) the numerator is $(s-4)$. This may be re-written as $(s+2) - 6$ without changing its value. Equation (vii) becomes

$$F(s) = -\frac{1}{8}\left[\frac{(s+2)}{(s+2)^2 + 2^2} - \frac{6}{(s+2)^2 + 2^2}\right]$$

$$= -\frac{1}{8}\left[\frac{(s+2)}{(s+2)^2+2^2} - 3\left(\frac{2}{(s+2)^2+2^2}\right)\right].$$

Therefore, $f(t) = -\frac{1}{8}(e^{-2t}\cos 2t - 3e^{-2t}\sin 2t)$ \hfill (viii)

Bringing this together with the first term (equation (v)) gives

$$f(t) = \frac{1}{8}(1 - e^{-2t}\cos 2t + 3e^{-2t}\sin 2t).$$

Similar methods apply to types 5, 6, 7 and 8 following (only the partial fractions are quoted).

5. $F(s) = \dfrac{E}{s(s+a)}$. The partial fractions are: $\dfrac{A}{s} + \dfrac{B}{(s+a)}$.

6. $F(s) = \dfrac{1}{(s+p)^2(s^2+q)}$. Partial fractions:
$\dfrac{A}{s+p} + \dfrac{B}{(s+p)^2} + \dfrac{Cs+D}{(s^2+q)}$.

7. $F(s) = \dfrac{1}{s^2(xs+y)}$. Partial fractions: $\dfrac{A}{s} + \dfrac{B}{s^2} + \dfrac{C}{xs+y}$.

8. $F(s) = \dfrac{E}{(s^2+p^2)(s+a)}$. Partial fractions: $\dfrac{A}{s+a} + \dfrac{Bs+C}{s^2+p^2}$.

10.5 CIRCUIT TRANSFORMS

10.5.1 Resistance

$v_R = iR$.

Taking Laplace transforms

$$\mathcal{L}v_R = \mathcal{L}iR$$

$$R = \frac{\overline{v_R}}{\overline{i}}$$

where $\overline{v_R} = \mathcal{L}v_R$ and $\overline{i} = \mathcal{L}i$ \hspace{1em} (Fig. 10.4(a)).

LAPLACE TRANSFORMS AND THE SOLUTION OF CIRCUIT TRANSIENTS

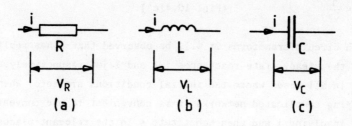

Fig. 10.4

10.5.2 Pure inductance

$$v_L = L \frac{di}{dt}.$$

Taking Laplace transforms

$$\mathcal{L}v_L = L(s\bar{i} - i_{(o)}) \quad \text{(Laplace pair no. 9)}$$

again writing $\bar{i}$ for $\mathcal{L}i = F(s)$. $i_{(o)}$ is the value of current at $t = 0$. If the current is zero at $t = 0$ then

$$\mathcal{L}v_L = Ls\,\bar{i}$$

$$\frac{\bar{v_L}}{\bar{i}} = sL \quad \text{(Fig. 10.4(b))}.$$

10.5.3 Pure capacitance

$$i = C\frac{dv}{dt}.$$

$$\mathcal{L}i = C(s\bar{v_C} - v_{(o)}) \quad \text{(Laplace pair no. 9)}$$

writing $\bar{v_C}$ for $\mathcal{L}v_C = F(s)$.
If the capacitor is initially uncharged, i.e. $v_{(o)} = 0$, then

$$\bar{i} = Cs\bar{v_C}$$

ELECTRICAL PRINCIPLES FOR HIGHER TEC

$$\frac{\bar{v}_C}{\bar{i}} = \frac{1}{Cs}$$ (Fig. 10.4(c)).

With circuit transforms it will be observed that s has replaced (jω) in the steady-state reactances jωL and 1/jωC respectively. This is true in all cases where the initial conditions are zero. When simplifying complicated networks it is convenient to use conventional methods involving j and then substitute s in the relevant places in the final expression for impedance when a transient solution is required.

When solving equations, $\bar{i}$ or $\bar{v}$ is written to indicate the value of F(s) being sought as it is more convenient than using $\mathcal{L}i$ or $\mathcal{L}v$.

<u>Worked example 10.1</u>   Derive expressions for the current and capacitor voltage in an RC circuit suddenly connected to a direct source of V volts.

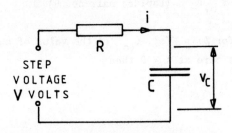

Fig. 10.5

$V = iR + v_C$.

Now $dq = C\, dv_C$, and transposing $dv_C = \frac{dq}{C} = \frac{i\, dt}{C}$. The voltage V is applied to the circuit at t = 0. The capacitor is initially uncharged.

After time t, capacitor voltage, $v_C = \int dv_C = \int_0^t \frac{i\, dt}{C}$.

Therefore, $V = iR + \int_0^t \frac{i\, dt}{C}$.

Taking Laplace transforms.

LAPLACE TRANSFORMS AND THE SOLUTION OF CIRCUIT TRANSIENTS

Laplace pair 1: $\mathcal{L}V = \dfrac{V}{s}$. Laplace pair 11: $\mathcal{L}\left[\displaystyle\int_0^t i\, dt\right] = \dfrac{\bar{i}}{s}$

(writing $\bar{i}$ for $\mathcal{L}i$).

$$\dfrac{V}{s} = \bar{i}R + \dfrac{\bar{i}}{Cs}$$

$$= \bar{i}\left(R + \dfrac{1}{Cs}\right).$$

This result could have been obtained directly by at first writing the solution in steady-state form

$$V = IR + \dfrac{I}{j\omega C} = I\left(R + \dfrac{1}{j\omega C}\right).$$

Then substituting the voltage and circuit transforms

$$\dfrac{V}{s} = \bar{i}\left(R + \dfrac{1}{Cs}\right).$$

Transposing to obtain $\bar{i}$ gives

$$\bar{i} = \dfrac{\dfrac{V}{s}}{R + \dfrac{1}{Cs}}$$

$$= \dfrac{\dfrac{V}{s}}{\dfrac{RCs + 1}{Cs}}$$

$$= \dfrac{VCs}{s(RCs + 1)} = \dfrac{VC}{(RCs + 1)}.$$

Inspecting the Laplace transform pairs will show that no transform exists with other than unit s or s^2. Divide the inside of the denominator bracket by RC and multiply the outside by RC to leave the value unchanged

$$\bar{i} = \dfrac{VC}{RC\left(s + \dfrac{1}{RC}\right)} = \dfrac{V}{R}\dfrac{1}{\left(s + \dfrac{1}{RC}\right)}.$$

Compare this with Laplace pair 3: $\mathcal{L}e^{-at} = \dfrac{1}{s + a}$, by letting $a = \frac{1}{RC}$, $i = $ inverse of $\left\{\frac{V}{R} \dfrac{1}{\left(s + \frac{1}{RC}\right)}\right\} = \frac{V}{R} e^{-t/RC}$.

But $v_C = \int_0^t \frac{i\,dt}{C}$. Substituting the value for i gives

$$v_C = \int_0^t \frac{\frac{V}{R} e^{-t/RC}}{C}\,dt$$

$$= \left[\frac{\frac{V}{R} e^{-t/RC}}{-C \times \frac{1}{RC}}\right]_0^t$$

$$= \left(-V e^{-t/RC} + V e^{-0}\right)$$

$$v_C = V\left(1 - e^{-t/RC}\right) \text{ volts.}$$

At $t = \infty$, $e^{-t/RC} = 0$ and $v_C = V$. The value of V is the final steady-state value of capacitor voltage. The steady-state solution is $v_C = V$, the supply voltage.

At any time t, v_C = steady-state voltage $- \left(V e^{-t/RC}\right)$.

Generally it may be stated that
condition at time t = final steady-state condition + transient response.

<u>Worked example 10.2</u>   A capacitor C carrying a charge q_o and with a standing voltage V volts is suddenly connected to a resistance R Ω at t = 0. Using Laplace transforms develop an expression for the current in the resistor at any time t. (The instantaneous capacitor voltage is v_C volts.)

At $t = 0$, $q_o = CV$, or transposing: $\dfrac{q_o}{C} = V$. \hfill (i)

Charge leaves the capacitor, flowing through the resistor.

LAPLACE TRANSFORMS AND THE SOLUTION OF CIRCUIT TRANSIENTS

Charge leaving = $\int_0^t i\, dt$ coulombs

After time t, charge remaining on the capacitor, $q = q_0 - \int_0^t i\, dt$ C (ii)

Since $q = Cv_C$, $v_C = \dfrac{q}{C}$. (iii)

Substituting (ii) in (iii)

$$v_C = \dfrac{q_0 - \int_0^t i\, dt}{C} = \dfrac{q_0}{C} - \int_0^t \dfrac{i\, dt}{C} \,.$$ (iv)

v_C is the potential difference at the capacitor terminals, and since it is connected to a resistor, it must be the p.d. across the resistor. Also,

p.d. across the resistor = iR volts

Therefore, $v_C = iR$. (v)

Substituting (v) and (i) in (iv) gives

$$iR = V - \int_0^t \dfrac{i\, dt}{C} \,.$$

Taking Laplace transforms of both sides (the individual transforms are as used in worked example 10.1)

$$\overline{iR} = \dfrac{V}{s} - \dfrac{\overline{i}}{Cs}$$

$$\dfrac{V}{s} = \overline{i}\left(R + \dfrac{1}{Cs}\right)$$

$$\overline{i} = \dfrac{V}{s\left(R + \dfrac{1}{Cs}\right)} \,.$$

This is identical to the expression found for the charging current in worked example 10.1 and the solution is

$$i = \dfrac{V}{R} e^{-t/RC} \text{ A.}$$

ELECTRICAL PRINCIPLES FOR HIGHER TEC

The final steady-state current is zero when the capacitor is fully discharged.

Current at time t = steady-state current + transient response.

<u>Self-assessment example 10.3</u> A coil with resistance R Ω and inductance L henry is suddenly connected to a direct voltage of V volts at time $t = 0$. Deduce (a) that the Laplace transform for the current at time t is given by the expression $\bar{i} = \dfrac{V}{(R + sL)s}$,

(b) that $i = \dfrac{V}{R}\left(1 - e^{-Rt/L}\right)$ A.

(Use partial fractions to find the inverse of $\bar{i}$, see section 10.4, equation type 5.)

<u>Worked example 10.4</u> A steady current I amperes is established in a coil with resistance R Ω and inductance L henry. At $t = 0$, the supply is disconnected and the input terminals are simultaneously short-circuited forming the circuit shown in Fig. 10.6. Deduce an expression for the current at any time t seconds after the terminals are short-circuited.

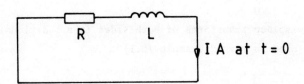

Fig. 10.6

When the coil is short-circuited, the current cannot suddenly reduce to zero. As it falls there will be an induced voltage given by

$$e = L\frac{di}{dt} \text{ V}$$

which by Lenz's law will attempt to maintain the current. At any instant, this driving voltage will be balanced by the iR voltage drop

LAPLACE TRANSFORMS AND THE SOLUTION OF CIRCUIT TRANSIENTS

in the resistance, Hence

$$L\frac{di}{dt} = iR.$$

Using Laplace pair 9: $\mathcal{L}\frac{di}{dt} = s\bar{i} - i_{(o)}$

$i_{(o)}$ is the value of current at $t = 0$. $i_{(o)} = IA$. Therefore

$$\mathcal{L}L\frac{di}{dt} = L(s\bar{i} - I).$$

Hence, $L(s\bar{i} - I) = \bar{i}R.$

Transposing

$$\bar{i}(sL - R) = IL$$

$$\bar{i} = \frac{IL}{sL - R} = \frac{I}{\left(s - \frac{R}{L}\right)}.$$

Taking inverses (Laplace pair 3): $i = I\, e^{-Rt/L}$.
The final steady-state current is zero.
The general forms for currents and voltages in inductive and capacitive circuits dealt with in examples 10.1 to 10.4 are shown in Fig. 10.7.

<u>Worked example 10.5</u>   Determine an expression for the current at time t seconds when an RL circuit is fed with a ramp voltage given by $v = kt$ volts. The initial conditions are zero.

Initially the circuit carries no current so the circuit transforms can be used directly.

Steady-state: $V = I(R + j\omega L)$ and $V = kt$.

$$\mathcal{L}kt = \frac{k}{s^2} \quad \text{(Laplace pair 4)}.$$

Therefore $\dfrac{k}{s^2} = \bar{i}(R + sL)$

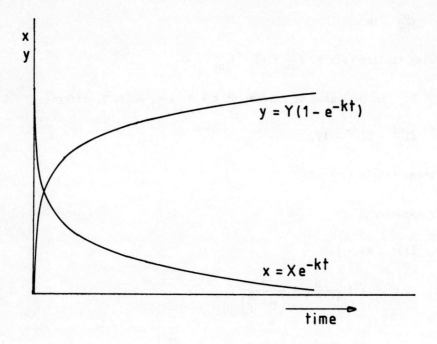

Fig. 10.7

$$\bar{i} = \frac{k}{s^2(R + sL)}.$$

Using partial fractions (section 10.4, type 7)

$$\bar{i} = \frac{k}{s^2(R + sL)} = \frac{A}{s} + \frac{B}{s^2} + \frac{C}{R + sL} \qquad \text{(i)}$$

$$= \frac{As(R + sL) + B(R + sL) + Cs^2}{s^2(R + sL)}.$$

Hence $k = ARs + ALs^2 + BR + BLs + Cs^2$.

Equating values of s^2: $\quad 0 = AL + C$ (ii)
Equating values of s: $\quad 0 = AR + BL$ (iii)
Equating numbers: $\quad k = BR$.

Therefore $B = \dfrac{k}{R}$. (iv)

Substituting (iv) in (iii): $0 = AR + \dfrac{kL}{R}$

hence $A = -\dfrac{kL}{R^2}$. (v)

Substituting (v) in (ii): $0 = -\dfrac{kL^2}{R^2} + C$

hence $C = \dfrac{kL^2}{R^2}$. (vi)

Substituting (iv), (v) and (vi) in (i)

$$i = \dfrac{-\dfrac{kL}{R^2}}{s} + \dfrac{\dfrac{k}{R}}{s^2} + \dfrac{\dfrac{kL^2}{R^2}}{(R + sL)}.$$

Taking inverses: $i = -\dfrac{kL}{R^2} + \dfrac{k}{R}t + \dfrac{kL}{R^2}e^{-Rt/L}$ amperes.

The form of the current is shown in Fig. 10.8.

<u>Self-assessment example 10.6</u>   Determine an expression for the input current to the parallel circuit shown in Fig. 10.9. The input voltage is a ramp given by $v = kt$. The initial conditions are zero.

There are two methods, both of which could be attempted to advantage.
(i) From worked example 10.5 the current in R and L is known for a ramp input. Determine an expression for the current in C alone and then add this to the former value.
(ii) The impedance of a parallel circuit is given by

$$Z_{eq} = \dfrac{Z_1 Z_2}{Z_1 + Z_2}.$$

Writing the impedances in, in their steady-state form, and then substituting s for (jω) gives

$$\mathcal{L}Z_{eq} = \dfrac{(R + sL)\dfrac{1}{Cs}}{R + sL + \dfrac{1}{Cs}} \quad \text{and} \quad \mathcal{L}Z = \mathcal{L}\dfrac{v}{i}.$$

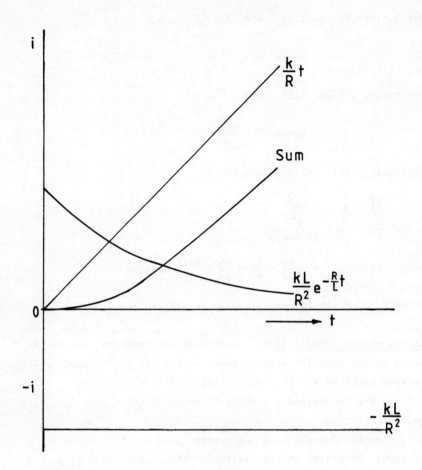

Fig. 10.8

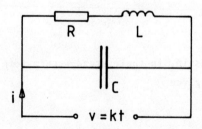

Fig. 10.9

After simplification and the use of partial fractions the following result is achieved

$$i = kC - \frac{kL}{R^2} + \frac{k}{R}t + \frac{kL}{R^2} e^{-Rt/L} \text{ amperes.}$$

10.6 THE DOUBLE-ENERGY SERIES CIRCUIT

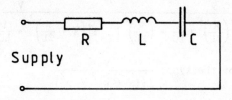

Fig. 10.10

A series circuit comprising R, L and C has two elements in which energy is stored and such a circuit can cause oscillations due to energy transfer from the capacitance to the inductance, and vice-versa. If there were no resistance, such oscillations, once created, would continue for ever. This condition has been achieved in super-conducting circuits.

The effect will now be examined of applying a step input voltage V to the circuit in Fig. 10.10 which is originally at zero conditions. Using the voltage and circuit transforms direct

$$\frac{V}{s} = \bar{i}\left(R + sL + \frac{1}{Cs}\right)$$

$$\bar{i} = \frac{V}{s\left(R + sL + \frac{1}{Cs}\right)}.$$

Taking L outside the denominator bracket and s inside the bracket gives:

$$\bar{i} = \frac{V}{L\left(s\frac{R}{L} + s^2 + \frac{1}{LC}\right)} = \frac{V}{L\left(s^2 + \frac{R}{L}s + \frac{1}{LC}\right)}.$$

By adding $\left(\frac{R}{2L}\right)^2$ to $\left(s^2 + \frac{R}{L}s\right)$, a perfect square is created since

$$\left[s^2 + \frac{R}{L}s + \left(\frac{R}{2L}\right)^2\right] = \left(s + \frac{R}{2L}\right)^2.$$

To maintain the value of the denominator the value $\left(\frac{R}{2L}\right)^2$ must be subtracted from $\frac{1}{LC}$, hence

$$\bar{i} = \frac{V}{L\left[s^2 + \frac{R}{L}s + \left(\frac{R}{2L}\right)^2 + \frac{1}{LC} - \left(\frac{R}{2L}\right)^2\right]} = \frac{V}{L\left[\left(s + \frac{R}{2L}\right)^2 + \frac{1}{LC} - \left(\frac{R}{2L}\right)^2\right]}.$$

This transform is of the type: $\dfrac{V}{L[(s+k)^2 + p^2]}$

where $p^2 = \dfrac{1}{LC} - \left(\dfrac{R}{2L}\right)^2$ or $p = \sqrt{\left[\dfrac{1}{LC} - \left(\dfrac{R}{2L}\right)^2\right]}$.

Therefore, $\bar{i} = \dfrac{V}{L\left[\left(s + \dfrac{R}{2L}\right)^2 + \left\{\sqrt{\left[\dfrac{1}{LC} - \left(\dfrac{R}{2L}\right)^2\right]}\right\}^2\right]}$. (10.2)

Equation 10.2 has four solutions

1. Undamped, when $R = 0$.
2. Underdamped when $\left(\dfrac{R}{2L}\right)^2$ is less than $\dfrac{1}{LC}$.
3. Critically damped when $\left(\dfrac{R}{2L}\right)^2 = \dfrac{1}{LC}$, or by transposing, $R = 2\sqrt{\left(\dfrac{L}{C}\right)}$.
4. Overdamped, when $\left(\dfrac{R}{2L}\right)^2$ is greater than $\dfrac{1}{LC}$.

Each of these solutions will be examined in turn.

1. Undamped. With $R = 0$ in equation 10.2,

$$\bar{i} = \frac{V}{L\left(s^2 + \dfrac{1}{LC}\right)}$$

from work on series resonance

LAPLACE TRANSFORMS AND THE SOLUTION OF CIRCUIT TRANSIENTS

$\omega_o = \sqrt{\left(\frac{1}{LC}\right)}$ (equation 9.1), hence $\frac{1}{LC} = \omega_o^2$

and $\bar{i} = \frac{V}{L(s^2 + \omega_o^2)}$.

To make a perfect inverse, ω_o is needed in the numerator. Multiply and divide by ω_o

$$\bar{i} = \frac{V}{\omega_o L}\left(\frac{\omega_o}{s^2 + \omega_o^2}\right).$$

From Laplace pair 5

$$i = \frac{V}{\omega_o L} \sin \omega_o t \text{ amperes.} \qquad (10.3)$$

This is a sine wave of amplitude $\frac{V}{\omega_o L}$ amperes and angular velocity ω_o rad/s (see Fig. 10.11).

2. Underdamped. $\left(\frac{R}{2L}\right)^2$ less than $\frac{1}{LC}$.

Equation 10.2 stands unchanged, and it may be seen to be in the form

$$\bar{i} = \frac{V}{L[(s + k)^2 + \omega^2]} \quad \text{(Laplace pair 7)}$$

$k = \frac{R}{2L}$ and $\omega = \sqrt{\left[\frac{1}{LC} - \left(\frac{R}{2L}\right)^2\right]}$.

It requires ω in the numerator for a perfect inverse. Multiply and divide by ω

$$\bar{i} = \frac{V}{\omega L}\left[\frac{\omega}{(s + k)^2 + \omega^2}\right].$$

$$i = \frac{V}{\omega L} e^{-kt} \sin \omega t.$$

Substituting for k and ω

$$i = \frac{V}{L\sqrt{\left[\frac{1}{LC} - \left(\frac{R}{2L}\right)^2\right]}} e^{-Rt/2L} \sin\sqrt{\left[\frac{1}{LC} - \left(\frac{R}{2L}\right)^2\right]} t \quad \text{A.} \tag{10.4}$$

This is an oscillatory current with angular velocity ω, which is decaying exponentially. The rate of decay is determined by the ratio of R to L (see Fig. 10.11).

3. Critically damped. $\left(\frac{R}{2L}\right)^2 = \frac{1}{LC}$.

Equation 10.2 reduces to

$$\bar{i} = \frac{V}{L\left(s + \frac{R}{2L}\right)^2}.$$

The inverse of $\frac{1}{s^2} = t$ (Laplace pair 4).

Adding R/2L to the s involves the exponential shift theorem (section 10.3) and this multiplies t by $e^{-Rt/2L}$. Therefore

$$\text{inverse of } \left\{\frac{1}{\left(s + \frac{R}{2L}\right)^2}\right\} = te^{-Rt/2L}$$

and inverse of $\left[\dfrac{V}{L\left(s + \dfrac{R}{2L}\right)^2}\right] = \dfrac{V}{L} te^{-Rt/2L}$ A. (10.5)

This is non-oscillatory (see Fig. 10.11).

4. Overdamped, $\left(\frac{R}{2L}\right)^2$ greater than $\frac{1}{LC}$.

In equation 10.2, $\frac{1}{LC} - \left(\frac{R}{2L}\right)^2$ will be negative and it is therefore necessary to write the second term in the denominator as

$$-\left\{\sqrt{\left[\left(\frac{R}{2L}\right)^2 - \frac{1}{LC}\right]}\right\}^2.$$

LAPLACE TRANSFORMS AND THE SOLUTION OF CIRCUIT TRANSIENTS

$$\bar{i} = \frac{V}{L\left[\left(s + \frac{R}{2L}\right)^2 - \left\{\sqrt{\left[\left(\frac{R}{2L}\right)^2 - \frac{1}{LC}\right]}\right\}^2\right]}$$

This is of the form $\dfrac{V}{L[(s+k)^2 - a^2]}$ (Laplace pair 14).

However, an a is needed in the numerator to make a perfect inverse. Multiplying and dividing by a gives

$$\bar{i} = \frac{Va}{aL[(s+k)^2 + a^2]}$$

where $a = \sqrt{\left[\left(\dfrac{R}{2L}\right)^2 - \dfrac{1}{LC}\right]}$ and $k = \dfrac{R}{2L}$.

Taking the inverse: $i = \dfrac{V}{aL} e^{-kt} \sinh at$

and substituting for a and k gives

$$i = \frac{V}{L\sqrt{\left[\left(\frac{R}{2L}\right)^2 - \frac{1}{LC}\right]}} e^{-Rt/2L} \sinh \sqrt{\left[\left(\frac{R}{2L}\right)^2 - \frac{1}{LC}\right]} \, t \text{ A}. \qquad (10.6)$$

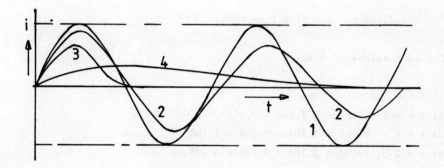

Fig. 10.11

Figure 10.11 shows typical response curves for the four conditions covered in this section (labelled 1 to 4).

373

ELECTRICAL PRINCIPLES FOR HIGHER TEC

Worked example 10.7   A circuit comprising a coil with resistance 1 Ω and inductance 0.5 H in series with a capacitor with capacitance 2000 μF is connected at t = 0 to a direct voltage of 50 V. Determine an expression for the current at time t and sketch the waveform over the first 2 seconds.

Firstly it is necessary to determine the state of damping of the circuit.

Evaluate $\left(\frac{R}{2L}\right)^2$: $\left(\frac{1}{2 \times 0.5}\right)^2 = 1$

Evaluate $\frac{1}{LC}$: $\frac{1}{0.5 \times 2000 \times 10^{-6}} = 1000$

Hence $\left(\frac{R}{2L}\right)^2$ is less than $\frac{1}{LC}$ and the circuit is underdamped.

For the underdamped circuit, the solution is given in equation 10.4

$$i = \frac{V}{L\sqrt{\left[\frac{1}{LC} - \left(\frac{R}{2L}\right)^2\right]}} e^{-Rt/2L} \sin\sqrt{\left[\frac{1}{LC} - \left(\frac{R}{2L}\right)^2\right]}t$$

$$= \frac{50}{0.5\sqrt{(1000 - 1)}} e^{-t} \sin\sqrt{(1000 - 1)}t$$

i = 3.164 e^{-t} sin 31.61t amperes.

Consider values of 3.164 e^{-t}:

at t = 0, value = 3.164
at t = 0.5 s, value = 3.164 × 0.6065 = 1.92
at t = 1 s, value = 3.164 × 0.368 = 1.164
at t = 2 s, value = 3.164 × 0.135 = 0.428.

During positive half-cycles, sin ωt is positive and the product (3.164 e^{-t} sin 31.61t) is positive. During negative half-cycles the product is negative. The current wave therefore lies within an envelope with height +3.164 e^{-t} (above the time axis) and -3.164 e^{-t}

(below the time axis).

At 31.61 rad/s, f = 5.031 Hz. The periodic time = 0.198 s. One quarter cycle takes 0.0497 s.

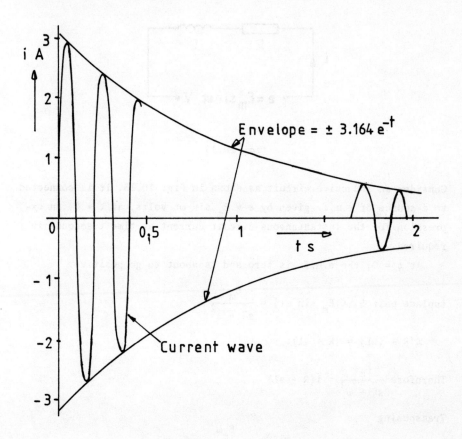

Fig. 10.12

The resultant current wave is shown in Fig. 10.12.

<u>Self-assessment example 10.8</u>   To the circuit used in worked example 10.7 is added a variable resistor connected in series with the coil.

Determine the value to which the resistor must be set to give critical damping.

Sketch the waveform of the current with this value of resistance in circuit when the supply is at a steady 25 V and is suddenly connected to the circuit at t = 0.

10.7 INDUCTIVE CIRCUITS SUPPLIED WITH A SINUSOIDAL VOLTAGE

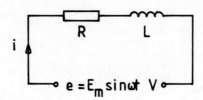

Fig. 10.13

Consider an inductive circuit as shown in Fig. 10.13. It is connected to a source of e.m.f. given by $e = E_m \sin \omega t$ volts, at $t = 0$. An expression for the instantaneous circuit current at time t seconds is required.

At $t = 0$, the e.m.f. is zero and is about to go positive.

Laplace pair 5: $\mathcal{L}[E_m \sin \omega t] = \dfrac{E_m \omega}{s^2 + \omega^2}$.

$\mathcal{L}[R + j\omega L] = (R + sL)$.

Therefore $\dfrac{E_m \omega}{s^2 + \omega^2} = \bar{i}(R + sL)$.

Transposing

$$\bar{i} = \dfrac{E_m \omega}{(s^2 + \omega^2)(R + sL)} = \dfrac{\frac{E_m \omega}{L}}{(s^2 + \omega^2)(s + \frac{R}{L})}.$$

Using partial fractions (section 10.4, type 8)

$$\dfrac{\frac{E_m \omega}{L}}{(s^2 + \omega^2)(s + \frac{R}{L})} = \dfrac{A}{s + \frac{R}{L}} + \dfrac{Bs + C}{s^2 + \omega^2} \quad \text{(i)}$$

$$\dfrac{E_m \omega}{L} = As^2 + A\omega^2 + Bs^2 + \dfrac{BR}{L}s + Cs + C\dfrac{R}{L}.$$

Equating values of s^2: $A + B = 0$, therefore $A = (-B)$. (ii)

Equating values of s: $\dfrac{BR}{L} + C = 0$ (iii)

Substituting (ii) in (iii): $-\dfrac{AR}{L} + C = 0$, $C = \dfrac{AR}{L}$. (iv)

Equating numbers: $A\omega^2 + C\dfrac{R}{L} = \dfrac{E_m \omega}{L}$. (v)

Substituting (iv) in (v): $A\omega^2 + \dfrac{AR^2}{L^2} = \dfrac{E_m \omega}{L}$

$$A = \dfrac{E_m \omega}{L\left(\omega^2 + \dfrac{R^2}{L^2}\right)} = \dfrac{E_m \omega}{\dfrac{R^2}{L} + \omega^2 L} = \dfrac{E_m \omega L}{R^2 + \omega^2 L^2}.$$

From (ii): $B = \dfrac{-E_m \omega L}{R^2 + \omega^2 L^2}$.

Substituting the value of B in (iii)

$$\left(\dfrac{-E_m L}{R^2 + \omega^2 L^2}\right)\dfrac{R}{L} + C = 0$$

$$C = \dfrac{E_m \omega R}{R^2 + \omega^2 L^2}$$

Substituting values for A, B and C in (i)

$$\bar{i} = \dfrac{\dfrac{E_m \omega L}{R^2 + \omega^2 L^2}}{s + \dfrac{R}{L}} - \dfrac{\left(\dfrac{E_m \omega L}{R^2 + \omega^2 L^2}\right)s + \dfrac{E_m \omega R}{R^2 + \omega^2 L^2}}{s^2 + \omega^2}$$

$$= \dfrac{E_m \omega}{R^2 + \omega^2 L^2}\left(\dfrac{L}{s + \dfrac{R}{L}} - \dfrac{sL}{s^2 + \omega^2} + \dfrac{R}{s^2 + \omega^2}\right).$$

Taking inverses:

$$i = \frac{E_m \omega}{R^2 + \omega^2 L^2}\left(Le^{-Rt/L} - L\cos\omega t + \frac{R}{\omega}\sin\omega t\right)$$

$$= \frac{E_m \omega}{R^2 + \omega^2 L^2} Le^{-Rt/L} + \frac{E_m \omega}{R^2 + \omega^2 L^2}\left(-L\cos\omega t + \frac{R}{\omega}\sin\omega t\right).$$

The term in brackets is the rectangular form of the phasor shown in Fig. 10.14.

The resultant $= \sqrt{\left[L^2 + \left(\frac{R}{\omega}\right)^2\right]} = \frac{1}{\omega}\sqrt{\left(\omega^2 L^2 + R^2\right)}$ and $\tan\phi = \frac{\omega L}{R}$.

$$i = \frac{E_m \omega}{R^2 + \omega^2 L^2} Le^{-Rt/L} + \frac{E_m \omega}{R^2 + \omega^2 L^2} \times \frac{1}{\omega}\sqrt{\left(\omega^2 L^2 + R^2\right)}\sin(\omega t - \phi) \text{ A}$$

$$= \frac{E_m \omega}{R^2 + \omega^2 L^2} Le^{-Rt/L} + \frac{E_m}{\sqrt{\left(R^2 + \omega^2 L^2\right)}}\sin(\omega t - \phi) \text{ A}. \qquad (10.7)$$

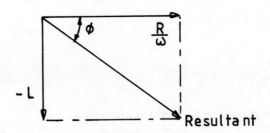

Fig. 10.14

<u>Worked example 10.9</u> A coil with resistance 1 Ω and inductance 15.9 mH is connected at time t = 0 to a voltage expressed by the equation v = 50 sin 314 t volts. Draw to scale the first two cycles of input current. The secondary is open-circuited (it has no effect).

$X_L = 2\pi fL = 314 \times 15.9 \times 10^{-3} = 5$ Ω. Therefore, $\omega^2 L^2 = 25$.

LAPLACE TRANSFORMS AND THE SOLUTION OF CIRCUIT TRANSIENTS

Substituting values in equation 10.7

$$i = \frac{50 \times 314}{1 + 25} \times 0.0159 \, e^{-1t/0.0159} + \frac{50}{\sqrt{(1^2 + 25)}} \sin(314t - \phi)$$

$$= 9.6 \, e^{-62.89t} + 9.805 \sin(314t - \phi) \text{ A}.$$

$$\phi = \arctan\left(\frac{314 \times 0.0159}{1}\right) = 78.67° = 1.373 \text{ rad}.$$

The following table can be drawn up

t	$9.6 \, e^{-62.89t}$	$9.805 \sin(314t - 1.373)$	i (A)
0	9.6	-9.6	0
0.005	7	1.923	8.923
0.0094	5.32	9.805	15.12 (peak value)
0.01	5.118	9.614	14.73
0.015	3.737	-1.922	1.814
0.02	2.73	-9.61	-6.88
0.025	2	1.923	3.923
0.03	1.455	9.614	11.069
0.035	1.06	-1.922	-0.862
0.04	0.77	-9.61	-8.8

From these results Fig. 10.15 is drawn.

After four or five cycles the transient disappears and the final steady-state sine wave current is left. It should be observed that the right-hand term in equation 10.7 is the normal steady-state solution for this circuit.

When an inductive circuit is energised from a sinusoidal supply, at the instant the voltage wave passes through a zero, the peak value of supply current is far greater than its final steady-state value. In the case of large power transformers this current inrush must be designed for, otherwise the mechanical forces involved will be disruptive.

Clearly the point in the voltage waveform at which the switch is

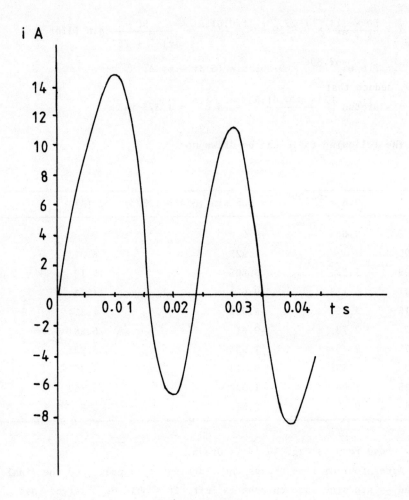

Fig. 10.15

closed is a matter of chance, so that it is worth examining what would happen if the switch were closed when the voltage wave was at a maximum value.

Self-assessment example 10.10 A resistive inductive circuit as in Fig. 10.14 is energised from a sinusoidal alternating supply at the instant the voltage wave is at a maximum value. Therefore at $t = 0$, $e = E_m$. Hence $e = E_m \cos \omega t$ volts. Deduce the following expressions:

(a) $\bar{i} = \dfrac{\dfrac{E_m s}{L}}{(s^2 + \omega^2)(s + \dfrac{R}{L})}$.

(b) Simplifying (a) using partial fractions as in worked example 10.9, deduce that

$$\bar{i} = \dfrac{\dfrac{-E_m R}{\omega^2 L^2 + R^2}}{s + \dfrac{R}{L}} + \dfrac{\dfrac{E_m R s}{\omega^2 L^2 + R^2} + E_m \left(\dfrac{\omega^2 L}{\omega^2 L^2 + R^2}\right)}{s^2 + \omega^2}$$

(c) $i = \dfrac{-E_m R}{\omega^2 L^2 + R^2} e^{-Rt/L} + \dfrac{E_m R}{\omega^2 L^2 + R^2} \cos \omega t + \dfrac{E_m \omega L}{\omega^2 L^2 + R^2} \sin \omega t$

$= \dfrac{-E_m R}{\omega^2 L^2 + R^2} e^{-Rt/L} + \dfrac{E_m}{\sqrt{(R^2 + \omega^2 L^2)}} \sin(\omega t + \phi)$

where $\phi = \arctan \dfrac{R}{\omega L}$.

The form of the resulting current calculated for the circuit parameters in worked example 10.9 is shown in Fig. 10.16. It will be noted that almost perfect symmetry of current waveform is achieved immediately and therefore the ideal time at which to energise an inductive circuit is as the applied voltage passes through a maximum value.

The instant of closing a switch is arbitrary so that conditions vary between the best and worst. When energising a transformer a transient hum is often heard for the first 5–10 cycles indicating that the instant of closing the switch was not ideal. When energising a three-phase transformer, since the voltage waveforms are displaced by 120°, at least two phases will be energised at such an instant.

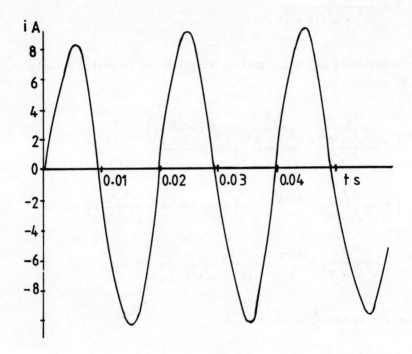

Fig. 10.16

Table for self-assessment example 10.10

$$i = \frac{-50}{1 + 25} e^{-62.89t} + \frac{50}{\sqrt{(1 + 25)}} \sin(\omega t + 0.197)$$

$$= 1.92 \, e^{-62.89t} + 9.805 \sin(314t + 0.197).$$

t	$-1.92 \, e^{-62.89t}$	$9.805 \sin(314t + 0.197)$	i (A)
0	-1.92	1.92	0
0.0044	-1.46	9.805	8.335
0.005	-1.402	9.614	8.216
0.01	-1.024	-1.92	-2.946
0.015	-0.747	-9.614	-10.36
0.02			1.375
0.025			9.216

LAPLACE TRANSFORMS AND THE SOLUTION OF CIRCUIT TRANSIENTS

Table for self-assessment example 10.10 (cont.)

t	$-1.92\, e^{-62.89t}$	$9.805 \sin(314t + 0.197)$	i (A)
0.03			-2.21
0.035			-9.826
0.04			1.765
0.045			9.5
0.05			-2.0
0.055			-9.67

10.8 TRANSFER FUNCTION

When a network has a pair of input terminals and a pair of output terminals, its transfer function is defined as the ratio of output to input expressed in Laplace form, for a particular loading.

The open-circuit voltage transfer function = $\dfrac{\overline{v_{out}}}{\overline{v_{in}}}$ with the output open-circuited.

The short-circuit current transfer function = $\dfrac{\overline{i_{out}}}{\overline{i_{in}}}$ with the output short-circuited.

<u>Worked example 10.11</u>   Determine the open-circuit voltage transfer function for the network shown in Fig. 10.17.

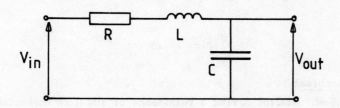

Fig. 10.17

$$\mathcal{L}V_{in} = \mathcal{L}iZ = \overline{i}\left(R + sL + \frac{1}{Cs}\right)$$

$$\mathcal{L}V_{out} = \mathcal{L}iC = \bar{i} \times \frac{1}{Cs} .$$

$$\text{Transfer function} = \mathcal{L}\left[\frac{v_{out}}{v_{in}}\right] = \frac{\bar{i}\frac{1}{Cs}}{\bar{i}\left(R + sL + \frac{1}{Cs}\right)}$$

$$= \frac{1}{Cs\left(R + sL + \frac{1}{Cs}\right)}$$

$$= \frac{1}{s^2LC + CRs + 1}$$

$$= \frac{1}{LC\left(s^2 + \frac{R}{L}s + \frac{1}{LC}\right)} .$$

<u>Self-assessment example 10.12</u> Show that the open circuit voltage transfer function of the filter network shown in Fig. 10.18 is given by

$$\frac{1}{LC\left(s^2 + \frac{1}{LC}\right)} .$$

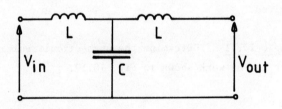

Fig. 10.18

Further problems

10.13. A 6 mH inductor having a resistance of 100 Ω is connected in series with a 0.05 μF capacitor. The circuit so formed is suddenly connected to a direct source of 20 V at t = 0. Given that the initial conditions are zero, show that the current in the circuit t s after connecting the supply, can be expressed in the form

$$i = I\, e^{-kt} \sin \omega t \text{ A.}$$

Calculate the values of I, k and ω.

10.14. A capacitor with capacitance 0.15 µF in series with a 600 Ω resistor is switched on to a voltage given by v = 100t at t = 0. The capacitor is initially uncharged. Deduce an expression for the current at time t seconds after switching on. Using the expression for current, determine an expression for the capacitor voltage at time t seconds.

What is the value of circuit current at t = 15 µs?

10.15. A capacitor is charged from a direct source to a voltage of V_o volts. It is then discharged into a resistive coil. Using Laplace transforms deduce a general expression for the circuit current at any time t seconds after the capacitor has been connected to the coil.

Given that C = 0.2 µF, L = 0.5 H and R = 5 Ω, evaluate the general expression in terms of V_o and the circuit parameters.

10.16. A ramp voltage v = kt is applied at t = 0 to the circuit shown in Fig. 10.19. Given that the capacitor is initially uncharged and the inductor current is zero at t = 0, evaluate the value of the input current at t = 0.04 second given that R_1 = 10 Ω, R_2 = 1000 Ω, L = 0.35 H, C = 0.1 µF and k = 100.

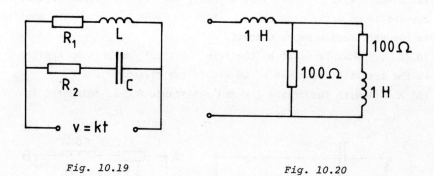

Fig. 10.19   Fig. 10.20

10.17. A series circuit consisting of a resistor with value 1 Ω and an inductor with inductance 20 mH is connected to a 240 V r.m.s., 50 Hz, sinusoidal supply at the instant the voltage wave is passing through a zero and is about to become positive.

Derive an expression for the instantaneous current and sketch the form of the first five cycles of the current.

10.18. The circuit shown in Fig. 10.20 is connected at t = 0 to a

10 volt direct supply. Deduce an expression for the input current at any time t seconds after connection.

10.19. A coil has resistance R Ω and inductance L H. A step voltage V volts is applied to it.

Establish and solve the equation for the current at any instant t seconds after the voltage is applied.

If R = 5 Ω, L = 5.5 H and V = 10 V, calculate (a) the value of current 0.6 seconds after the voltage is applied, (b) the time taken for the current to reach 1.86 A, (c) the time taken for the current to decay to 0.25 A if, after some considerable time, the coil is shorted-out using a 10 Ω resistor and the supply is simultaneously removed.

10.20. A 50 μF capacitor is charged to 500 V and then connected to a coil of inductance 1 H and resistance 50 Ω. Derive an expression for the current at time t seconds after the connection. Sketch the form of the current wave.

10.21. (a) Derive an expression for the open-circuit voltage transfer function for the circuit shown in Fig. 10.21.

(b) Working from first principles determine (i) the value of resistance which will give critical damping, and (ii) the value of circuit current 10 ms after a constant voltage v_{in} = 50 V is suddenly applied to the critically-damped circuit.

10.22. (a) What is meant by the term 'critical damping' as applied to the transient response of an electrical circuit?

(b) A coil with inductance L H and resistance R Ω is connected in

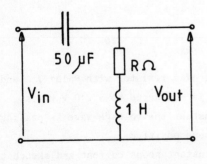

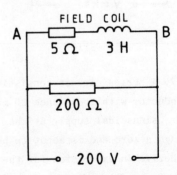

Fig. 10.21   Fig. 10.22

series with a capacitor of capacitance C F and a variable resistor R_X Ω. Deduce an expression for the value of R_X to give critical damping. Calculate the value of R_X given that L = 1.5 H, C = 50 μF and R = 50 Ω.

10.23. The field coil of a d.c. motor is shunted by a 200 Ω arc-quenching resistor as shown in Fig. 10.22. Determine (a) an expression for the current in the field coil at time t seconds after connection to the 200 V supply, (b) the total current taken by the parallel components 0.5 seconds after switching on, (c) the peak voltage V_{AB} at the instant the supply is switched off, (d) the time taken for the coil current to fall to 10% of its normal running value after the supply is switched off; both (c) and (d) being considered when steady-state conditions have been achieved after switching on.

10.24. A transformer primary winding with resistance 2 Ω and inductance 6.6 H is connected to a 50 Hz, 440 V, sinusoidal supply at the instant the voltage wave is a maximum. Determine the value of the first current peak in the primary winding.

10.25. The transformer in problem 10.24 is energised as the voltage wave passes through a zero. Determine the value of the first current peak in the primary winding.

Table of a selection of Laplace transforms

Laplace Pair Text reference number	f(t)	Form of f(t)	$\mathcal{L}f(t) = F(s)$
1	V		$\dfrac{V}{s}$
2	e^{at}		$\dfrac{1}{s-a}$
3	e^{-at}		$\dfrac{1}{s+a}$
4	kt		$\dfrac{k}{s^2}$
5	$\sin \omega t$		$\dfrac{\omega}{s^2 + \omega^2}$
6	$\cos \omega t$		$\dfrac{s}{s^2 + \omega^2}$
7	$e^{-kt} \sin \omega t$		$\dfrac{\omega}{(s+k)^2 + \omega^2}$

LAPLACE TRANSFORMS AND THE SOLUTION OF CIRCUIT TRANSIENTS

Laplace Pair Text reference number	$f(t)$	Form of $f(t)$	$\mathcal{L}f(t) = F(s)$
8	$e^{-kt} \cos \omega t$	(damped cosine waveform)	$\dfrac{s + k}{(s + k)^2 + \omega^2}$
9	$\dfrac{df(t)}{dt}$		$sF(s) - f_{(o)}$
10	$\dfrac{d^2 f(t)}{dt^2}$		$s^2 F(s) - sf_{(o)} - f_{(o)}$
11	$\int_0^t f(t)dt$		$\dfrac{F(s)}{s}$
12	$\sinh at$		$\dfrac{a}{s^2 - a^2}$
13	$\cosh at$		$\dfrac{s}{s^2 - a^2}$
14	$e^{-kt} \sinh at$		$\dfrac{a}{(s + k)^2 - a^2}$
15	$e^{-kt} \cosh at$		$\dfrac{s + k}{(s + k)^2 - a^2}$

CHAPTER 10   ANSWERS

10.8. R_{total} = 31.6 Ω, add 30.6 Ω. Current: t = 0, i = 0; t = 0.02 s, i = 0.53 A; t = 0.05 s, i = 0.514 A; t = 0.1 s, i = 0.211 A; t = 0.2 s, i = 0.018 A; t greater than 0.5 s, i = 0

10.13. I = 0.058 A; k = 8333; ω = 57.1 rad/s

10.14. $i = 0.15 \times 10^{-4} (1 - e^{-11111t})$ A; $v_C = 100(t + 9 \times 10^{-5}(e^{-11111t} - 1))$ V; i = 2.3 μA

10.15. General expression as equation 10.2. Circuit underdamped. Required answer as equation 10.4 with values:

$$i = \frac{V_o}{1.58 \times 10^3} e^{-5t} \sin 3162t \text{ A}$$

10.16. Inductive branch: $i = 100\left(\frac{-0.35}{10^2} + \frac{0.04}{10} + \frac{0.35}{100} e^{-10/0.35 \times 0.04}\right)$

$= 1.62 \times 10^{-3}$ A

Capacitive branch: $i = 10(1 - e^{-10^4 \times 0.04}) = 10$ μA

Total current 1.63 mA

10.17. $i = 53.5 \sin(\omega t - 80.95°) + 52.96 e^{-50t}$ A

10.18. $i = 0.2 - 0.2 e^{-150t} \cosh 111.8t - \frac{20}{111.8} \sinh 111.8t$.

This simplifies to: $i = 0.2 - 0.1895 e^{-38.2t} - 0.0105 e^{-261.8t}$ A

10.19. (a) 0.84 A   (b) 2.925 s   (c) 0.76 s (don't forget the additional resistance)

10.20. $i = 3.59 e^{-25t} \sin 139.2t$ A

10.21. (a) $\dfrac{R + sL}{R + sL + \dfrac{1}{Cs}}$ (b) (i) 282.8 Ω (ii) 0.1215 A

10.22. (b) 296.4 Ω

10.23. (a) $\dfrac{V}{R}(1 - e^{-Rt/L})$ (b) 12.3 A   (c) 4000 V   (d) 0.034 s

10.24. 0.3 A (closely)

10.25. 0.6 A (closely)

11 Transmission Lines

11.1 POWER TRANSMISSION

11.1.1 Short lines

A three-phase transmission or distribution system is fed at its sending end by a star-connected transformer or generator, the star point of either being connected to earth. Power is transmitted on three wires.

For a star-connected system, the phase and line currents are identical.

$$I_{ph} = I_L$$

The phase voltage is equal to the line voltage divided by the square root of three.

$$V_{ph} = \frac{V_L}{\sqrt{3}}$$

The values of capacitance between lines and between each line and earth of a three-phase overhead line are small per unit length since the spacings between lines are large and so, up to a few kilometres in length, these may be ignored. Only the series resistance and inductance are considered. Calculations are carried out per phase considering the system to be symmetrical so that the line currents in each phase are identical as are the phase (and line) voltages. For asymmetric loading a more complicated analysis is involved using what are called 'symmetrical components'.

One phase of a three-phase system is shown in Fig. 11.1.

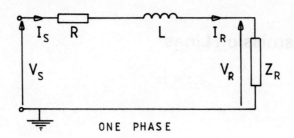

Fig. 11.1

V_S and V_R = phase voltages at the sending end and receiving end, respectively.
I_S and I_R = phase (and line) currents at the sending end and receiving end, respectively.
ϕ_S = phase angle at the sending end.
Z_R and ϕ_R = impedance and phase angle of the load.
R = resistance per phase of the transmission line.
L = series inductance per phase of the transmission line.

Referring to chapter 8:

equation 8.1: $V_S = AV_R + BI_R$,   equation 8.2: $I_S = CV_R + DI_R$.

For a network with series impedance only:

equations 8.3 and 8.4: A = 1, B = Z, C = 0 and D = 1.

<u>Worked example 11.1</u>   A three-phase, 50 Hz, 11 kV distribution line 10 km in length has a conductor resistance of 0.155 Ω/km/phase and a series inductance of 1.2 mH/km/phase. A load of 2 MW at 0.75 power factor lagging is supplied at the receiving end of the line at 11 kV. Calculate (a) the sending-end voltage, (b) the sending end power factor, and (c) the transmission efficiency.

The phase voltage at 11 kV line = $\dfrac{11000}{\sqrt{3}}$ = 6350.8 volts (= V_R).

TRANSMISSION LINES

Power in a three-phase system = $\sqrt{3} V_L I_L \cos\phi$ watts.

Considering the receiving end

$$\sqrt{3} \times 11000 \times I_L \times 0.75 = 2 \times 10^6.$$

Therefore, $I_L \; (= I_R) = \dfrac{2 \times 10^6}{\sqrt{3} \times 11000 \times 0.75} = 140$ A.

Taking the receive-end voltage as reference: $V_R = 6350.8 \underline{/0°}$ volts, I_R lags V_R by an angle whose cosine is 0.75. Hence $\phi_R = -41.4°$.

$$I_R = 140\underline{/-41.4°} \text{ A.}$$

The four-terminal network parameters are now required.

$A = D = 1$
$C = 0$
$B = Z$

Now, the impedance per kilometre = $(0.155 + j\, 2\pi\, 50 \times 1.2 \times 10^{-3})$
$= (0.155 + j0.377)\; \Omega.$

Therefore, for a length of 10 km, $B = 10(0.155 + j0.377)$
$= 1.55 + j3.77 = 4.076\underline{/67.65°}\; \Omega.$

(a) $V_S = AV_R + BI_R = 1 \times 6350.8\underline{/0°} + 4.076\underline{/67.65°} \times 140\underline{/-41.4°}$
$= 6350.8 + 570.64\underline{/26.25°}$
$= 6350.8 + 511.79 + j252.38$
$= 6862.6 + j252.38 = 6867.2\underline{/2.1°}$ V (phase)

Sending-end line voltage = $\sqrt{3} \times 6867.2 = 11894$ V.

(b) Since there are no shunt components, $I_S = I_R$. By inspection of Fig. 11.2, it is seen that the angle between I_S and V_S is $(41.4 + 2.1)° = 43.5°$. Therefore, sending-end power factor = $\cos 43.5° = 0.725$.

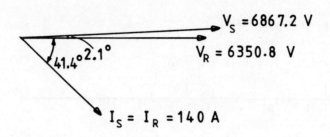

Fig. 11.2

(c) Sending-end power = $\sqrt{3} V_L I_L \cos\phi$ watts (using sending-end quantities).
$\qquad\qquad\qquad\;\; = \sqrt{3} \times 11894 \times 140 \times 0.725$
$\qquad\qquad\qquad\;\; = 2.091$ MW.

Efficiency = $\dfrac{\text{output}}{\text{input}} = \dfrac{2 \text{ MW}}{2.091 \text{ MW}} = 0.956$ p.u.

<u>Self-assessment example 11.2</u>   A three-phase, 50 Hz transmission line 18 km long has a conductor resistance of 0.2 Ω/km/phase and a series inductance of 1.47 mH/km/phase. Determine the sending-end voltage and its phase with respect to the receiving-end voltage when the line supplies a load of 10 MW at 33 kV and unity power factor.

11.1.2 Medium length lines

Increasing the length of an overhead transmission line increases the overall capacitance and also, since for larger distances higher voltages are used, the capacitive or charging current becomes significant and has to be considered. In addition, at higher voltages the leakage current across the insulators may also become significant.

For underground power cables of whatever length, since the cores are generally close together the capacitance is large and must be considered. The leakage current through the insulation may also have an appreciable effect. (See chapter 5, sections 5.13.2 and 5.13.3 on dielectric losses.)

For overhead power lines up to a few hundred kilometres in length and short lengths of interconnecting underground cables, calculations are generally carried out with reasonable accuracy (within 10%) by

TRANSMISSION LINES

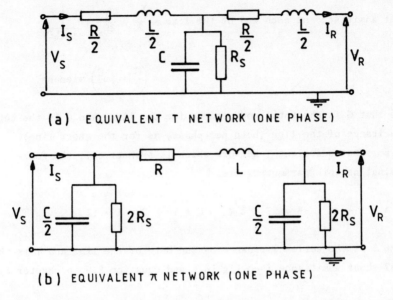

Fig. 11.3

considering line models made up with 'lumped' constants.

In Fig. 11.3, R and L, V_R, V_S, I_R and I_S are as defined for the short line.

C = total effective capacitance to earth of one line (see also, chapter 2, worked example 2.13).
R_s = total shunt resistance to earth of one line.

In the equivalent T-network (Fig. 11.3(a)), the total capacitance and shunt resistance are considered to act midway along the length of the line. On either side of these two components is one half of the line resistance and one half of the line inductance.

Shunt admittance, $Y_s = \dfrac{1}{R_s} + \dfrac{1}{\dfrac{1}{j\omega C}} = (G + j\omega C)$ siemens. (11.1)

In the equivalent π-network, one half of the total capacitance is considered to act at each end of the line. The leakage resistance associated with it is doubled so giving one half of the total leakage current at each end of the line. This is shown in Fig. 11.3(b).

Shunt admittance at each end of the line = $\dfrac{Y_s}{2} = \dfrac{1}{2R_s} + \dfrac{1}{\dfrac{1}{j\omega(\tfrac12 C)}}$

$$= \left(\dfrac{G}{2} + j\omega\dfrac{C}{2}\right) \text{ siemens.}$$

Note that G is the total conductance for the line, and C is the total capacitance of the line (both per-phase, as for the short line).

For the equivalent T-network which is symmetrical, the four-terminal network parameters are

$$A = 1 + YZ \qquad B = 2Z + Z^2 Y \qquad C = Y \qquad D = 1 + YZ.$$

Where Z is the series impedance of ONE HALF of the line and Y is the total shunt admittance (see Fig. 8.6 and equation 8.9 in chapter 8).

<u>Worked example 11.3</u>   The per-phase resistance, reactance and capacitive susceptance of a three-phase transmission line are 20 Ω, 50 Ω and $4 \times 10^{-4} \underline{/90°}$ S respectively. The load at the receiving end of the line is 60 MW at 132 kV line and 0.85 power factor lagging. Considering each phase as a T-network, determine the four-terminal network parameters of the line and hence determine the sending-end power input.

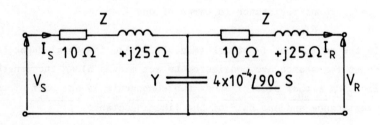

Fig. 11.4

$V_R = \dfrac{132000}{\sqrt{3}} = 76210$ V.

TRANSMISSION LINES

Receiving-end power = $\sqrt{3} V_L I_L \cos\phi$ watts.
Using receiving-end values

$$60 \times 10^6 = \sqrt{3} \times 132000 \times I_R \times 0.85$$
$$I_R = 308.7 \text{ A}.$$

Using V_R as reference, $V_R = 76210\underline{/0^\circ}$ V.
$$I_R = 308.7\underline{/\arccos 0.85} = 308.7\underline{/-31.79^\circ} \text{ A}$$

(angle is negative since current lags voltage).

To determine the four-terminal network parameters

$$A = D = 1 + YZ = 1 + 4 \times 10^{-4}\underline{/90^\circ} \times (10 + j25)$$
$$= 1 + 4 \times 10^{-4}\underline{/90^\circ} \times 26.93\underline{/68.2^\circ}$$
$$= 1 + 0.01077\underline{/158.2^\circ}$$
$$= 1 - 0.01 + j4 \times 10^{-3}$$
$$A = D = 0.99 + j4 \times 10^{-3}$$

$$B = 2Z + Z^2 Y = 2(10 + j25) + (10 + j25)^2 \times 4 \times 10^{-4}\underline{/90^\circ}$$
$$= (20 + j50) + (26.93\underline{/68.2^\circ})^2 \times 4 \times 10^{-4}\underline{/90^\circ}$$
$$= (20 + j50) + 0.29\underline{/226.4^\circ}$$
$$= 20 + j50 - 0.2 - j0.21$$
$$B = 19.8 + j49.79$$

$$C = Y = 4 \times 10^{-4}\underline{/90^\circ}$$

Substituting these values in $V_S = AV_R + BI_R$ gives

$$V_S = (0.99 + j4 \times 10^{-3})(76210\underline{/0^\circ}) + (19.8 + j49.79)(308.7\underline{/-31.79^\circ})$$
$$= 0.99\underline{/0.23^\circ} \times 76210\underline{/0^\circ} + 53.58\underline{/68.31^\circ} \times 308.7\underline{/-31.79^\circ}$$
$$= 75447.9 + j302.9 + 13292.4 + j9844.5$$
$$= 88740 + j10147.4 = 89318.3\underline{/6.52^\circ} \text{ V } (\times \sqrt{3} = 154703 \text{ V line}).$$

Substituting these values in $I_S = CV_R + DI_R$ gives

$$I_S = 4 \times 10^{-4}\underline{/90^\circ} \times 76210\underline{/0^\circ} + 0.99 + j4 \times 10^{-3} \times 308.7\underline{/-31.79^\circ}$$
$$= j30.48 + 260.4 - j159.95$$
$$= 290.8\underline{/-26.44^\circ} \text{ A}.$$

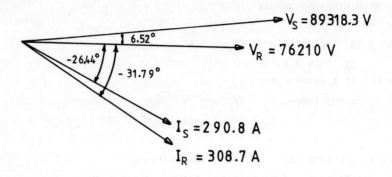

Fig. 11.5

The various quantities are shown in Fig. 11.5.

Angle between V_S and I_S = $26.44°$ + $6.52°$ = $32.96°$.

Input power = $\sqrt{3} V_L I_L \cos \phi$ watts (using input values)
= $\sqrt{3} \times 154703 \times 290.8 \times \cos 32.96°$
= 65.38 MW.

<u>Self-assessment example 11.4</u>   (a) Deduce the values of the four-terminal network parameters A, B, C and D for a symmetrical π-configuration with a total series impedance Z Ω and two shunt admittances each equal to Y/2 siemens.

(b) A 160 km length of 132 kV, 3-phase, 50 Hz transmission line has total conductor resistance of 29.5 Ω/phase, inductive reactance of 66.5 Ω/phase and capacitance of 1.44 μF/phase. Leakage current may be ignored. Using the π-representation, determine the A, B, C and D parameters of the line. Hence calculate the required value of the sending-end voltage for a receiving-end power of 50 MVA at 132 kV and 0.8 power factor lagging.

11.2   TRANSMISSION LINES FOR HIGH FREQUENCIES
From worked example 11.3 and self-assessment example 11.4 it should be observed that there is a phase change between the sending-end voltage V_S, and the receiving-end voltage V_R. In an overhead power

line this is generally small, an extremely long length being required to produce a change of $90°$. According to spacing this length lies between 2500 km and 6000 km.

The length of line which would be required to cause the receiving-end voltage to be exactly one complete cycle behind the input voltage is known as a wavelength.

Where high frequencies are involved a wavelength may only be a fraction of a metre and in problems where lines have lengths which are significant compared with a wavelength and especially where small powers are involved so that the criteria for maximum power transfer must be considered, the lumped-parameter approach is not sufficiently accurate. The resistance, reactance and capacitance must be taken as being uniformly distributed along the line.

11.2.1 The transmission line equations

Consider a transmission line as shown in Fig. 11.6. It has series impedance $(R + j\omega L)$ Ω per unit length and shunt resistance and capacitance R_s and C respectively per unit length. It is supplied with voltage V_S from an alternating source and the input current is I_S. The receiving-end voltage and current are V_R and I_R respectively.

At a distance x from the sending end an element of line δx in length is considered in which the current is I amperes. At this point the potential difference between line and earth is V volts.

Figure 11.6(a) shows the line complete, length ℓ m. Figure 11.6(b) shows the element of line δx in length. It is represented as a T-equivalent. The impedance of the line, $Z = (R + j\omega L)$ Ω per unit length, so that for length δx, the series impedance = $(R + j\omega L)\delta x$ Ω = $Z\delta x$ Ω. Similarly the shunt components are $R_s \delta x$ and $C\delta x$ for the element.

An expression for the characteristic impedance Z_o of the line
(See chapter 8, section 8.8, for a definition of Z_o.)
Consider the small element of line δx in Fig. 11.6(b). Using the iterative method of determination, looking in at terminals AC, the impedance must be equal to Z_o when terminals BD are connected to an impedance Z_o. Hence, in Fig. 11.7, $\frac{Z\delta x}{2}$ is in series with a parallel combination of the shunt impedance $\frac{1}{Y_s \delta x}$ and $\left(\frac{Z\delta x}{2} + Z_o\right)$ Ω.

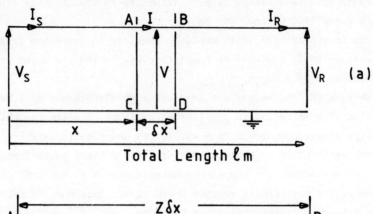

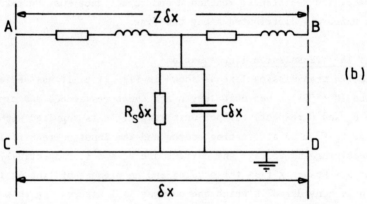

Fig. 11.6

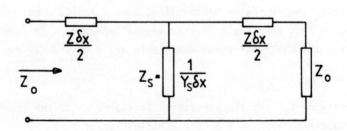

Fig. 11.7

$$Z_o = \frac{Z\,\delta x}{2} = \frac{1}{Y_s\,\delta x}\left[\frac{\frac{Z\,\delta x}{2} + Z_o}{\frac{1}{Y_s\,\delta x} + \frac{Z\,\delta x}{2} + Z_o}\right].$$

Multiplying through by $\left(\frac{1}{Y_s\,\delta x} + \frac{Z\,\delta x}{2} + Z_o\right)$ and expanding the bracket gives

$$\cancel{\frac{Z}{Y_s\,\delta x}} + \cancel{\frac{Z_o Z\,\delta x}{2}} + Z_o^2 = \frac{Z\,\delta x}{2Y_s\,\delta x} + \frac{Z^2(\delta x)^2}{4} + \cancel{\frac{Z Z_o \delta x}{2}} + \frac{Z\,\delta x}{2Y_s\,\delta x} + \cancel{\frac{Z_o}{Y_s\,\delta x}}$$

$$Z_o^2 = \frac{Z\,\delta x}{Y_s\,\delta x} + \frac{Z^2(\delta x)^2}{4}.$$

As $\delta x \to 0$, $(\delta x)^2$ may be ignored and $Z_o^2 = \frac{Z}{Y_s}$.

But $Z = (R + j\omega L)$ Ω and $Y_s = (G + j\omega C)$ S (section 11.1.2, equation 11.1). Therefore

$$Z_o = \sqrt{\left[\frac{(R + j\omega L)}{(G + j\omega C)}\right]} \qquad (11.2)$$

Now, returning to Fig. 11.6,

current drawn by the shunt components in the element δx

$$= VY_s\,\delta x = V(G + j\omega C)\delta x \text{ A}.$$

The current leaving the element of line is less than that entering it by this amount. The change in current through the element equals $-\delta I$, hence

$$-\delta I = V(G + j\omega C)\delta x \text{ A}.$$

In the limit, as $\delta x \to 0$

$$-\frac{\delta I}{\delta x} = V(G + j\omega C). \qquad (11.3)$$

Due to the current I A in the series impedance, there will be a fall in line potential $-\delta V$ volts with respect to earth across the element of line.

$$-\delta V = IZ\,\delta x = I(R + j\omega L)\delta x.$$

In the limit, as $\delta x \to 0$

$$-\frac{\delta V}{\delta x} = I(R + j\omega L). \tag{11.4}$$

Partial derivatives are used since both voltage and current are also time-varying.

Differentiating 11.3 with respect to x gives

$$-\frac{\delta^2 I}{\delta x^2} = \frac{\delta V}{\delta x}(G + j\omega C). \tag{i}$$

Differentiating 11.4 with respect to x gives

$$-\frac{\delta^2 V}{\delta x^2} = \frac{\delta I}{\delta x}(R + j\omega L). \tag{ii}$$

Substituting 11.4 in (i)

$$\frac{\delta^2 I}{\delta x^2} = I(R + j\omega L)(G + j\omega C). \tag{11.5}$$

Substituting 11.3 in (ii)

$$\frac{\delta^2 V}{\delta x^2} = V(R + j\omega L)(G + j\omega C). \tag{11.6}$$

These are the two equations for a transmission line with uniformly distributed parameters.

11.2.2 A solution of the transmission line equations for a line terminated in its characteristic impedance

Equation 11.6: $\dfrac{\delta^2 V}{\delta x^2} = V(R + j\omega L)(G + j\omega C)$.

To simplify the expression, write P^2 for $(R + j\omega L)(G + j\omega C)$ or $P = \sqrt{[(R + j\omega L)(G + j\omega C)]}$. Then equation 11.4 becomes

TRANSMISSION LINES

$$\frac{\delta^2 V}{\delta x^2} = P^2 V.$$

Taking Laplace transforms of both sides (chapter 10, Laplace pair 10)

$$s^2 \overline{V} - s V_{(o)} - V'_{(o)} = P^2 \overline{V}$$

$$\overline{V}(s^2 - P^2) = s V_{(o)} + V'_{(o)}$$

$$\overline{V} = \frac{s V_{(o)}}{s^2 - P^2} + \frac{V'_{(o)}}{s^2 - P^2}$$

Taking inverses (Laplace pairs 12 and 13, chapter 10)

$$V = V_{(o)} \cosh Px + \frac{V'_{(o)}}{P} \sinh Px. \tag{i}$$

Now $V'_{(o)}$ is the value of $\frac{\delta V}{\delta x}$ at $x = 0$.

From equation 11.4, $\frac{\delta V}{\delta x} = -I(R + j\omega L)$, and as $x = 0$ is at the sending end of the line, $I = I_S$, the total input current. It is only as progress is made along the line that the value of the current changes due to the shunt components. Therefore

$$V'_{(o)} = -I_S (R + j\omega L).$$

Also in equation (i) above, $V_{(o)}$ is the value of voltage at $x = 0$, which is the sending-end voltage V_S. Therefore $V_{(o)} = V_S$. Writing these values into (i)

$$V = V_S \cosh Px - \frac{I_S (R + j\omega L)}{P} \sinh Px \tag{ii}$$

But $P = \sqrt{[(R + j\omega L)(G + j\omega C)]}$, therefore equation (ii) becomes

$$V = V_S \cosh Px - \frac{I_S (R + j\omega L)}{\sqrt{[(R + j\omega L)(G + j\omega C)]}} \sinh Px$$

$$= V_S \cosh Px - I_S \sqrt{\left[\frac{(R + j\omega L)}{(G + j\omega C)}\right]} \sinh Px.$$

Substituting equation 11.2 for Z_o

$$V = V_S \cosh Px - I_S Z_o \sinh Px. \tag{11.7}$$

For a correctly terminated line, the input impedance is Z_o so that $I_S Z_o = V_S$.

Now, $\cosh Px = \dfrac{e^{Px} + e^{-Px}}{2}$ and $\sinh Px = \dfrac{e^{Px} - e^{-Px}}{2}$.

Equation 11.7 can therefore be written as

$$V = V_S \left(\frac{e^{Px} + e^{-Px}}{2}\right) - V_S \left(\frac{e^{Px} - e^{-Px}}{2}\right)$$

$$= \frac{V_S}{2}(e^{Px} + e^{-Px} - e^{Px} + e^{-Px})$$

$$V = V_S e^{-Px} \text{ volts} \tag{11.8}$$

V is the line voltage at any distance x from the sending end. Transposing equation 11.8

$$\frac{V_S}{V} = e^{+Px} \quad \text{or} \quad Px = \log_e\left(\frac{V_S}{V}\right).$$

In chapter 8, section 8.12.2, the propagation coefficient of a network γ is defined as

$$\gamma = \log_e \frac{I_1}{I_2} = \log_e \frac{V_1}{V_2}$$

where subscripts 1 and 2 denote input and output conditions respectively.

Therefore the quantity 'P' is in fact the propagation coefficient of the transmission line per unit length, and equation 11.8 may be written as

$$V = V_S e^{-\gamma x} \text{ volts.} \tag{11.9}$$

And from equation 8.38, equation 11.9 becomes

TRANSMISSION LINES

$$V = V_S e^{-\alpha x} \underline{/-\beta x} \text{ volts} \tag{11.10}$$

α = the attenuation coefficient in nepers per unit length
β = the phase-change coefficient per unit length
γ = propagation coefficient per unit length = $\sqrt{[(R + j\omega L)(G + j\omega C)]}$.

At the end of the line, $x = \ell$, and equation 11.10 becomes

$$V_R = V_S e^{-\alpha \ell} \underline{/-\beta \ell}.$$

In exactly the same manner, writing I for V starting from equation 11.5, a solution may be arrived at for the current I at any distance x from the sending end. An outline proof is given here.

$$\frac{\delta^2 I}{\delta x^2} = P^2 I.$$

Using Laplace transforms

$$I = I_{(o)} \cosh Px + \frac{I'_{(o)}}{P} \sinh Px$$

$I_{(o)}$ = current when x is zero = I_S.

$I'_{(o)} = \frac{\delta I}{\delta x}$ (at $x = 0$) = $-V_Z(G + j\omega C)$ \hfill (equation 11.3).

Therefore, $I = I_S \cosh Px - \dfrac{V_S(G + j\omega C)}{\sqrt{[(R + j\omega L)(G + j\omega C)]}} \sinh Px$

$$= I_S \cosh Px - \frac{V_S}{Z_o} \sinh Px. \tag{11.11}$$

For a correctly terminated line, $I_S = \dfrac{V_S}{Z_o}$.

Therefore, $I = \dfrac{I_S}{2}(e^{Px} + e^{-Px} - e^{Px} + e^{-Px})$

$$= I_S e^{-Px}$$

$$I = I_S e^{-\alpha x} \underline{/-\beta x} \text{ A}. \tag{11.12}$$

At the receiving end of the line, $x = \ell$, and $I_R = I_S e^{-\alpha \ell} \underline{/-\beta \ell}$.

11.2.3 The four-terminal network parameters of a transmission line with uniformly distributed constants

From equations 8.13 and 8.14 and Fig. 8.8(b)

$$V_2 = DV_1 - BI_1 \quad \text{and} \quad -I_2 = CV_1 - AI_1.$$

Treating V_1 as the input voltage V_S, and I_1 as the input current I_S, V_2 and I_2 become the voltage and current respectively, at any point on the line (V and I in section 11.2.2). Hence

$$V = DV_S - BI_S.$$

From equation 11.7

$$V = V_S \cosh \gamma x - I_S Z_0 \sinh \gamma x$$

it is seen that $D = \cosh \gamma x$, and $B = Z_0 \sinh \gamma x$. Also

$$-I = CV_S - AI_S$$
$$I = -CV_S + AI_S.$$

From equation 11.11

$$I = -\frac{V_S}{Z_0} \sinh \gamma x + I_S \cosh \gamma x$$

it is seen that $C = \dfrac{\sinh \gamma x}{Z_0}$, and $A = \cosh \gamma x$.

These A, B, C and D parameters are for a length x of line. The parameters for the whole line are obtained by writing ℓ for x.

<u>Worked example 11.5</u> Determine the values of (a) Z_0, (b) α, and (c) β, for a cable which has the following primary constants per kilometre length.

$R = 46.8 \: \Omega$, $G = 38 \: \mu S$, $L = 1.1$ mH, $C = 0.06 \: \mu F$ and $\omega = 10000$ rad/s.

$(R + j\omega L) = 46.8 + j10000 \times 1.1 \times 10^{-3} = 46.8 + j11 = 48.075 \underline{/13.2^\circ} \: \Omega$.

$(G + j\omega C) = 38 \times 10^{-6} + j10000 \times 0.06 \times 10^{-6}$

$= 38 \times 10^{-6} + j6 \times 10^{-4} = 6.01 \times 10^{-4} \underline{/86.37^o}$ S.

(a) $Z_o = \sqrt{\left(\dfrac{R + j\omega L}{G + j\omega C}\right)} = \sqrt{\left(\dfrac{48.075\underline{/13.2^o}}{6.01 \times 10^{-4}\underline{/86.37^o}}\right)} = \sqrt{(79991.7\underline{/-73.17^o})}$

$= 282.8\underline{/-36.58^o}\ \Omega.$

(b) and (c) $\gamma = \sqrt{[(R + j\omega L)(G + j\omega C)]}$

$= \sqrt{(48.075\underline{/13.2^o} \times 6.01 \times 10^{-4}\underline{/86.37^o})}$

$= \sqrt{(0.02889\underline{/99.57^o})} = 0.17\underline{/49.78^o} = 0.109 + j0.13.$

$\alpha = 0.109$ nepers per km, $\qquad \beta = 0.13$ radians per km.

11.2.4 The logarithmic spiral

Using the data from worked example 11.5 equation 11.12 may be written in the form

$I = I_S e^{-0.109x}\underline{/-0.13x}$ A (where x is measured in km).

Or using degrees instead of radians

$I = I_S e^{-0.109x}\underline{/-7.45x^o}$ A.

It follows that for increasing values of x the modulus of the current will be decreasing while its phase with respect to I_S will be increasingly more lagging. This is best demonstrated using specific values.

Let us consider an applied voltage of $282.8\underline{/-36.58^o}$ V $(= Z_o)$. Then

$I_S = \dfrac{V_S}{Z_o} = \dfrac{282.8\underline{/-36.58^o}}{282.8\underline{/-36.58^o}} = 1\underline{/0^o}$ A.

At x = 1 km

$I = 1e^{-0.109 \times 1}\underline{/-7.45^o \times 1}$

$= 0.897\underline{/-7.45^o}$ A.

The current, after a length of 1 km, has a modulus of 0.897 A and is lagging I_S by $7.45°$. For 2 km of line

$$I = 1e^{-0.109 \times 2}\underline{/-7.45° \times 2}$$
$$= 0.804\underline{/-14.9°} \text{ A.}$$

After 2 km of line the modulus of the current is 0.804 A and is lagging I_S by $14.9°$.

Similar calculations may be made for any length of line required. The following table is drawn up for lengths up to 20 km.

Length (km)	Current (A)	Phase angle (degrees)
1	0.897	-7.45
2	0.804	-14.9
3	0.72	-22.35
4	0.646	-29.8
5	0.58	-37.25
10	0.336	-74.5
20	0.113	-149

These results are plotted in Fig. 11.8. The line joining the ends of the phasors is known as the logarithmic spiral.

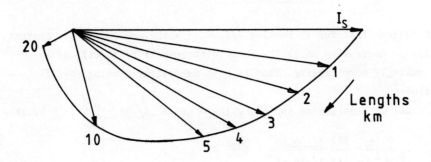

Fig. 11.8

A similar spiral may be drawn for the voltage using the equation

TRANSMISSION LINES

$$V = V_S e^{-0.109x} \underline{/-7.45x^\circ} \text{ V} \qquad \text{(equation 11.10)}.$$

At any point on the line $V/I = Z_o$, and the phase angle between V and I will be that of Z_o. In the example used this is 36.58°.

11.2.5 Wavelength (λ)

That length of line which will cause the voltage or current phasor to be retarded by one complete revolution is known as the wavelength.

Since β is the phase change in radians per unit length and one revolution equals 2π radians, it will require $2\pi/\beta$ unit lengths to give the required phase change.

$$\lambda = \frac{2\pi}{\beta}. \qquad (11.13)$$

11.2.6 Phase velocity (V_p)

Consider a cosine wave input to a transmission line expressed by the equation, $v = V_m \cos \omega t$ volts. It is switched on to the line at $t = 0$, i.e. when $v = V_m$ (Fig. 11.9(a)). The sending end of the line is at V_m volts at $t = 0$.

It takes a finite time for the effect of the applied voltage to be felt further along the line so that after time interval t_1 the effect is felt at x_1 (Fig. 11.9(b)), by which time the input wave has fallen from its peak value. As the voltage wave travels along the line it is attenuated according to equation 11.10. At $t_2 = \pi/2\omega$

$$v = V_m \cos \omega \left(\frac{\pi}{2\omega}\right) = 0 \text{ volts}.$$

The wave has travelled a distance x_2 (Fig. 11.9(c)).

Progressing through the cycle to $t_3 = 2\pi/\omega$

$$v = V_m \cos \omega \left(\frac{2\pi}{\omega}\right) = V_m$$

the input is again at value V_m and the effect has been felt at a distance x_3 along the line (Fig. 11.9(d)). At this point the effect is being felt of a wave which was impressed on the line one cycle beforehand, i.e. the effect has been retarded by exactly one complete cycle. The length x_3 is therefore one wavelength (section 11.2.5).

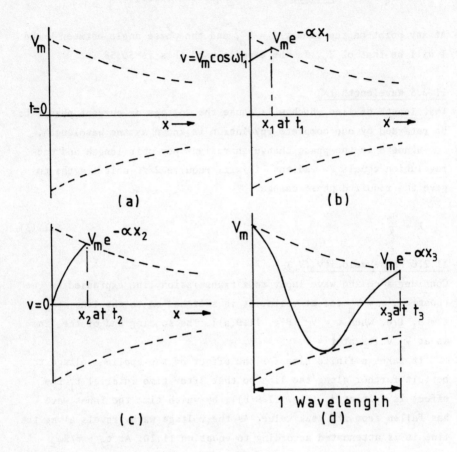

Fig. 11.9

$x_3 = \lambda$.

How long has it taken for the effect to be felt? It has taken the periodic time of the wave, namely 1/f seconds. Now

$$\text{velocity} = \frac{\text{distance}}{\text{time}} = \frac{\lambda}{\frac{1}{f}} = \lambda f \text{ s}. \quad (11.14)$$

Equation 11.14 defines the phase velocity or velocity of propagation.

<u>Worked example 11.6</u> An underground cable is 20 km long and has the following primary constants per kilometre length

TRANSMISSION LINES

$R = 56 \ \Omega$, $L = 0.625$ mH, $C = 0.04 \ \mu F$ and $G = 1 \ \mu S$.

The line is terminated in its own characteristic impedance and has an input power of 1 watt at a frequency of 1 kHz.

Determine (a) the characteristic impedance of the line, (b) the attenuation coefficient, (c) the phase-change coefficient, (d) the receiving-end power, (e) the phase change between I_R and I_S, (f) the wavelength, (g) the velocity of propagation, and (h) the time delay introduced by the line.

(a) $Z_o = \sqrt{\left(\dfrac{R + j\omega L}{G + j\omega C}\right)} = \sqrt{\left(\dfrac{56 + j2\pi \times 1000 \times 0.625 \times 10^{-3}}{1 \times 10^{-6} + j2\pi \times 1000 \times 0.04 \times 10^{-6}}\right)}$

$= \sqrt{\left(\dfrac{56.14 \underline{/4.01^\circ}}{2.513 \times 10^{-4} \underline{/89.77^\circ}}\right)} = 472.65 \underline{/-42.88^\circ} = (346.35 - j321.6) \ \Omega.$

$\gamma = \sqrt{[(R + j\omega L)(G + j\omega C)]} = \sqrt{(56.14\underline{/4.01^\circ} \times 2.513\underline{/89.77^\circ})}$

$= 0.119\underline{/46.89^\circ} = 0.081 + j0.0867.$

Answers

(b) $\alpha = 0.081$ N/km.
(c) $\beta = 0.0867$ rad/km.
(d) The sending-end power = 1 watt

$= V_S I_S \cos \phi_S.$

Since the line is correctly terminated, signals at the input see only the characteristic impedance, Z_o. Therefore

$|I_S| = \dfrac{|V_S|}{|Z_o|}$ and $\phi_S = \phi$, the angle of Z_o, i.e. -42.88°.

Power $= V_S \times \dfrac{V_S}{Z_o} \cos \phi_0$

$1 = \dfrac{V_S^2}{Z_o} \cos \phi_0.$

Substituting values and transposing

$$V_S^2 = \frac{472.65 \times 1}{\cos(-42.88°)}$$

$V_S = 25.4$ V.

Again, power = $V_S I_S \cos \phi$ watts. Hence

$1 = 25.4 \times I_S \times \cos(-42.88°)$.

Transposing:

$$I_S = \frac{1}{25.4 \times 0.733} = 0.0537 \text{ A}.$$

Now $V_R = V_S e^{-\alpha \ell}$ (equation 11.10)

$$V_R = 25.4 e^{-0.081 \times 20} = 5.027 \text{ V}.$$

And $I_R = I_S e^{-\alpha \ell}$ (equation 11.12)

$$I_R = 0.0537 e^{-0.081 \times 20} = 0.0106 \text{ A}.$$

The phase angle between V_R and I_R is that of Z_o since the termination is in Z_o. Therefore

receiving-end power = $5.027 \times 0.0106 \times \cos(-42.88°)$
= 0.039 W.

Once the method is understood, it will be seen that the receiving-end power can more easily be calculated as

$$P_R = \text{sending power} \times (e^{-\alpha \ell})^2 \text{ watts}.$$

(e) Phase change = 0.0867 rad/km. Therefore over 20 km, phase change is $20 \times 0.0867 = 1.734$ rad (99.35°).

(f) Wavelength, $\lambda = \frac{2\pi}{\beta}$ (equation 11.13)

$$= \frac{2\pi}{0.0867} = 72.47 \text{ km}.$$

(g) Phase velocity, $V_p = f\lambda$ (equation 11.14)
$$= 1000 \times 72.47 = 72470 \text{ km/s}.$$

(Since the speed of light is 3×10^8 m/s, the phase velocity in this line is 0.24 times that of light.)

(h) To travel 20 km at 72470 km/s takes

$$\frac{20}{72470} = 276 \text{ μs}.$$

This is the time delay introduced by the line.

Self-assessment example 11.7   A cable 8 km in length has the following primary constants per 100 m length

$R = 4.5 \, \Omega$, $L = 0.035$ mH, $C = 0.013$ μF; G may be ignored.

The line is correctly terminated in its characteristic impedance and is supplied at 50 V and 10 kHz at the sending end.

Determine (a) the value of Z_o, (b) the receiving-end current, (c) the wavelength, and (d) the velocity of propagation.

11.2.7 The incorrectly terminated line

It is not the intention in this book to deal in depth with the incorrectly terminated line since how to deal with the problems involved and artificial matching methods is a complete study in itself. The topic is introduced so that the desirability of correct matching is appreciated.

From equation 11.7

$$V = V_S \cosh \gamma x - I_S Z_o \sinh \gamma x \text{ volts} \qquad (i)$$

where V is the line voltage at any distance x from the sending end. For the correctly-terminated line this reduces to

$$V = V_S e^{-\alpha x} \underline{/-\beta x} \text{ V.}$$

For any other terminating impedance than Z_o, $V_S/I_S \neq Z_o$, and (i) above becomes

$$V = V_S \left(\frac{e^{\gamma x} + e^{-\gamma x}}{2}\right) - I_S Z_o \left(\frac{e^{\gamma x} - e^{-\gamma x}}{2}\right)$$

$$= \frac{e^{\gamma x}}{2}(V_S - I_S Z_o) + \frac{e^{-\gamma x}}{2}(V_S + I_S Z_o) .$$

In the work on the logarithmic spiral, an expression containing $e^{-\gamma x}$ reduces in size as it travels in the x-direction, i.e. from the sending end towards the receiving end. An expression containing $e^{\gamma x}$ is one which increases in size as it travels in the x-direction. This appears to be impossible since taken to the limit at $x = \infty$, $V = \infty$.

In reality, what is happening is that the original wave travels along the line until it reaches the termination, where, since this is not correct, not all the power is absorbed. Some of it is re-transmitted towards the sending end, being attenuated as it travels along the line. The returning wave is decreasing in magnitude as it moves in the (-x)-direction.

$$e^{-\gamma x}\left(\frac{V_S + I_S Z_o}{2}\right)$$

is a wave travelling from the sending end towards the receiving end, being attenuated as it progresses along the line.

$$e^{\gamma x}\left(\frac{V_S - I_S Z_o}{2}\right)$$

is a wave travelling from the receiving end towards the sending end being attenuated as it progresses along the line. If it could be measured separately it would be increasing in size at increasing distances from the sending end.

Although the line is not terminated in Z_o the line still has its characteristic impedance and it is in this that the current flows.

In a low-loss line there will be several reflections before the magnitude is reduced to zero. This is one way in which several images successively to the right can be produced in a television receiver.

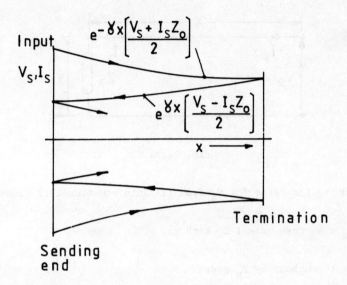

Fig. 11.10

The receiver needs to be matched to the co-axial down lead which in turn must be matched to the aerial. Failure to do so results in reflections from the receiver to the aerial and then back to the receiver possibly several times.

The same phenomenon may be demonstrated by considering the current equation 11.11.

One further effect of incorrect termination is that it does not simply change the input impedance to another (constant) value but makes this a function of the length of line. Adjusting the length of line, while keeping the terminating impedance constant can have a considerable effect on the input impedance.

Consider a line terminated in an impedance Z_R as shown in Fig. 11.11. From equation 11.7

$$V_R = V_S \cosh \gamma x - I_S Z_o \sinh \gamma x .$$

Now, $I_R = \dfrac{V_R}{Z_R}$ or $V_R = I_R Z_R$. (i)

415

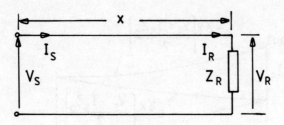

Fig. 11.11

Substituting the value for V_R from (i) into equation 11.7 gives

$$I_R Z_R = V_S \cosh \gamma x - I_S Z_o \sinh \gamma x.$$

Dividing throughout by Z_R gives

$$I_R = \frac{V_S}{Z_R} \cosh \gamma x - I_S \frac{Z_o}{Z_R} \sinh \gamma x. \qquad \text{(ii)}$$

But equation 11.11 is an equation for I_R

$$I_R = I_S \cosh \gamma x - \frac{V_S}{Z_o} \sinh \gamma x .$$

Equating 11.11 and (ii) gives

$$I_S \cosh \gamma x - \frac{V_S}{Z_o} \sinh \gamma x = \frac{V_S}{Z_R} \cosh \gamma x - I_S \frac{Z_o}{Z_R} \sinh \gamma x$$

$$I_S \left(\cosh \gamma x + \frac{Z_o}{Z_R} \sinh \gamma x \right) = V_S \left(\frac{1}{Z_R} \cosh \gamma x + \frac{1}{Z_o} \sinh \gamma x \right)$$

$$\frac{V_S}{I_S} = Z_{input} = \frac{\cosh \gamma x + \frac{Z_o}{Z_R} \sinh \gamma x}{\frac{1}{Z_R} \cosh \gamma x + \frac{1}{Z_o} \sinh \gamma x} . \qquad (11.15)$$

Equation 11.15 shows the input impedance to be dependent not only upon the value of Z_R, the terminating impedance, but upon x, the length of the line.

Further problems

11.8. The per-phase resistance, reactance and capacitive susceptance of a three-phase transmission line are 15 Ω, 45 Ω, and 4×10^{-4} S respectively. Calculate the values of the four-terminal network parameters for the line considering it as a T-equivalent.

11.9. The four-terminal network parameters for each phase of a 132 kV, 3-phase, 50 Hz overhead line are as follows

$$A = D = 0.9975 \underline{/0.6^\circ}, \quad B = 47.3 \underline{/72^\circ} \quad \text{and} \quad C = 4 \times 10^{-4} \underline{/90^\circ}.$$

The line delivers a load of 50 MVA at 0.8 power factor lagging at 132 kV to a factory sub-station.

Calculate (a) the sending-end voltage, (b) the sending-end current, and (c) the transmission efficiency.

11.10. A 160 km length of 3-phase, 132 kV, 50 Hz transmission line has conductor resistance 0.184 Ω/km/phase, inductance 1.32 mH/km/phase and capacitance 9 nF/km/phase. Leakage current may be ignored. Using a T-representation, calculate for a receiving-end load of 40 MW at 0.8 power factor lagging and 132 kV (a) the sending-end voltage, (b) the sending-end current, and (c) the transmission efficiency.

11.11. A transmission line 16 km in length has the following primary constants per kilometre length

$$R = 31.25 \text{ Ω}, \quad L = 0.625 \text{ mH}, \quad C = 37.5 \text{ nF}, \quad G = 0.$$

(a) Calculate the value of Z_o for ω = 5000 rad/s.
(b) When the line is terminated in its characteristic impedance and the sending-end voltage equals 2.5 V at 5000 rad/s, calculate (i) the current in the terminating impedance, (ii) the wavelength, (iii) the phase velocity V_p.

11.12. A transmission line 5 km long has a characteristic impedance of 600 $\underline{/0^\circ}$ Ω and a propagation coefficient (0.025 + j0.18) per km at a frequency of 1591.55 Hz.

Given that the line is terminated in its characteristic impedance and has a p.d. of 10 $\underline{/0^\circ}$ V at the above frequency applied to the sending end, calculate (a) the magnitude of the receiving-end current, (b) the angle through which the receiving-end voltage has been retarded with respect to the sending end, and (c) the magnitude of the line voltage at a point 2.8 km from the sending end.

11.13. A co-axial cable has inductance 171 nH/m, resistance 0.1 Ω/m and a capacitance of 65 pF/m. Zero conductance may be assumed. A signal of 0.2 mV at 2 MHz is applied to a length of the cable terminated in its characteristic impedance Z_o.

Calculate (a) the value of Z_o, (b) the length of the cable if the time delay introduced by it is 0.05 s, and (c) the voltage at the termination.

11.14. A co-axial cable 15 m in length has inductance 250 nH/m, resistance 0.3 Ω/m and capacitance 51.02 pF/m. It has an input signal of 5 mV at 28 MHz and is terminated in an impedance designed to prevent reflections.

Calculate (a) the characteristic impedance Z_o, (b) the phase shift in the line, (c) the time delay introduced by the line, and (d) the power dissipated in the line termination (Z_o).

11.15. The primary constants of a telephone pair per metre length are

$$R = 0.01875 \text{ Ω}, L = 0.0125 \text{ mH}, C = 37.5 \text{ μμF}, G = 5 \times 10^{-9}.$$

The line is 10 km in length and the frequency is 3400 Hz. An oscillator with an internal resistance of 600 Ω which generates an e.m.f. of $6\underline{/0^o}$ V at 3400 Hz is connected to the sending-end of the line. Assuming correct termination, calculate (a) the value of Z_o, (b) the input voltage to the line, (c) the power input to the line, (d) the power at the termination, and (e) the time delay introduced by the line.

CHAPTER 11 ANSWERS

11.2. $34.18\underline{/+4.23^o}$ kV

11.4. $A = D = 1 + \frac{YZ}{2} = 0.986\underline{/0.39^o}$, $B = Z = 72.75\underline{/66.07^o}$

$C = Y + \left(\frac{Y}{2}\right)^2 Z = 4.49 \times 10^{-4}\underline{/90.2^o}$ $I_R = 218.7\underline{/-36.9^o}$ A

(Observe: 50 MVA not MW)

418

TRANSMISSION LINES

$V_S = 89400 \underline{/5.3^\circ}$ V (phase) 154.8 kV line. (Values vary slightly with degrees of accuracy in using A, B, C, and D.)

11.7. (a) $Z_o = 78.3 \underline{/-31.98^\circ}$ Ω (b) $0.638 e^{-2.71} = 0.0423$ A
(c) 11.6 km (d) 115600 km/s

11.8. $A = D = 0.991 + j0.003$; $B = 14.9 + j44.8$; $C = j0.0004$

11.9. $I_R = 218 \underline{/-36.8^\circ}$ A (a) $V_S = 85000 \underline{/4.6^\circ}$ V (phase) (b) $I_S = 199.4 \underline{/-28.7^\circ}$ A (c) 0.938 p.u.

11.10. (a) $89200 \underline{/5.3^\circ}$ V (phase) (b) $196.9 \underline{/-28.4^\circ}$ A (c) 0.91
(This is the same line as in self-assessment example 11.4 for comparison of π with T)

11.11. (a) $409.3 \underline{/-42.15^\circ}$ Ω (b) (i) 2.68×10^{-3} A (ii) 110.5 km
(iii) 87933 km/s ($\gamma = (0.0515 + j0.0569)$ per km)

11.12. (a) 14.72 mA (b) 0.9 rad (51.6°) (c) 9.32 V

11.13. (a) $51.3 \underline{/-1.33^\circ}$ Ω; $\gamma = (9.798 \times 10^{-4} + j4.189 \times 10^{-2})$ per metre;
$V_p = 3 \times 10^8$ m/s (b) 15 m (c) 0.197 mV

11.14. (a) 70 Ω (b) 9.42 rad (539.7°) (c) 53.57 ns
(d) 0.356 W

11.15. (a) $578 \underline{/-1.83^\circ}$ Ω (b) 2.944 V (c) 14.98 mW (d) 12.56 mW
(e) 216.6 μs

Appendix 1

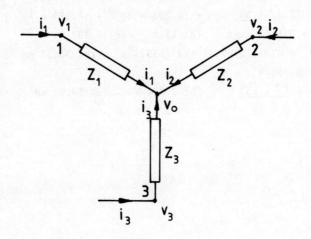

Fig. A1.1

Firstly consider the star shown in Fig. A1.1.

Currents i_1, i_2 and i_3 flow towards the star point. Voltages v_1, v_2, v_3 and v_0 are measured with respect to a datum point, usually earth.

From Kirchhoff's first law, $i_1 + i_2 + i_3 = 0$

$$i_1 = \frac{v_1 - v_0}{Z_1} \; ; \quad i_2 = \frac{v_2 - v_0}{Z_2} \; ; \quad i_3 = \frac{v_3 - v_0}{Z_3}$$

each being the potential difference across an impedance divided by

THE STAR/DELTA TRANSFORMATION

the value of that impedance. Adding the three currents

$$\frac{v_1 - v_0}{Z_1} + \frac{v_2 - v_0}{Z_2} + \frac{v_3 - v_0}{Z_3} = 0.$$

Transposing to get all the v_0 terms on the right-hand side

$$\frac{v_1}{Z_1} + \frac{v_2}{Z_2} + \frac{v_3}{Z_3} = \frac{v_0}{Z_1} + \frac{v_0}{Z_2} + \frac{v_0}{Z_3}$$

$$\frac{v_1}{Z_1} + \frac{v_2}{Z_2} + \frac{v_3}{Z_3} = v_0\left(\frac{1}{Z_1} + \frac{1}{Z_2} + \frac{1}{Z_3}\right)$$

$$v_0 = \frac{\frac{v_1}{Z_1} + \frac{v_2}{Z_2} + \frac{v_3}{Z_3}}{\frac{1}{Z_1} + \frac{1}{Z_2} + \frac{1}{Z_3}}$$

Substituting for v_0 in any of the current equations (selecting the equation for i_1)

$$i_1 = \frac{v_1 - v_0}{Z_1} = v_1 - \frac{\left[\frac{\frac{v_1}{Z_1} + \frac{v_2}{Z_2} + \frac{v_3}{Z_3}}{\frac{1}{Z_1} + \frac{1}{Z_2} + \frac{1}{Z_3}}\right]}{Z_1}.$$

To bring v_1 over the common denominator $\left(\frac{1}{Z_1} + \frac{1}{Z_2} + \frac{1}{Z_3}\right)$, multiply it by this quantity.

$$i_1 = \frac{\frac{v_1}{Z_1} + \frac{v_1}{Z_2} + \frac{v_1}{Z_3} - \left(\frac{v_1}{Z_1} + \frac{v_2}{Z_2} + \frac{v_3}{Z_3}\right)}{Z_1\left(\frac{1}{Z_1} + \frac{1}{Z_2} + \frac{1}{Z_3}\right)}.$$

The $\frac{v_1}{Z_1}$ terms disappear.

$$i_1 = \frac{\frac{v_1}{Z_2} + \frac{v_1}{Z_3} - \frac{v_2}{Z_2} - \frac{v_3}{Z_3}}{Z_1\left(\frac{1}{Z_1} + \frac{1}{Z_2} + \frac{1}{Z_3}\right)} = \frac{v_1 - v_2}{Z_1 Z_2\left(\frac{1}{Z_1} + \frac{1}{Z_2} + \frac{1}{Z_3}\right)} + \frac{v_1 - v_3}{Z_1 Z_3\left(\frac{1}{Z_1} + \frac{1}{Z_2} + \frac{1}{Z_3}\right)} \quad (i)$$

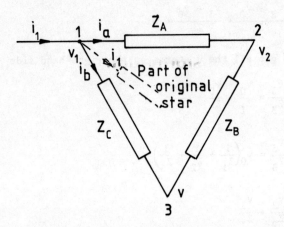

Fig. A1.2

Now consider the equivalent delta as in Fig. A1.2.

The input current from terminal 1 was i_1 A in the original star. If the delta is to be exactly equivalent, the current entering terminal 1 must still be i_1 A. In Fig. 2.30, $i_1 = i_a + i_b$

$$i_a = \frac{v_1 - v_2}{Z_A} \quad \text{and} \quad i_b = \frac{v_1 - v_3}{Z_C}$$

Hence, $i_1 = \dfrac{v_1 - v_2}{Z_A} + \dfrac{v_1 - v_3}{Z_C}$. \hfill (ii)

Comparing equations (i) and (ii) it is seen that for equivalence

$$Z_A = Z_1 Z_2 \left(\frac{1}{Z_1} + \frac{1}{Z_2} + \frac{1}{Z_3}\right) \quad \text{and} \quad Z_C = Z_1 Z_3 \left(\frac{1}{Z_1} + \frac{1}{Z_2} + \frac{1}{Z_3}\right).$$

By a similar process it may be deduced that $Z_B = Z_2 Z_3 \left(\dfrac{1}{Z_1} + \dfrac{1}{Z_2} + \dfrac{1}{Z_3}\right).$

Appendix 2
The Delta/Star Transformation

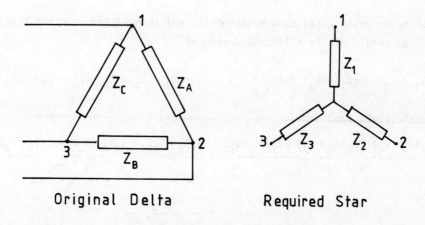

Fig. A2.1

The values of the impedances in the equivalent star are determined by considering the impedance between each pair of terminals in both networks. Consider the impedance from terminal 1 to terminal 2 in the delta network. An electrical path exists through Z_A direct. Another parallel path exists through Z_C and Z_B in series.

Equivalent impedance of these two parallel paths = $\dfrac{\text{product of impedances}}{\text{sum of impedances}}$

$$= \frac{Z_A(Z_B + Z_C)}{Z_A + Z_B + Z_C} \,.$$

In the star, between terminals 1 and 2, impedance = $Z_1 + Z_2$. For equivalence

$$Z_1 + Z_2 = \frac{Z_A(Z_B + Z_C)}{Z_A + Z_B + Z_C} \quad . \tag{i}$$

Considering terminals 2 and 3, in the delta, Z_B is in parallel with $(Z_A + Z_C)$. In the star between terminals 2 and 3 the impedance = $Z_2 + Z_3$. For equivalence

$$Z_2 + Z_3 = \frac{Z_B(Z_A + Z_C)}{Z_A + Z_B + Z_C} \quad . \tag{ii}$$

Finally, between terminals 3 and 1

$$Z_1 + Z_3 = \frac{Z_C(Z_B + Z_A)}{Z_A + Z_B + Z_C} \quad . \tag{iii}$$

Taking equation (iii) from equation (i) and then adding equation (ii) to the result yields a left-hand side of

$$(Z_1 + Z_2) - (Z_1 + Z_3) + (Z_2 + Z_3) = Z_1 + Z_2 - Z_1 - Z_3 + Z_2 + Z_3 = 2Z_2.$$

On the right-hand side

$$\frac{Z_A(Z_B + Z_C) - Z_C(Z_B + Z_A) + Z_B(Z_A + Z_C)}{Z_A + Z_B + Z_C} = \frac{Z_A Z_B + Z_A Z_C - Z_C Z_B - Z_C Z_A + Z_B Z_A + Z_B Z_C}{Z_A + Z_B + Z_C}$$

$$= \frac{2Z_A Z_B}{Z_A + Z_B + Z_C} \quad .$$

Equating left- and right-hand sides gives

$$2Z_2 = \frac{2Z_A Z_B}{Z_A + Z_B + Z_C}$$

$$Z_2 = \frac{Z_A Z_B}{Z_A + Z_B + Z_C} \quad .$$

Note that this is similar in form to the equations for the star/delta transformation.

By a similar method it is found that

THE DELTA/STAR TRANSFORMATION

$$Z_1 = \frac{Z_A Z_C}{Z_A + Z_B + Z_C} \quad \text{and} \quad Z_3 = \frac{Z_B Z_C}{Z_A + Z_B + Z_C}.$$

Appendix 3
Mutual Inductance

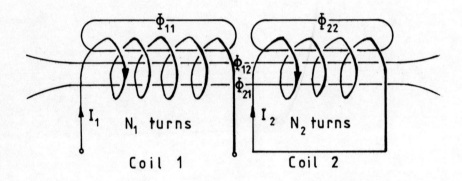

Fig. A3.1

Figure A3.1 shows two coils in close proximity. A current I_1 is supplied to coil 1 from an external source of e.m.f.

Φ_{11} is the flux set up in coil 1 which only links with coil 1.
Φ_{22} is the flux set up in coil 2 which only links with coil 2.
Φ_{12} is the flux set up in coil 1 which links with coil 2.
Φ_{21} is the flux set up in coil 2 which links with coil 1.

Assume that the flux produced is proportional to the current.
 An expression for self-inductance from equation 6.13 is

$$L = \frac{N \, d\Phi}{di} \quad \text{H}.$$

MUTUAL INDUCTANCE

For self-inductance the flux linking with the turns creating it is involved.

Assume that initially the circuit is not energised. All fluxes and currents are zero. Establishing a current in coil 1 creates a flux $\Phi_{12} + \Phi_{11}$, that is, the flux which links with coil 2 plus the flux which links only coil 1 or part of it.

di is the change in current from zero to its final value I_1

$$di = I_1.$$

The change in flux is from zero to $(\Phi_{12} + \Phi_{11})$

$$d\Phi = (\Phi_{12} + \Phi_{11}).$$

Therefore, using equation 6.13

$$L_1 = \frac{N_1(\Phi_{12} + \Phi_{11})}{I_1} \text{ H}.$$

Multiplying out the bracket gives two terms, each of which is a self-inductance.

$$L_1 = \frac{N_1\Phi_{12}}{I_1} + \frac{N_1\Phi_{11}}{I_1} \tag{ii}$$

Dealing with the second term

$$\frac{N_1\Phi_{11}}{I_1}$$

is an inductance present due to the imperfection in coupling between the coils. If perfect coupling were present all the flux from coil 1 would link with coil 2 and Φ_{11} would be zero.

$$\frac{N_1\Phi_{11}}{I_1} = L_{11}$$

where L_{11} is the leakage inductance of coil 1.

Let $\dfrac{N_1\Phi_{12}}{I_1} = L_A$

where L_A is an inductance due to the flux which does link the two coils.

Equation (ii) may therefore be written as $L_1 = L_{11} + L_A$.

The same reasoning applies to the secondary. Due to the changing flux, an e.m.f. is induced which drives a current I_2 if the circuit is closed. The current produces two fluxes, Φ_{21} and Φ_{22}. Hence $L_2 = L_{22} + L_B$, where L_2 = total inductance, L_{22} = leakage inductance.

The two coils may be represented schematically as in Fig. A3.2.

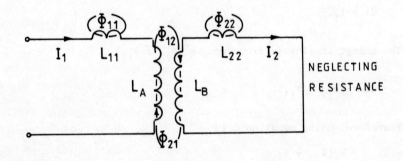

Fig. A3.2

Now if there were no leakage

$$L_1 = L_A = \frac{N_1 \Phi_{12}}{I_1}$$

and $L_2 = L_B = \dfrac{N_2 \Phi_{21}}{I_2}$.

Hence, $L_1 L_2 = \dfrac{N_1 \Phi_{12}}{I_1} \times \dfrac{N_2 \Phi_{21}}{I_2}$

or $\quad L_1 L_2 = \dfrac{N_1 \Phi_{21}}{I_2} \times \dfrac{N_2 \Phi_{12}}{I_1}$. \hfill (iii)

Now $N_1 \Phi_{21}$ are flux linkages of secondary-produced flux by I_2 with the primary coil of N_1 turns. Just as self-inductance is defined in equation 6.13 as flux linkages with a coil created by a current in that coil, mutual inductance may be defined as flux linkages with a second coil created by a current in the first coil or vice-versa. Hence

MUTUAL INCTANCE

$$\frac{N_1 \Phi_{21}}{I_2} = M$$

where M is the mutual inductance between the coils. Similarly, $N_2 \Phi_{12}$ are flux linkages of primary-produced flux by current I_1 with the secondary of N_2 turns. Hence

$$\frac{N_2 \Phi_{12}}{I_1} = M.$$

Therefore equation (iii) becomes

$$L_1 L_2 = M \times M = M^2. \tag{iv}$$

This result is achieved assuming perfect coupling, i.e. no leakage flux and so the mutual inductance is at its maximum possible value, M_{max}.

Taking the square root of both sides of equation (iv) gives

$$M_{max} = \sqrt{(L_1 L_2)} \text{ H}. \tag{9.9}$$

If indeed there is leakage flux then the effective linked inductances of coil 1 and coil 2 become L_A and L_B respectively when

$$M = \sqrt{(L_A L_B)} \text{ H}.$$

This value will be less than M_{max}.

$$\frac{M}{M_{max}} = k$$

where k is the coupling coefficient. Transposing

$$M = kM_{max} = k\sqrt{(L_1 L_2)}. \tag{9.10}$$

Index

Admittance, 28
Ampere, definition, 209
Attenuation coefficient, 287
Attenuation
    decibels, 283
    nepers, 284

Balancing current, 332
Bandwidth, 323
Bridges, 131 et seq.
    Anderson, 146
    Hay, 138
    Maxwell comparison, 133
    Maxwell Wien, 136
    Nul detectors for, 132
    Owen, 135
    Schering, 140
    Wien, 143

Capacitance
    of concentric cable, 172, 195
    of parallel-plate capacitor 170, 191
    of twin line, 176
Cascaded networks, 280
Characteristic impedance, 278, 399
Charge density, 163
Complex waves, 88 et seq.
Conductance, 28
Coulomb's law, 161
Coupling coefficient, 312
Critical coupling, 319
Curvi-linear squares, 193

Decibel, 283
Delta Star transformation, 73
Demagnetisation curve, 246
Diamagnetism, 206
Dielectrics
    imperfections in, 184

Dielectrics (*cont.*)
    properties of, 187
    strength, 183
Domain theory, 210
Dynamic impedance, 302

Eddy current losses, 226
Electric field plotting, 188
Electric force, 164
Electric potential, 166
Electromagnetism, 204 et seq.
Energy
    density in electrostatic field, 182
    in electromagnetic field, 217
    in electrostatic field, 180
Equipotential surfaces, 168

Ferrites, 240
Force
    between current-carrying conductors, 209
    on charge in electrostatic field, 164, 170
Forced magnetisation of iron, 97
Four terminal parameters
    admittance, 162
    high-frequency transmission line, 406
    series impedance, 261
    T network, 262
    π network, 262
Free magnetisation of iron, 95

Harmonics, 88
    content of wave, 106
    errors in measurements due to, 121
    in single-phase circuits, 109
    production of, 94
Hysteresis
    loop, 213
    loss, 212, 214

Image impedance, 268, 271
Induced e.m.f., 207
Inductance
    coaxial cable, 221
    defining equations, 216, 217
    parallel pair of conductors, 223
    toroid, 217
Insertion-loss ratio, 285
Iron loss separation, 231
Iterative impedance, 274
Iterative transfer coefficient, 286

j-operator, 4
    significance in a.c. circuits, 17 et seq.

INDEX

Laplace transforms, 346 et seq.
    circuit transforms, 358
    definition of, 347
    derivations of, 350
    of derivatives, 350
    double-energy circuit, 369
    exponential factor (shifting theorem), 354
    inductive circuits fed from an alternating supply, 376
    table of transforms, 388, 389
    transfer function, 383
Loss angle, 186

Magnetising force, 206
Magnetism, definitions, 205
Materials for transformer cores, 239
Maximum power and characteristic impedance, 279
Maximum power transfer theorems, 77
Mutual inductance, 310
    bandwidth, 323
    critical coupling, 319
    effect of secondary load on input impedance, 312
    with tuned secondary, 317

Networks in cascade, 280
Neper, 284
Norton's theorem, 63
Nul detectors (bridges), 132

Open-circuit test (two port network), 260, 271
Optimum operating point, 249

Parallel resonance, 300
Paramagnetism, 206
Permanent magnets, 245 et seq.
Permeability, 206
Permittivity, 166
Phase-change coefficient, 287
Phase velocity, 409
Polar form, 8
    division using, 15
    multiplication using, 13
Pole shoes, 253
Potential gradient, 164
Potentiometer
    a.c. polar type, 150
    d.c. type, 149
    measurements using, 153
    rectangular (cartesian co-ordinate type), 152
Power in a.c. circuits, 32 et seq.
Power conveyed by complex wave, 103
Power factor, 38, 105
    of capacitor, 142
Power transformer, 329
    balancing current, 332
    core loss, 333

Power transformer (*cont.*)
  e.m.f. equation, 208, 336
  referring values of resistance and reactance, 335
Power transmission, 391
  equivalent π circuit, 398
  equivalent T circuit, 395
Propagation coefficient, 287

Q-factor, 302
  and bandwidth, 303, 307

Rectangular coordinates, 6
  division using, 14
  multiplication using, 12
Reluctance, 205
Resonance
  parallel, 300
  series, 298
R.M.S. value of complex wave, 99

Selective resonance, 114
Series resonance, 298
Short-circuit test, two-port network, 261
Standards, 130
Star delta transformation, 68
Steinmetz index, 226, 236
Stress in cable dielectric, 173
Superposition theorem, 50
Susceptance, 28

Thevenin's theorem, 54
Transfer coefficient (iterative), 286
Transfer function, 383
Transformer e.m.f. equation, 208, 336
Transmission line (high frequency)
  characteristic impedance, 399
  equations, 399, 402
  four-terminal parameters, 406
  incorrectly terminated, 413
  logarithmic spiral, 407
  phase velocity, 409
  wavelength, 409
Transmission line (power *see under* power transmission)

Volt-amperes reactive, 37

Wavelength, 409